FUNGICIDES FOR PLANT AND ANIMAL DISEASES

Edited by **D. Dhanasekaran, N. Thajuddin** and **A. Panneerselvam**

INTECHOPEN.COM

Fungicides for Plant and Animal Diseases
Edited by D. Dhanasekaran, N. Thajuddin and A. Panneerselvam

Published by InTech

Janeza Trdine 9, 51000 Rijeka, Croatia

Edition 2014

Publishing Process Manager Marko Rebrovic

Technical Editor Teodora Smiljanic

Cover Designer InTech Design Team

Additional hard copies can be obtained from orders@intechopen.com

Fungicides for Plant and Animal Diseases,
Edited by D. Dhanasekaran, N. Thajuddin and A. Panneerselvam
p. cm.
ISBN 978-953-307-804-5

INTECH

open science | open minds

Contents

Preface

A fungicide is a chemical pesticide compound that kills or inhibits the growth of fungi. In agriculture, fungicide is used to control fungi that threaten to destroy or compromise crops. Gardeners use fungicide as a household pesticide to protect plants from potential destruction. In medicine, fungicide is used to kill fungal infections. The drugs used to kill these infections are referred to as antifungal drugs.

Government agencies monitor the use of fungicide in agriculture and medicine. Fungicide is classified as a pesticide when used on plants and is subject to Environmental Protection Agency (EPA) standards. Most fungicides are toxic to humans and can present both acute and chronic problems if absorbed into food eaten by people. There are also environmental issues, that the EPA must address, resulting from the use of fungicide in agriculture, including runoff into streams and lakes. In medicine, infections caused by fungus, such as ringworm and yeast infections, must be treated with an antifungal medication. Antifungal medications can be orally or topically administered, and many fungal infections are treated with both methods.

We are thankful to all the authors who have contributed to this book. We wish to record our appreciation of the high standard of work put up by the InTech Open Access Publisher, committed to making peer reviewed scientific literature free for the world. This book comprises of 14 chapters devoted to the application of fungicides in the control and management of fungal diseases, which should be very helpful to the undergraduate and postgraduate students, researchers, teachers of microbiology, biotechnology, agriculture and horticulture.

Dr. D. Dhanasekaran,
Assistant Professor, Department of Microbiology, School of Life Sciences,
Bharathidasan University, India

Dr. N. Thajuddin,
Associate Professor and Head, Department of Microbiology,
Bharathidasan University, India

Dr. A. Panneerselvam,
Associate Professor, PG and Research Dept. of Microbiology and Botany,
AVVM Sri Pushpam College (Autonomous), India

Part 1

Fungicidal Compounds

1

Natural Fungicides Obtained from Plants

Juan A. Martínez
Universidad Politécnica de Cartagena
Spain

1. Introduction

Fungicides belong to a group of pesticides which inhibited fungal growth either causing damage to the cells or preventing the fungal development. As pesticides, they offer great economic and social benefits through the protection and preservation of materials, food and the prevention of diseases. Since pesticides are designed specifically to fight harmful or even dangerous life forms and therefore are toxic to them, they may present hazards to the environment by their potential effect upon non-target organisms, including humans, particularly when misused. The need to balance these benefits against the risks presents a challenge to the EPA (Environmental Protection Agency) unlike other chemicals.

The widespread use of synthetic fungicides (chemical fungicides) in agriculture is a relatively recent phenomenon, and most of the major developments have taken place during the last 60 years. It has been the major way of fungal disease control in the world during the past decades and nowadays it plays an important role in crop protection. However, the chemical residues are liable to remain on the plant or within its tissues following fungicidal treatment. Fungicide residues in plants and their fruits pose a great health risk to the consumer, led to the search for safe alternatives to synthetic fungicides. In days gone by, farmers often ignored or did not recognize the effect that fungal pathogens had on the yield and quality of their crops. Nowadays, however, the losses are unacceptable, and farmers rely on the use of chemicals to control fungal diseases. As a consequence, commercial fungicides have become an important component of the total agrochemical business, with world-wide sales in 1996 of about \$5.9 billion, equivalent to 18.9% of the total agrochemical market (Clough & Godfrey, 1998; Wood Mackenzie, 1997).

Economic losses caused by plant diseases are one of the main problems in crop management and postharvest storage. Some chemical fungicides are able to present curative and preventive action at the same time. Generally, they have a high persistent action; therefore they can develop a preventive action for long time. In spite of their effective action depend on dose and mode of plant application; it is known that they present a high effectively. This fact together to the simple application, handle and low cost, these chemical products have been used from approximately 1950. However, the massive and continue use of these products, together to the lack of controlled and adequate conditions for using them it have generated numerous problems such as new fungal pathogen strains resistant to fungicides and the increase of waste residues and the toxic effects for humans and animals.

The use of synthetic fungicides has been the major commercial means of postharvest decay control for several decades. However, the chemical residues that are liable to remain on the fruit or within its tissues following fungicidal treatment and the 1986 report from the US National Academic of Sciences (Research Council, Board of Agriculture, 1987) indicating that fungicide residues on food pose a great health risk to the consumer, led to the search for safe alternatives to synthetic fungicides. The fact that the effectiveness of synthetic fungicides has been reduced by the frequent development of resistance by the pathogens further highlighted the need for new substances and methods for the control of storage diseases. Naturally occurring plant products are important sources of antifungal compounds with low toxicity to mammals and safe to the environment which may serve as substitutes for synthetically produced fungicides. It was later suggested that these compounds might be developed either as products by themselves or used as starting point for synthesis (Knight et al., 1997).

2. Chemicals alternative to conventional fungicides

The continuing development of fungicide resistance in plant and human pathogens necessitates the discovery and development of new fungicides. Several preharvest and postharvest pathogens have developed resistance to commonly used fungicides. Hence, a wide range of chemicals has been evaluated for their potential for use as alternative to the current fungicides, e.g. plant extracts and some compounds obtained from plants. Discovery and evaluation of natural product based fungicides for agriculture is largely dependent upon the availability of miniaturized antifungal bioassay. Essentials for natural product bioassays include sensitivity to very small quantities, selectivity to determine optimum target pathogens, and adaptability assays should be relevant to potential pathogen target sites in the natural infection process of the host and applicable to the agrochemical industry. Bioassays should take advantage of current high throughput technology available to evaluate dose-response relationships, commercial fungicides standards, modes of action, and structure activity studies (Wedge & Smith, 2006).

Since the early 1970s, agriculture worldwide has struggled with the evolution of pathogen resistance to disease control agents. Increased necessities for repeated chemical applications, development of pesticide cross-resistance and disease resistance management strategies have characterized the use of agricultural chemicals to-date. As a consequence, producers are currently attempting to control agricultural pests with a decreasing arsenal of effective crop protection chemicals. In addition, the desire for safer pesticides with less environmental impact has become a major public concern. Particularly desirable is the discovery of novel pesticide agents from new chemical classes that are able to operate using different modes of action and, consequently, against plant pathogens with resistance to currently used chemistries. In this regard, evaluating natural products and extracts as a source of new pesticides is one strategy for the discovery of new chemical moieties that have not previously been created by synthetic chemists (Wedge & Smith, 2006).

In January 2009, European Parliament agreed the text of a Regulation to replace 91/414/EC which Plant Protection Products have been regulated up to now. The new Regulation, originally proposed in July 2006, maintained risk assessment but also introduced the concept of hazard assessment or 'cut-off criteria' for the approval of active substances. In addition it proposed the concept of comparative assessment and substitution for those active substances meeting certain 'less stringent' hazard criteria. According to this Regulation, the potential impact in terms of percentage reductions in available active fungicides from 31 to 43% not approved (Richardson, 2009).

Under this situation there have been investigated new methods to control plant diseases as an alternative way to chemical fungicide application either eliminating these synthetic compounds from agriculture or meanly controlling their use together with natural fungicide substances in an unique strategy plan called Integrated Plant Management (IPM).

3. Plants contain naturally fungicides

Plants possess a range of tools for combating fungal infections. Defence mechanisms include physical barriers and chemical defence includes many naturally occurring compounds and systems to prevent microbial attacks by using several types of pre-formed compounds and induced response plant defence compounds.

Plants have defensive response by means of some formed or pre-formed substances belong to the commonly called plant secondary compounds (secondary metabolism). Secondary metabolism is known as all those compounds that they are not involved in primary metabolic processes such as respiration and photosynthesis. However, many compounds do have roles intermediate between primary and secondary metabolism. For instance, some plant secondary compounds like plant growth substances are associated with plant growth and development (Seigler, 1998). In fact, secondary metabolites are compounds with a restricted occurrence in taxonomic groups that are not necessary for a cell (organism) to live with its environment, ensuring the survival of the organism in its ecosystem (Verpoorte, 2000).

There has been much speculation over many years about why organisms take the trouble to produce secondary metabolites or natural products which often have considerable structural complexity. All such structures serve the producing organisms by improving their survival fitness (Williams et al., 1989). In this case, it seems likely that the fungi and bacteria which produce strobilurins, oudemansins, myxothiazols and/or melithiazols, which are substances with antifungal properties synthesized by fungi or bacteria give themselves an advantage by doing so, presumably by suppressing other organisms which compete for nutrients in the same environment (Williams et al., 1989).

Secondary metabolites are now thought to mediate chemical defence mechanisms by providing chemical barriers against animal and microbial predators. Plants produce numerous chemicals for defence and communication, and can elicit their own form of offensive chemical warfare by targeting the proliferation of pathogens. These chemicals may have general or specific activity against key target sites in bacteria, fungi and viruses.

These substances are classified into different ways: based on chemical characteristics, plant origin or biosynthetic origin. From a chemical point of view, the compounds can be divided into a number of groups based on typical characteristics such as alkaloids and phenolic compounds, principally, Based on biosynthetic origin, there are terpenoids, phenylpropanoids, polyketides and other compounds belong to other minor groups must be considered. Therefore, the total number of these compounds is very elevated. A great number of those are used to prevent animal consumption. However, many of these defensive substances are toxic for the plants themselves and they are inactivated as glycosides or by polymeration or they are confined in intercellular spaces.

4. Classification of fungicides in plants

Natural compounds with antifungal activity belong to the secondary metabolism in plants as above mentioned. This type of metabolism can be classified into three important groups:

1. Preformed compounds (preformed resistance in plants)

Substances naturally forming part of the plant cell composition have antifungal properties. Phytonzide is normally called to all the substances naturally occurring in plants which have antimicrobial properties inhibiting the growth and development of bacteria and fungi. These compounds are present in the plants in their biologically active form. These include phenolics, terpenoids and steroids. Some preformed compounds are directly toxic, while others exist as conjugates such as glycosides that are not directly toxic but become toxic following disruption of the conjugate. On the other hand, some plant preformed compounds are toxic as glycosides, but lose toxicity when deglycosylated.

2. Inducible preformed compounds (inducible preformed resistance in plants)

Some substances which are normally present in healthy tissues but they can be further induced in the host in response to pathogen attack or presence, as well as to other stresses. On the other hand, there are compounds that occur as inactive precursors and are activated in response to pathogen attack. These preformed compounds or phytoanticipins differ from the inducible phytoalexins that are synthesized from remote precursors in response to pathogen attack (Prusky, 1997).

3. Phytoalexins - Induced inhibitory compounds

Inducible resistance mechanisms are active, energy-requiring systems typified by specific recognition of an invader that ultimately leads to the production of proteins or metabolites that are antagonist to the invader. Such active resistance mechanisms are usually referred as the hypersensitive response (HR).

In this group, we can include other induced inhibitory compound like pathogenesis-related proteins (PR), active oxygen species and lectins, although they are not phytoalexins. Therefore they are in a chapter apart (see below).

4.1 Preformed compounds (preformed resistance in plants)

Preformed resistance involves the presence in healthy tissues of biologically active, low-molecular-weight compounds whose activity affords protection against infection (Barkai-Golan, 2001). These compounds belong to the secondary metabolism which includes all the compounds belong to the three categories described above. It could also be included other defensive physical structures (physical resistance) like structural barriers (cuticle, epicuticular wax, cell wall and processes like lignification and suberization of cells at the wound areas).

4.1.1 Plant extracts

Plant extracts is a group of substances extracted from different parts of plants which contain a great many of compounds with antimicrobial properties. These compounds can be obtained from roots, barks, seeds, buds, leaves and fruits. Aromatic plants are especially rich in those substances and they have been used in foods as flavouring agents (Davidson, 1997). Most of these compounds are terpenes and have fungicide properties are described below.

Plant extracts can be directly used or substances responsible for the antimicrobial properties can be isolated. A new extract with potent antifungal properties is the extract obtained from *Aloe vera*, which has been found to have antifungal activity against four common postharvest pathogens: *Penicillium digitatum*, *P. expansum*, *Botrytis cinerea* and *Alternaria alternata* (Barkai-Golan, 2001).

Among the compounds belonging to plant extracts, some important examples have been selected and explained below. They can be classified in other groups, mainly as phenolic compounds or essential oils.

Few studies have focused on the mechanism by which plant extracts and their essential oils inhibit microorganisms. The terpenes present of the essential oils are the primary antimicrobials. Many of the most active terpenes, for example, eugenol, thymol and carvacrol, are phenolic in nature. Therefore, it would seem reasonable that their modes of action might be related to those of other phenolic compounds. Essential oils may inhibit enzyme systems in yeasts, including those involved in energy production in cells and synthesis of their structural components.

Eugenol [2-methoxy-4-(2-propenyl)-phenol] is a substance obtained from clove (*Syzygium aromaticum* L.), which contain 95% of eugenol as the main volatile oils. Cinnamic aldehyde [3-phenyl-2-propenal] is another antimicrobial substance obtained from cinnamon (*Cinnamomum zeylanicum* J Presl.). This plant also contains eugenol (only 8% of volatile oil and 75% of the first one). Both eugenol and cinnamon inhibits spores of *Bacillus anthracis*, while eugenol and aqueous clove infusions inhibited outgrowth of germinated spores of *B. subtilis* in nutrient agar (Davidson, 1997). In addition, clove extracts has antifungal properties. Thus, it inhibited growth initiation for over 21 days at 25°C of *Aspergillus* and *Penicillium* species. Cinnamon was the next most effective spice, inhibiting three *Penicillium* species for over 21 days (Davidson, 1997).

The genus *Allium* contains various compounds with resistance to fungal disease. Some are constitutive inhibitors, such as the phenolic compounds catechol that is present in the outer bulb layers of pigmented *Allium cepa* (onion) cultivars were it confers resistance to *Colletotrichum* (Link & Walker, 1933). Onion also produces a class of cyclopentane phytoalexin upon pathogen infection, designated tsibulins (Dmitriev et al., 1990), which accumulate in bulb scales at infection sites during incompatible interactions with *B. cinerea*. Tsibulins inhibited spore germination and germ tube elongation of *B. cinerea in vitro*. The ED_{50} values were lower than the actual phytoalexin content in bulb scale spots, were *B. cinerea* lesion formation was restricted (Dmitriev et al., 1990). Only little accumulation of cyclopentane phytoalexins was observed in onion bulb scales during infection by the specialized pathogen *B. allii* (Dmitriev et al., 1990). *B. allii* seems to either suppress tsibulin accumulation; analogous to the interaction of *B. narcissicola* and its host narcissus of actively degrade tsibulins, as discussed for other antifungal compounds above.

Isothiocyanates derive from glucosinolates in cells of plants of the Cruciferae, or mustard family (cabbage, kohirabi, Brussels sprout, cauliflower, broccoli, kale, horseradish, mustard, turnips, rutabaga) are potent antifungal and antimicrobial agents (Davidson, 1997). These compounds are formed from the action of the enzyme myrosinase (thioglucoside glucohydrolase; EC 3.2.3.1) on the glucosinolates when plant tissues are injured or disrupted. Common isothiocyanates include allyl, ethyl, methyl, benzyl and phenyl. The compounds are inhibitory to fungi, yeasts and bacteria in the range from 0.016 to 0.062 μg mL^{-1} in the vapour phase or 10 to 600 μg mL^{-1} in liquid media (Davidson, 1997). The mechanism by which isothiocyanates inhibit cells may involve enzymes by direct reaction with disulfide bond or through thiocyanate anion reaction to inactive sulphydryl enzymes. The isothiocyanates may act as uncouplers of oxidative phosphorylation (Davidson, 1997).

Purified metabolic extract of carrots and found it to be antifungal activity against the yeast *Candida lambica* at concentrations ranging from 55 to 220 mg mL^{-1} and control some bacteria as well (Davidson, 1997). This inhibition was due to the compounds dodecanoic (monolaurin) and pentadanoic acids. An extract of carrots inhibited sporulation and aflatoxin production by *Aspergillus parasiticus* (Davidson, 1997). The inhibitor was not identified but was determined not to be 6-methoxymellein, *p*-hydroxybenzoic acid, or falcarindiol but rather to be part of the volatile carrot seed oil, a mixture of terpenoid compounds (Davidson, 1997).

Extracts of onions (Allium cepa L.) and garlic (Allium sativum L.) also contain volatile antimicrobial substances which inhibit Aspergillus flavus, A. parasiticus, Candida albicans, Cryptococcus, Penicillium, Rhodotorula, Saccharomyces, Torulopsis and Trichosporon species. The main compounds obtained from this extract is allicin (diallylthiosulfinate; thio-2-propene-1-sulfinic acid-S-allyl ester). This substance is produced by tissues when the cells are damaged and the enzyme (alliin-lyase; E.C. 4.4.1.4) act on the substrate (S-allyl-L-cysteine sulphoxide) because of enzyme and substrate are separated by cellular membranes inside the cells (Davidson, 1997). This has been suggested as a plant fungicide which undergoes thioldisulphide exchange reactions with free thiol groups in proteins and it is thought that this is the basis of its antimicrobial action. At 50 μg mL^{-1}, allicin in garlic juice inhibited the germination of sporangia and cysts and subsequent germ tube growth by Phytophthora infestans both in vitro and in vivo on the leaf surface. Similarly, in growth room experiments at concentrations from 50–1,000 μg mL^{-1}, allicin reduced the severity of cucumber downy mildew caused by Pseudoperonospora cubensis by approximately 50–100% (Davidson, 1997). The antimicrobial sulfur compounds of onions contain phenolic compounds protocatechuic acid and catechol, which contribute to their antimicrobial activities. Garlic oil inhibits ethanol production by Saccharomyces cerevisiae and delay sporulation of Hansenula anomala and Loderomyces elongisporus.

Resin from the flowers of the hop vine (*Humulus lupulus* L.) is used in the brewing industry for imparting a desirable bitter flavour to beer and about 3 to 12% of the resin is composed of α-bitter acids, including humulone (humulon), cohumulone and adhumulone and also β-bitter acids, including lupulone (lupulon), colupulone, xanthohumol and adlupulone (Davidson, 1997). Both types of bitter acids possess antimicrobial activity against gram-positive bacteria and fungi. These acids are considered to be the primary quality feature of hops, because they are a measure of the bitter potential and due to their antimicrobial and preservative properties.

4.1.2 Essential oils

Essential oils are highly volatile substances synthesized and stored in glandular trichomes of odiferous plants. The term essential oil refers not only to the complex oils isolated from the plant, but also to their constituent compounds. Volatile components of essential oils are lipophilic molecules that volatilize at low temperatures.

Culinary aspects of many plants are based on essential oils, although other uses have been attributed (e.g., medicinal practices). Plants of the family *Laminaceae* (mints) present typically these oils in their composition (e.g., *Mentha* sp., *Monarda* sp., *Origanum* sp., *Rosmarinus* sp., *Thymus* sp.). Many of these essential oils isolated from mints are terpenes. However, other plants belonging to other families present essential oils as well, e.g. *Rutaceae*. They are mainly terpenes as well. The antimicrobial effects of two important plants: oregano (*Origanum vulgare* L.) and thyme (*Thymus* sp.) have been attributed to their essential oils which contain the terpenes carvacrol [2-methyl-5-(1-methylethyl) phenol] and thymol [5-methyl-2-(1-methylethyl)-phenol]. Firstly, the antimicrobial activities of oregano and thyme were proposed as antibacterial compounds. However, in recent works they have been proposed as antifungal properties as well. Essential oils obtained from thyme (*Thymus vulgaris* L.) and juniper (*Juniperus communis* L.) have been proved as fungicides against *P. digitatum* and *P. italicum* (Martínez & Hernández, 2007). They are actives against a great number of bacteria, moulds and yeasts.

In general, essential oils and plant extracts are sources of antifugal activity against a wide range of fungi (Grange & Ahmed, 1988). A rapid assay to determine antifungal activity in both plant extracts and essential oils have been shown by Barkai-Golan (2001) concluding that among 345 plant extracts analyzed, 13 showed high levels of activity against *B. cinerea*, which served as a test fungus. Garlic (*Allium* sp.) and pepper (*Capsicum* sp.) seem to be the major plants showing the highest persistent antifungal activity (Barkai-Golan, 2001). In addition, among the 49 essential oils tested, those of palmarosa (*Cymbopogon martinii* Roxb.) and red thyme (*Thymus zygis* L.) showed the greatest inhibitory effect on *B. cinerea* spore germination at the lowest concentration. The next best inhibitors were essential oils of clove buds (*Syzygium aromaticum* L.) and cinnamon leaf (*Cinnamomun zeylanicum* J Presl.). The most frequently occurring constituents in essential oils showing high antifungal activity were: *D*-limonene, cineole, α-pinene, β-pinene, β-myrcene and camphor. The fungicidal activity of the individual components, singly and in combination, is being studied (Barkai-Golan, 2001). Essential oil derived from another species of *Thymus, T. capitatus* L., markedly reduced development of *B. cinerea* in inoculated mandarin fruit when applied as a vapour. Scanning electron microscope observations indicated a direct damaging effect of the thyme oil on fungal hyphae (Arras & Piga, 1994).

In the family *Rutaceae,* several natural antifungal compounds have been isolated from lemon peel and have been identified. Among these are citral (lemonal), limetin, 5-geranosy-7-methoxycoumarin and isopimpeneyin. The antifungal activity of citral (terpenoids) has been exhibited against various fungi (Barkai-Golan, 2001). Introducing *Penicillum digitatum* spores into the oil glands of the peel of young lemon fruits revealed that citral is the main factor within the glands responsible for the inhibition of the pathogen growth (Barkai-Golan, 2001). In postharvest studies in lemon fruits it have been concluded that the changes in citral concentration in the lemon peel may determine the fruit sensitivity to postharvest decay. In parallel with citral decline the flavedo of yellow lemon exhibited an increased level of neryl-acetate, which is a monoterpene ester which exhibits no inhibitory activity against *Penicillium digitatum*, and which, in low concentration (less than 500 ppm) may even stimulate development of the pathogen. The increase in the level of monoterpenes in the peel may explain, at least in part, the stimulatory effect on *P. digitatum* and *P. italicum* development, of the etheric oil derived from stored citrus fruit. The resistance of citrus peel to inoculations applied between the oil glands, found in early studies by Schiffmann-Nadel and Littauer (1956), was attributed to another factor, which does not change with fruit ripening and may be related to its chemical composition or its anatomic structure (Barkai-Golan, 2001).

Tulip bulbs and pistils contain high concentration of fungitoxic compounds, identified as lactones and termed tulipalins (van Baarlen et al., 2004). These compounds are sesquiterpenes, an important essential oils group. Tulipalin A is found in bulbs, whereas aerial plant parts contain mainly tulipalin B. Tulipalins A and B are stored in a glycosylates form, named tuliposides A and B. Tuliposides are less toxic to fungi than the corresponding unglycosylated lactones and are lactonized into tulipalins when the pH exceeds 5. All *Botrytis* species tested by Schönbeck & Schroeder, (1972) so far are sensitive to pure tulipalin A, except for the tulip-specific pathogen *B. tulipae*. Higher doses (7-10 μM, more than 30 μM for *B. tulipae*) are lethal to *Botrytis* spp. Conidia are at least three times more sensitive to tulipalin A than mycelium.

Among 22 naturally occurring monoterpenoids screened for activity against *Botrytis cinerea* and *Monilia fructicola* causing grey mould and brown rot diseases, the phenolic monoterpenoids carvacrol and thymol were the most effective in inhibiting spore germination and mycelial growth of the pathogens (Narayanasamy, 2006; Rong & Ting, 2000).

Rhizoctonia solani is sensitive to mint oils. Growth of *R. solani* was reduced when grown on culture medium amended with oils of *Monarda didyma* or when exposed to the vapours in a microatmosphere test (Gwinn et al., 2010). When dispersed in fungal culture media, oils from thyme (*Thymus vulgaris* L.), *Salvia fruticosa* Mill. and peppermint (*Mentha* x *piperita* L.), also reduced growth of *R. solani*. Growth of *R. solani* was completely suppressed by the phenolic monoterpene, carvacrol at a concentration of 100 mg mL^{-1} and by its structural isomer, thymol, at a slightly higher concentration (150 mg mL^{-1}) (Gwinn et al., 2010). Carvacrol was the major constituent of essential oils collected from *Thymbra spicata* L., a member of the mint family that grows wild in the eastern Mediterranean; oil extracted from *T. spicata* completely suppressed growth of *R. solani*. Fungicidal activity of thyme oils has been correlated with thymol content (Gwinn et al., 2010).

Damping-off of seedlings caused by species of *Pythium* and *Rhizoctonia* can reduce number and quality of tomato seedlings. Losses due to *Pythium myriotylum* were reduced when monarda herbage (ground, dried leaves) from some varieties of *Monarda didyma* L., *M. clinopodia* L. and *M. menthifolia* L. were added to greenhouse growing medium at a rate of 10% (v/v) (Gwinn et al., 2010). Thymol was the major essential oil component of 'Croftway Pink', the most effective monarda herbage treatment for reducing losses due to *P. myriotylum*.

Bettiol et al. (2011) studied the effect of several essential and fixed oils (a mixture of not volatile esters of fatty acids) on the control *in vitro* and *in vivo* of green mould of oranges caused by *Penicillium digitatum*. They tested oils extracted from *Pogostemon cablin* Benth., *Mentha arvensis* L., *Cymbopogon citratus* Otto Stapf., *Ocimum basilicum* L., *Rosmarinus officinalis* L., *Lippia sidoides* Cham., *Zingiber officinale* Rosc., *Citrus aurantifolia* L., *Piper aduncum* L., *Allium sativum* L., *Copaifera langsdorffii* Desf., *Eucaliptus* spp. and *Azadirachta indica* A. Juss. The oils at 10,000 and 100,000 µL L^{-1} controlled green mould and inhibited spores germination and mycelial growth in a similar level as compared to the fungicide treatment. However, treatment with oil in concentrations higher than 10,000 µL L^{-1} caused ring damage and changed fruits flavour, which makes its implementation impractical in high concentrations.

4.1.3 Phenolic compounds

Phenolic compounds are those substances that possess an aromatic ring with one or more hydroxyl groups and can include functional derivates. Some of the important phenolic compounds include alkyl esters of parabens, phenolic antioxidants (e.g., BHA and TBHQ) and certain of the terpene fraction of the essential oil (e.g., thymol, carvacrol, eugenol and vanillin). Sesquiterpenes, with monoterpenes, are an important constituent of essential oils in plants. Simple phenolic compounds include monophenols (e.g.., cresol), diphenols (e.g., hydroquinone) and triphenols (e.g., gallic acid). Gallic acid occurs in plants as quinic acid esters or hydrolyzable tannins (tannic acid) (Davidson, 1997). These compounds are naturally present in plants. They have antibacterial properties but some of them have antifungal properties as well, for example against *Penicillium* sp., *Rhizopus* sp. and *Geotrichum candidum*.

The mechanism of phenolic compounds centres on their effects on cellular membranes. Simple phenols disrupt the cytoplasmic membrane and cause leakage of cells. Phenolics may also inhibit cellular proteins directly. However, some researchers have concluded that phenolic compounds may have a great many of mechanisms of action and that there may be several targets which lead to inhibition of microorganisms (Davidson, 1997).

Phenolic compounds have long been implicated in disease resistance in many horticultural crops (Barkai-Golan, 2001). Some occur constitutively and are considered to function as preformed or passive inhibitors, while others are formed in response to the ingress of pathogen and their appearance is considered as part of an active defence response (Barkai-Golan, 2001). They contribute to resistance through their antimicrobial properties; with elicit direct effects on the pathogen, or by affecting pathogenicity factors of the pathogen. However, they may also enhance resistance by contributing to the healing of wounds via lignifications of cell walls around wound zones. The cells surrounding the wound can produce and deposit lignin and suberin in their walls (Eckert, 1978). This compound protects the host from pathogen penetration or from the action of cell-wall degrading enzyme produced by the pathogen. As a result of wounding, the production of antimicrobial polyphenolic compounds can also contribute to wound protection. Phytoalexins are other toxic compounds can be formed at the wound area following inducement by initial infection. In this way, the inoculation of potatoes tubers with *Fusarium sambicinum*, the fungal pathogen of potato dry rot, resulted in an increase in phenolic acids suggesting that phenolic acid biosynthesis was induced. Following such inducement free phenolic acids are removed as they are converted into lignin or are joined onto cell walls (Barkai-Golan, 2001).

Studies with cultured carrot cells indicated that phenolic compounds with low molecular weight, which are a link in lignin biosynthesis and free radicals produced during its polymerization may take part in resistance inducement by damaging fungal cell membranes, fungal enzymes or toxins (Barkai-Golan, 2001). Accumulation of phenolic compounds and callose deposition in cell walls of young tomato fruits, following inoculation with *B. cinerea*, were found to arrest fungal development thus retarding or preventing decay (Barkai-Golan, 2001). The mechanism by which phenolic compounds accumulate in the host is not yet clear, but research carried out with wheat leaves suggested that the chitin in the fungal cell walls acts as a stimulator to lignifications in the leaves (Pearce & Ride, 1982, as cited in Barkai-Golan, 2001).

In vitro assays have shown that the phenolic compounds, chlorogenic acid and ferulic acid directly inhibited *Fusarium oxysporum* and *Sclerotinia sclerotiorum* respectively. Benzoic acid derivatives have been shown to be the best inhibitors of some of the major postharvest pathogen, such as *Alternaria* spp., *B. cinerea, Penicillium digitatum, S. sclerotiorum* and *F. oxysporum* (Latanzio et al., 1995, as cited in Barkai-Golan, 2001).

The principal phenol in the pear fruit epidermis and subtending cell layers are chlorogenic and caffeic acids. The concentration of these phenols decline as fruit mature, with a corresponding increase in fruit susceptibility to the brown rot fungus (*Monilinia fructicola*). In fact, fungal spore germination or mycelial growth were not inhibited by concentrations similar to or exceeding those that occur in the tissue of inmature, resistant fruit.

Tannins, which are polyphenols, have been described by Byrde et al. (1973, as cited in Barkai-Golan, 2001) as inhibitors of polygalacturonase (PG) activity of *Sclerotinia fructicola* (anamorph of *Monilinia fructicola*) in apples and other pathogen/host combinations. Tannins of young banana fruits (Green & Morales, 1967, as cited in Barkai-Golan, 2001) and benzylisothiocyanate in unripe papaya fruits (Patil et al., 1973, as cited in Barkai-Golan, 2001) are additional examples of in-fruit toxic compounds.

Proanthocyanidins are widely distributed in the plant Kingdom and are constitutive components in a number of discrete tissues in most plant organs. The chemical structure and composition of proanthocyanidins vary among plant species, organs and also with the stage of organ development. A special type of tannins, proanthocyanidins (condensed tannins) are polymeric flavonoids that results from the condensations of two or more derivates of flavan-3,4-diol. Plant proanthocyanidins maintain *B. cinerea* in a quiescent stage, leading to delayed development of symptoms. The transition from quiescence into expansion is triggered during host senescence or ripening and occurs at a less senescent or ripe stage in susceptible varieties. Prolonging the quiescence of *B. cinerea* infections by increasing the proanthocyanidin content would reduce losses to grey mould, especially after harvest. However, proanthocyanidin levels are constitutive and are not known to be subject to modulation by external elicitors. Moreover, knowledge is lacking on the genes and enzymes involved in the subtle modifications of proanthocyanidin structure that affect their biological activity. The use of proanthocyanidin content as an indicator of grey mould resistance for the selection of cultivars with improved shelf-life has been suggested for grape and strawberry (van Baarlen et al., 2004).

Oleuropein is another phenolic compound found in olive leaf from the olive tree. This substance inhibits *Rhizopus* sp. and *Geotrichum candidum* (Davidson, 1997).

Another class of inhibitors of cell wall-degrading enzymes comprises PG-inhibitory proteins present in both infected and uninfected plant tissue (Barkai-Golan, 2001). Research carried out with pepper fruit has shown that cell wall protein of the host inhibited pectolytic enzyme production by *Glomerella cingulata,* whereas the pectolytic activity of *Botrytis cinerea* was much less affected by these proteins (Brown & Adikaram, 1983, as cited in Barkai-Golan, 2001). The fact that *B. cinerea* can rot an immature pepper fruit whereas *G. cingulata* can attack only the ripened fruit, suggested that protein inhibitors might play a role in the quiescent infection of pepper fruit by *Glomerella.*

A new protein inhibitor that may be involved in the inhibition of enzymes necessary for microbial development was isolated from cabbages (Lorito et al., 1994, as cited in Barkai-Golan, 2001); it significantly inhibited the growth of *B. cinerea* by blocking chitin synthesis, so causing cytoplasmic leakage. Several studies supported the theory that natural protein compounds within the plant tissue may act as inhibitors of pathogen enzymes and that these inhibitors may be responsible for the low levels of PG and PL found in infected tissue (Barkai-Golan, 2001). Recent studies show a close correlation between the changes in the level of epicatechin in the peel of avocado fruit and the inhibition of pectolytic enzyme activity of *Colletotrichum gloesporioides* (Wattad et al., 1994, as cited in Barkai-Golan, 2001).

4.1.4 Hydroxycinnamic acids

Hydroxycinnamic acids can be considered as phenolic compounds and are a class of polyphenols which are hydroxyl derivates of cinnamic acid and include caffeic, chlorogenic, *p*-coumaric, ferulic and sinapinic acids. They occur frequently as esters and less often as glucosides. Many of the studies with hydroxycinnamic acids have involved their antifungal properties. It has been reported that 500 µg mL^{-1} of caffeic acid and 1,000 µg mL^{-1} of chlorogenic acid inhibit some species of *Fusarium* (Davidson, 1997). It has been shown that ferulic acid at 5.0 mg 26 mL^{-1} inhibits aflatoxin B_1 and G_1 production of *Aspergillus flavus* by approximately 50% and that of *A. parasiticus* by 75%. Salicylic and *trans*-cinnamic acids totally inhibit aflatoxin production at the same 5.0 mg 26 mL^{-1} (Davidson, 1997). After a study of the effects of caffeic, chlorogenic, *p*-coumaric and ferulic acids at pH 3.5 on the growth of *Saccharomyces cerevisiae* was concluded that caffeic and chlorogenic acid had little effect on the organism at 1,000 µg mL^{-1} (Davidson, 1997). In the presence of *p*-coumaric acid, however, the organism was completely inhibited by the same concentration. Ferulic acid was the most effective growth inhibitor tested. At 50 µg mL^{-1}; this compound extended the lag phase of *S. cerevisiae* and at 250 µg mL^{-1}, growth of the organism was completely inhibited. The degree of inhibition was inversely related to the polarity of the compounds.

4.1.5 Flavonoids

Flavonoids are a special group of phenolic compounds and some aspects of this group have been described above. The flavonoids consist of catechins and flavons, flavonols and their glycosides. Proanthocyanidins or condensed tannins are polymers of favan-3-ol and are found in apples, grapes, strawberries, plums, sorghum and barley (Davidson, 1997).

Benzoic acid, proanthocyanidins and flavonols account for 66% of cranberry microbial inhibition against the yeast *Saccharomyces bayanus*, with the latter two being the most important.

4.1.6 Plant growth substances and regulators

At the moment, some differences between plant growth substances and plant growth regulators can be considered (Arteca, 1996). Plant growth substances (PGS) or commonly called phytohormones are synthesized by plants whereas plant growth regulators (PGR) are those organic compounds other than nutrients (materials which supply either energy or essential mineral elements), which in small amounts promote, inhibit, or otherwise modify any physiological process in plants. Arteca (1996) used the term PGR to designate synthetic compounds and the term PGS for naturally occurring compounds produced by the plant.

Plant growth substances and regulators seem to be a group of important substances which they control the growth of plants and can have antifungal properties as well. It is very well known that phytohormones control all the physiological process in plants growth and development; therefore, they interfere with the influence of pathogens to attack plants. However, some of PGS and PGR have directly effect as fungicides. In this way, Martínez et al. (2011) reported that 100 mg of indole-3-acetic acid (IAA) delayed the *in vitro* mycelial growth of several *Botrytis cinerea* isolates obtained from potted plants in an isolate-dependent manner (Fig. 1).

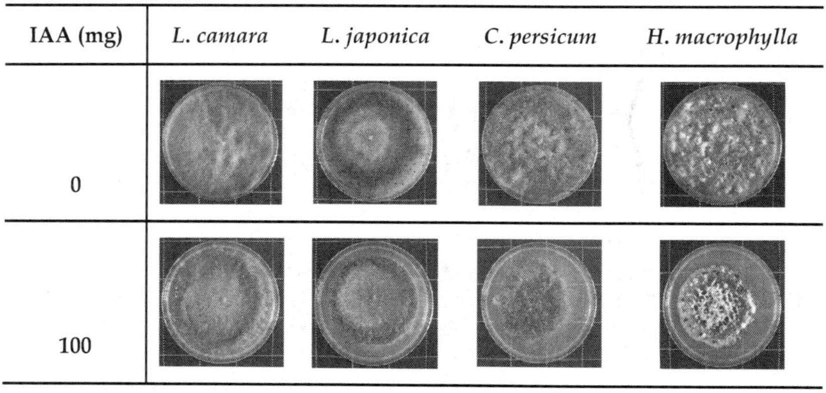

Fig. 1. Mycelia of four isolates of *B. cinerea* obtained from different potted plants (*Lantana camara, Lonicera japonica, Cyclamen persicum*, and *Hydrangea macrophylla*) grown *in vitro* on potato dextrose agar (PDA) at 26°C after 35 days with 0 mg or 100 mg of IAA (adapted from Martínez et al., 2011)

The synthesis of plant growth substances in many fungi has been demonstrated, but now, the synthesis pathways have been only established in a few cases. Moreover, new substances seem to be important effect as growth regulator in fungi. Until now, indol-3-acetic acid, gibberellic acid, abscisic acid and ethylene, important hormones in plants, have been discovered in fungi like *Botrytis cinerea* (Sharon et al., 2004).

Jasmonic acid and its derivates, mainly methyl-jasmonate are present in most of the plants playing a plant growth substance role (Arteca, 1996). Jasmonic acid and the corresponding methyl ester are fragrant constituents of the essential oils of *Jasminum* sp. as well as other perfumery plants. These plant growth substances are now under study for evaluating their effects on citrus fruit decay and the decrease of chilling injury in postharvest. These substances applied as vapour, drencher or bath in citrus packinghouse at very low concentrations could be considered as an alternative to control decay in citrus industry in Spain (Monterde et al., 2002). Droby et al. (1999) found that methyl-jasmonate had antifungal activity against *Penicillium digitatum*, the principal fungus causing decay in citrus fruits.

The mode of action consists of that in response to wounding or pathogen attack, fatty acids of the jasmonate cascade are formed from membrane-bound linolenic acid by lipoxygenase-mediated peroxidation (Vick & Zimmerman, 1984). Linolenic acid is thought to participate in a lipid-based signalling system where jasmonates induce the synthesis of a family of wound-inducible defensive proteinase inhibitors (Farmer & Ryan, 1992) and low and high molecular weight phytoalexins such as flavonoids, alkaloids and terpenoids.

In relation with other PGS, treatment of celery prior to storage with gibberellic acid (GA_3) in juvenile plant tissue resulted in decay suppression during 1 month of storage at 2°C, although GA_3 does not have any effect on fungal growth *in vitro* (Barkai-Golan, 2001). It was suggested that the phytohormone retards celery decay during storage by slowing down the conversion of (+) marmesin to psoralens, thereby maintaining the high level of (+) marmesin and low levels of psoralens and, thus increasing celery resistance to storage pathogens (Barkai-Golan, 2001). Martínez & Bañón, (2007) and Martínez et al. (2007) demonstrated that GA_3 has some effects on growth and development of fungal structures of the *Botrytis cinerea* isolates obtained from potted plants, but this phytohormone either increased the fungal development or had no effect on the growth, depending on the isolate.

4.1.7 Acetaldehyde and other volatile compounds

Acetaldehyde is a natural volatile compound produced by various plant organs and accumulates in fruits during ripening. It has shown fungicidal properties against various postharvest pathogens (Barkai-Golan, 2001). It is capable of inhibiting both spore germination and mycelial growth of common storage fungi and the development of yeast species responsible for spoilage of concentrated fruit juices. It has been reported to inactivate ribonuclease and to bind other proteins but the mechanism of aldehyde toxicity to fungal spores is still unknown (Barkai-Golan, 2001).

Fumigation apples, strawberries with acetaldehyde reduced decay cause by *Penicillium expansum*, *Rhizopus stolonifer* and *Botrytis cinerea* (Barkai-Golan, 2001). Avvisar & Pesis (1991 as cited in Barkai-Golan, 2001) showed that 0.05% acetaldehyde applied for 24 h suppressed decay caused by *B. cinerea*, *R. stolonifer* and *Aspergillus niger*.

Injuries resulting from acetaldehyde vapours have been reported for various products, such as cultivars of apples, strawberries, grapes, lettuce and carrot tissue cultures (Barkai-Golan, 2001). The efficacy of acetaldehyde vapours and of a number of other aliphatic aldehydes, produced naturally by sweet cherry cv. Bing, was evaluated in *P. expansum* inoculated fruits

(Mattheis & Roberts, 1993, as cited in Barkai-Golan, 2001). High concentrations of acetaldehyde, propanal and butanal suppressed conidial germination but resulted in extensive stem browning and fruit phytoxicity, which increased with the aldehyde concentration. On the other hand, stem quality is less of a concern for fruits intended for processing and for this purpose aldehyde fumigation may present an alternative to the use of synthetic fungicides.

Various volatiles (benzaldehyde, methyl salicylate and ethyl benzoate) have been recorded as growth suppressors. Nine out of 16 volatile compounds occurring naturally in peach and plum fruits greatly inhibited spore germination of *B. cinerea* and *Monilinia fructicola* (Barkai-Golan, 2001). The volatiles most effective in inhibiting spore germination were benzaldehyde, benzyl alcohol, γ-caprolactone and γ-valerolactone. Of these, benzaldehyde was active at the lowest concentrations tested and completely inhibited germination of *B. cinerea* spores at concentrations of 25 $\mu L\ L^{-1}$ and germination of *M. fructicola* spores at 125 $\mu L\ L^{-1}$. Ethyl benzoate was fungicidal against *Monilinia* sp. and fungistatic against *Botrytis* sp.

4.1.8 Ethanol

Ethanol is a substance produced in fruits (Barkai-Golan, 2001). It has been tested for control of brown rot and Rhizopus rot in peach fruits with varying degrees of success. Recently, the effects of ethanol solutions, at concentration of 10-20%, were evaluated for the control of postharvest decay of citrus fruits, peaches and nectarines (Barkai-Golan, 2001).

4.1.9 Hinokitiol

Hinokitiol is a natural volatile extracted from the root of trunk of Japanese cypress (*Hiba arborvitae*) with outstanding antifungal properties (Barkai-Golan, 2001). Hinokitiol reduced spore germination in *Monilia fructicola, Rhizopus oryzae* and *Botrytis cinerea.* In parallel, this volatile prevented decay of commercially harvested peaches in which more than 40% of the treated fruit developed brown rot caused by *M. fructicola* (Sholberg & Shimizu, 1991, as cited in Barkai-Golan, 2001).

Today melons are coated with wax containing imazalil (200 ppm) and under the above conditions, this leaves fungicidal residue above the level approved in some countries (0.5 ppm). Introducing hinokitiol into wax was also found to control decay during cold storage, caused mainly by *Alternaria alternata* and *Fusarium* spp. without any phytotoxic effects.

4.1.10 Glucosinolates

Glucosinolates are a large class of compounds that are derived from glucose and an amino acid. They occur as secondary metabolites of almost all plants of the order Brassicales, especially are present in Cruciferae's family. The studies carry out at the moment have been done with encouraging results (Barkai-Golan, 2001). When cells of plant tissues that metabolize glucosinolates are damaged, these compounds come into contact with the enzyme myrosinase, which catalyzes hydrolysis. The antifungal activity of six isothiocyanates has been tested on several postharvest pathogens *in vitro* and *in vivo* on artificially inoculated pears with encouraging results.

4.1.11 Latex

Latex is a stable dispersion of naturally occurring polymer microparticles in an aqueous medium. It is found in 10% of all angiosperms. This complex emulsion consisting of alkaloids, starches, sugars, oils, tannins, resins and gums that coagulates on exposure to air. It is also rich in enzymes like proteases, glucosidases, chitinases and lipases. It has been demonstrated that this substance is a source of natural fungicides (Barkai-Golan, 2001) which is regarded as both safe and effective against various diseases of banana, papaya and other fruits. The water-soluble fraction of papaya latex can completely digest the conidia of many fungi, including important postharvest pathogens (Indrakeerthi & Adikaram, 1996). Other latex extracted from several plants showed a strong antifungal activity against *Botrytis cinerea, Fusarium* sp. and *Trichoderma* sp. (Barkai-Golan, 2001).

4.1.12 Steroids

Steroids are terpenes with a particular ring structure composed by a specific arrangement of four cycloalkane ring that are joined to each other. Saponins are plant steroids, often glycosylated.

The saponin, α-tomatine, is a secondary metabolite produced in tomato leaves unripe fruits (Friedman, 2002, as cited in van Baarlen et al., 2004). It is also present in high concentration in the peel of green tomatoes. It is a potent antifungal and insecticidal compound that interacts with sterols in membranes (van Baarlen et al., 2004), It Inhibit mycelial growth of *B. cinerea* while not affecting germination of conidia. This substance also affects other fungal pathogens and its involvement in the development of quiescent infection has been suggested (Verhoeff & Liem, 1975, as cited in van Baarlen et al., 2004). Tomatine presumably is toxic due to its ability to bind to 3-β-hydroxy sterols in fungal membranes (Steel & Drysdale, 1988, as cited in Barkai-Golan, 2001). Most tomato pathogens, on the other hand, can specifically degrade tomatine and detoxify its effects through the activity of tomatinase (Barkai-Golan, 2001).

4.2 Inducible preformed compounds (inducible preformed resistance in plants)

Over the last two decades several studies have indicated that preformed antifungal compounds, which are normally present in healthy plant tissues, can be further induce in the host in response to pathogen attack or presence, as well as to other stresses. Induction of existing preformed compounds can take place in the tissue in which they are already present, or in a different tissue (Prusky & Keen, 1995).

These inducible preformed compounds can be induced due to infection, after association with surface plant or under an abiotic stress. This abiotic stress can be induced with storage techniques of fruits and vegetables. For example, heating is a postharvest fruit technique which can be used to inactivate senescence enzymes for prolonging the shelf-life by controlling high temperature during some periods of time as a type of physical treatment. At the same time, heating allows to maintain or to prolong the fungicide activity of compounds present in citrus peel like citral or certain proteins like quitinase and β-1,3-glucanase (Barkai-Golan, 2001). It can also induce phytoalexins in superficial wounds inoculated with the pathogen. In the other hand, heating can also catalyze biosynthesis of lignin and other analogues compounds in wounds which act on as a physical barrier against hyphae penetration of pathogen.

Antifungal substances isolated from unripe avocado fruit peel include monoene and diene compounds, of which diene is the more important (Prusky & Keen, 1993). Diene compound (1-acetoxi-2-hydroxy-oxo-heneicosa 12,15-diene) which it is a hydrocarbon, inhibits spore germination and mycelial growth of *Colletotrichum gloeosporioides*, at concentrations lower than those present in the peel. Prusky et al. (1990, as cited in Barkai-Golan, 2001) found that inoculation of unharvested or freshly harvested avocado fruit with *C. gloeosporioides*, but not with the stem-end fungus *Diplodia natalensis*, resulted in a temporarily enhanced level of these compounds. The response to this challenge doubled the amount of the preformed diene after 1 day and the effect persisted for 3 days, suggesting persistence of the elicitor. On the other hand, wounding of freshly harvested fruit resulted in a temporarily enhanced diene accumulation in the fruit, inducement did not occur in fruit 3-4 days after harvest (Barkai-Golan, 2001).

γ irradiation is another abiotic factor capable of inducing diene accumulation and CO_2 treatment also increase it. An inducement of antifungal diene also followed a high-CO_2 application. Exposing freshly harvested avocado fruit to 30% CO_2 resulted in increased concentration of the diene upon removal from the controlled atmosphere storage.

Resorcinolic compounds (resorcinols) are also considered as inducible preformed compounds. These compounds have been described in mango fruit. A mixture of resorcinolic compounds normally occurs in fungitoxic concentrations (154-232 $\mu g\ mL^{-1}$ fresh weight) in the peel of unripe mangoes whereas only very low concentrations are present in the fresh of the fruit (Droby et al., 1986). These fungitoxic compounds showed antifungal activity against *Alternaria alternata,* the causal agent of black spot in citrus fruit. This enhancement was accompanied by an increase in fruit resistance to fungal attack.

Exposure of the fruit to a controlled atmosphere containing up to 75% CO_2 was found to enhance of level of resorcinols in the peel itself were they are normally present; this enhancement was accompanied by decay retardation, as indicated by a delay in the appearance of the symptoms of *Alternaria alternata* infection (Barkai-Golan, 2001).

Other compounds that increase the level after infection are the bioactive polyacetylenes, falcarinol and falcarindiol, present in carrots, celery, celeriac and other umbeliferous vegetables.

In carrot roots, high concentrations of the antifungal polyacetylene compound falcarindiol, were recorded. This compound is found in extracellular oil droplets within the root periderm and the pericyclic areas (Garrod & Lewis, 1979). The high concentrations of the antifungal compound were suggested to result from the continuous contact of the carrot with organisms in rhizosphere or with various pathogens. One of the important antifungal compounds in carrot roots is the polyacetylenic compound, falcarinol.

4.3 Phytoalexins – Induced inhibitory compounds

Phytolalexins are low-molecular-weight toxic compounds produced in the host tissue in response to initial infection by microorganisms, or to an attempt at infection. In other words, in other to overcome an attack by the pathogen, the host is induced by the pathogen to produce antifungal compounds that would prevent pathogen development. However, the accumulation of phytoalexins does not depend on infection only. Such compounds may be elicited by fungal bacterial or viral metabolites, by mechanical damage, by plant constituents released after injury, by a wide diversity of chemical compounds, or by low temperature irradiation and other stress conditions. Phytoalexins are thus considered to be general stress-response compounds, produced after biotic or abiotic stress. The most available evidence on the role of phytoalexins shows that disruption of cell membranes is a central factor in their toxicity (Barkai-Golan, 2001) and that the mechanism is consistent with the lipophilic properties of most phytoalexins.

The chemical composition of phytoalexins is elevated. Most important phytoalexins are terpenes and sesquiterpenes. The effects of these sesquiterpenoids – phytoalexins as well as non-phytoalexins were found to be much lower than the effect of the fungicide metalaxyl. In general, phytoalexins are not considered to be as potent as antibiotic compounds (Barkai-Golan, 2001).

An example of induction of phytoalexins by abiotic stress was reported by Kuc' (1972), as cited in Barkai-Golan, 2001) who observed that fruit peeling resulted in browning of the fresh accompanied by enhanced activity of phenylalanine ammonia lyase (PAL). There are indications that PAL activity is connected with productions of phytoalexins and other compounds involved in the defence mechanism of the plant. Radiation is also a cause of phytoalexins production; several studies with citrus fruits have also described γ-irradiation as a stress factor leading to the induction of antifungal phytoalexinic compounds in the treated fruit tissues.

Biosynthesis of toxic compounds as a result of wounding or other stress conditions is a ubiquitous phenomenon in various plant tissues. An example of such a synthesis is the production of the toxic compound 6-methoxymellein in carrot root in response to wounding or to ethylene application (Barkai-Golan, 2001); the application of *Botrytis cinerea* conidia and other fungal spores to the wounded area was found to stimulate the formation of this compound. A similar result is also achieved by the application of fungal produced pectinase, in spite of the fact that this enzyme does not affect cell vitality (Barkai-Golan, 2001). This toxic compound probably has an important role in the resistance of fresh carrots to infection. Carrots that have been stored for a long period at a low temperature lose the ability to produce this compound and, in parallel their susceptibility to pathogen increases. Enhanced resistance of carrots can also be induced by application of dead spores; carrot discs treated with *B. cinerea* spores which had previously been killed by heating developed a market resistance to living spores of the fungus, which was much greater than that of the control discs. The most effective inhibitor found in the tissues after the induction of resistance, as well as in the control tissue, were methoxymellein, *p*-hydroxybenzoic acid and polyacetylene falcarinol (Harding & Heale, 1980).

A sesquiterpenoid compound, rishitin, produced in potato tubers following infection by *Phythophthora infectans,* was first isolated by Tomiyama et al. (1968) from resistant potatoes that have been inoculated with the fungus. Rishitin and solavetivone have also been found to be induced in potato tuber discs 24 h after inoculation with *Fusarium sambucinum,* which causes dry rot in stored potatoes (Ray & Hammerschmidt, 1998, as cited in Barkai-Golan, 2001) and *Erwinia carotovora* (Coxon et al., 1974, as cited in Barkai-Golan, 2001). Other sesquiterpenoids that have been found in potatoes may also play a role in tubers disease resistance; they include rishitinol, lubimin, oxylubimin and others. The terpenoid phituberin was found to be constitutively present in tuber tissues at low levels, but it was further induced after inoculation with *F. sambucinum*. The phytoalexins, phytuberol and lubimin appeared in potato discs by 48 h after inoculation, while solavetivone was produced in very low quantities. At least eight additional terpenoid compounds were induced in potato tubers in response to inoculation with pathogenic strains of *F. sambucinum* and they appeared 48-70 h after inoculation. Rishitin suppressed mycelial growth of the potato pathogen *Phythophthora infectans* on a defined medium (Engström et al., 1999, as cited in Barkai-Golan, 2001). A similar effect, however, was recorded for the naturally occurring plant sesquiterpenoids abscisic acid, cedrol and farnesol, although these compounds are found in healthy in plant tissue and are not associated with post-infection responses.

Other phytoalexins compounds are next listed: several phytoalexinic compounds, such as umbelliferone, scopoletin and sculetin, are produced in sweet potato roots infected by de fungus *Ceratocistis fimbriata* (Minamikawa et al., 1963). In addition, the resistance of celery petioles to pathogens has been attributed over the years to psoralens, linear furanocoumarins with are considered to be phytoalexins. Another phytoalexin found in celery tissue columbianetin, which probably also plays a more important role than psoralens in celery resistance to decay (Barkai-Golan, 2001). On the other hand, the phytoalexin capsidiol is a sesquiterpenoid compound produced by pepper fruit in response to infection with arrange of fungi.

Benzoic acid is a phytoalexin produced in apples as a result of infection by *Nectria galligena* and other pathogens. This acid has proved to be toxic only as the undissociated molecule and it is expressed only at low pH values such as can be found in unripe apples were the initial development of the fungus was indeed halted. With ripening and the decline in fruit tissue acidity, in conjunction with increasing sugar levels, the benzoic acid is the composed by the pathogen, ultimately to CO_2 and the fungus can resume active growth (Swinburne, 1983, as cited in Barkai-Golan, 2001). The elicitor of benzoic acid synthesis was found to be a protease produced by the pathogen (Swinburne, 1975, as cited in Barkai-Golan, 2001). This protease a non-specific elicitor and a number of proteases from several sources may elicit the same response. On the other hand, *Penicillium expansum, B. cinerea, Sclerotinia fructigena, Aspergillus niger,* which do not produce protease in the infected tissue and do not induce the accumulation of benzoic acid, can rot immature fruit (Barkai-Golan, 2001).

Inoculating lemon fruit with *Penicillium digitatum,* the pathogen specific to citrus fruits, results in the accumulation of phytoalexin scoparone (6,7-dimethyloxycoumarin). The induced compound has a greater toxic effect than that of the preformed antifungal compound naturally found in the fruit tissue, such as citral and limetin, as indicated by the inhibition of *P. digitatum* spore germination (Ben-Yehosua et al., 1992). Scoparone production can also be induced in the peel of various citrus fruits by ultraviolet (UV) illumination (Rodov et al., 1992).

Stilbenoids (stilbenes) are other phytoalexins group. They are a group of secondary products of heartwood formation in trees. Plants, especially grapes, can produce resveratrol, that act directly in their defence by inhibiting pathogen proliferation, or indirectly by disrupting chemical signal processes related to growth and development of pathogens or herbivores (Wedge & Camper, 2000). *Trans*-resveratrol (3,5,4′-trihydroxystibene) is one of the simplest stilbenes. It is a product of the plant secondary phenolic metabolism by the action of resveratrol synthase on *p*-coumaroyl-CoA and malonyl-CoA. It occurs in unrelated groups of angiosperms (Morales et al., 2000, as cited in van Baarlen et al., 2004). Besides *trans*-resveratrol, numerous other stilbenes have been characterized in grapevine. These include a 3-*O*-β-glucoside of resveratrol called piceid that is formed by the action of a glycosyl transferase on reverastrol (van Baarlen et al., 2004) and a dimethylated derivate of reverastrol (3,5-dimethosy-4'hydroxystibene) named pterostilbene. This substance has the highest antifungal activity, but its concentration is less than 5 μg g^{-1} in leaves and fruit of various grapevine cultivars (van Baarlen et al., 2004). The potency of pterostilbene increases in the presence of glycolic acid and organic acid that accumulates to high concentration in mature grape berries. Pterostilbenes may thus act as constitutive defence component in berries (van Baarlen et al., 2004).

An early study on the antifungal activity of stilbenes revealed that they rapidly inhibit the respiration of fungal cells, probably by acting as uncoupling agents and by forming protein-phenol complexes (Hart, 1981, as cited in van Baarlen et al., 2004). Based on the structural similarity of hydroxystilbenes and aromatic hydrocarbons, it was inferred that their mode of action may involve lipid peroxidation by blocking cytochrome c reductase and monooxygenases (Pezet & Pont, 1995, as cited in van Baarlen et al., 2004).

Analogues of stilbenes, the hydroxystilbenes, are other potent phytoalexins. The most active on fungal respiration were pterostilbene and ε-viniferin with respective EC_{50} values of 20 μg mL^{-1}, and 37 μg mL^{-1} (van Baarlen et al., 2004).

4.4 Pathogenic-related proteins, active oxygen species and lectins

Other substances that are induced by several factors and have antifungal properties are pathogenic-related proteins (PR), active oxygen species (AOS) and lectins. They are briefly described below.

Pathogenic–related proteins (PR proteins) represent a large array of proteins code by the host plants that are co-ordinately expressed under pathological or related situations. They have been characterized in over 70 plant species and 13 plant families including mono- and dicotyledonous plants. They are extremely diverse in terms of enzymatic and biological activity and have been grouped into 13 protein families based on primary structure and serological relationships. They primarily accumulate in plant cell walls and vacuoles. At least, *B. cinerea* infection leads to PR protein induction in many plants (van Baarlen et al., 2004).

Peroxidases are the most important group of PR proteins whose activity has been correlated with plant resistance against pathogens. Plant peroxidases, which are glycoproteins that catalyze the oxidation by peroxide of many organic and inorganic substrates, have been implicated in a wide range of physiological processes, such as ethylene biosynthesis, auxin metabolism, respiration, lignin formation, suberization, growth and senescence.

The importance of peroxidase lies in the fact that the host cell wall constitutes one of the first lines of defence against pathogen and peroxidase is a key enzyme in the world-building processes. Such processes include the accumulation of lignin and phenolic compounds and suberization. However, the resistance against pathogen may also been related to the highly reactive oxygen species such as H_2O_2 or oxygenase, which are likely to be toxic to pathogens and which are formed by peroxidase activity during the deposition of cell wall compounds (Goodman & Novacky, 1994).

Several glucanohydrolases found in plants, such as chitinase and β-1,3-glucanase, have received considerable attention as they are considered to play a major role in constitutive and inducible resistance against pathogens (El Ghaouth, 1994, as cited in Barkai-Golan, 2001). These enzymes are low-molecular weight proteins, frequently referred to as pathogenesis related (PR) proteins. They hydrolyze the major components of fungal cell walls which results in the inhibition of fungal growth (Schlumbaum et al., 1986). The chitinases, which are ubiquitous enzymes of bacteria, fungi, plants and animals hydrolyze the β-1,4-linkage between the N-acetylglucosamine residues of chitin, a polyssacharide of the cell wall of many fungi (Neuhaus, 1999). The glucanases, which are abundant, highly regulated enzymes, widely distributed in cell-plant species, are able to catalyze endo-type hydrolytic cleavage of glucosidic linkages in β-1,3-glucans (Barkai-Golan, 2001). The chitinases and β-1,3-glucanases are stimulated by infection and in response to elicitors. It was thus suggested that the deliberate stimulation and activation of PR proteins in the fruit tissue might lead to disease suppression by enhancing host resistance to infection.

Postharvest treatment with this elicitor has been found to activate antifungal hydrolases in several fruits: treatments of strawberries, bell peppers and tomato fruits with chitosan induces the production of hydrolases, which remained elevated for up to 14 days after treatment and reduced lesion development by *Botrytis cinerea* (Barkai-Golan, 2001). When applied as a stem scar treatment to bell peppers, chitosan stimulated the activities of chitinase, chitosanase and β-1,3-glucanase. Being capable of degrading fungal cell walls, this antifungal hydrolases are considered to play a major role in disease resistance (Barkai-Golan, 2001).

Many of several PR protein families display some toxicity towards *B. cinerea in vitro*. For some of them, this may be caused by their potential to degrade chitin and β-glucan fragments of *B. cinerea* cell walls (van Baarlen et al., 2004). A grape PR-like protein (chitinase) has one of the highest botryticidal activities. It inhibits germination of conidia with an EC_{50} value of 7.5 μg mL^{-1} (Derckel et al., 1998, as cited in van Baarlen et al., 2004) and it restricts the elongation of hyphae. Despite their anti-microbial activity *in vitro*, there is little evidence to support a potential role of PR proteins in effective plant disease resistance to *B. cinerea*.

The commercial potential of plants exhibiting higher levels of PR proteins will be hampered by the fact that PR proteins are associated with several undesirable effects such as the formation of haze in grape juices (Waters et al., 1996, as cited in van Baarlen et al., 2004) and allergenic reactions (van Baarlen et al., 2004).

Active oxygen, produced by plant cells during interactions with potential pathogens and in response to elicitors, has recently been suggested to be involved in pathogenesis. In response to pathogens, plants are generally able to mount a spectrum of defence responses, often coinciding with an oxidative burst involving active oxygen species (AOS) that commonly confers resistance to a wide range of pathogens. Active oxygen species, including superoxide, hydrogen peroxide and hydroxyl radical, can affect many cellular processes involved in plant-pathogen interactions (Baker & Orlandi, 1995). The direct antimicrobial effect of active oxygen species has not yet been clarified, but they are considered to play a role in various defence mechanisms, including lignin production, lipid peroxidation, phytoalexin production and hypersensitive responses. However, active oxygen can be difficult to monitor in plant cells because many of the active oxygen species are short lived and are suggest to cellular antioxidant mechanisms such as superoxide dismutases, peroxidases, catalase and other factors (Baker & Orlandi, 1995).

A first report on the production of active oxygen in potato tubers undergoing a hypersensitive response was given by Doke (1983) who demonstrated that O_2 production occurred in potato tissues upon inoculation with and incompatible race of *Phytophthora infestans* (i.e.; a race causing a hypersensitive response).

Beno-Moualen & Prusky (2000) found that the level of reactive oxygen species in freshly harvested unripe avocado fruit, which is resistant to infection, was higher than that in the susceptible ripe fruit.

Lectins are a class of sugar-binding proteins that are widely distributed in nature and their occurrence in plants has been known since the end of the 20th century. However, the role of plant lectins is still not well defined and understood (Barkai-Golan, 2001). Now, it is considered that lectins act as recognition determinants in the formation of symbiotic relations between leguminous plants and nitrogen-fixing bacteria and, moreover, they can play a role in the defence of plants against various animals, as well as phytopathogenic fungi (Sharon, 1997), such as *Trichoderma viride, Phytophthora citrophthora, Geotrichum candidum, Botrytis cinerea, Furarium moniliforme* and other pathogens (Barkai-Golan, 2001)

5. Using plant fungicides for commercial purposes

The optimization of plant natural compounds fungicides against fungal diseases for agriculture is an important research because it would permit to search some important alternatives to the use of synthetic fungicides. At the same time, the study of the role of these compounds that they play in plant metabolism will permit to contribute to the knowledge of plant's metabolism. Most of these compounds present a weak antifungal activity, so additional studies are necessary in order to optimize the use of these compounds as fungicides.

A large number of fungicides are already available to the farmer; the Pesticide Manual (Tomlin, 1994) contains 158 different fungicidal active ingredients in current use. Nevertheless, further industrial research aimed at the discovery and development of new compounds is extremely intensive and this is due to a number of important factors. Firstly, the development of fungicides with novel modes of action remains an important strategy in the search for ways to overcome problems associated with resistance to established products. Secondly, it is becoming increasingly desirable (some would say essential) to replace certain existing products with compounds of lower toxicity to non-target species and acceptable levels of persistence in the environment. Finally, in the increasingly competitive world, agrochemical companies are forced to look for new compounds which show marketable technical advantages over their own and their competitors' products (Clough & Godfrey, 1998).

In principle, fungicidal natural products can either be used as fungicides in their own right, or may be exploited as leads for the design of other novel synthetic materials. In the former approach, purified natural products constitute the active ingredient of a formulated mixture, or are used in mixture with a synthetic material. However, the use of natural products by themselves as fungicides has not been particularly successful for a number of reasons. Firstly, natural products possessing marketable levels of activity against a broad spectrum of commercially important diseases have proved to be very hard to find. Furthermore, they are often inherently unstable (for example, to sunlight) and consequently are not sufficiently persistent in the field to deliver a useful effect. In addition, some lack selectivity of action and this can manifest itself in the form of toxicity to plants or mammals. Finally, many natural products derived from fermentation broths are present in low concentrations and are difficult to purify on a large scale. Some of these limitations can be overcome by making semisynthetic derivatives, but this inevitably adds to the overall cost (Clough & Godfrey, 1998).

For the reasons above, the agrochemical industry has largely focused on the second approach: the design of novel, fully synthetic compounds from a consideration of the structure of appropriate natural product leads. These synthetic compounds ideally possess optimized biological, physical and environmental properties and are often simpler in structure than their naturally occurring progenitors (Clough and Godfrey, 1998).

In spite of the arguments above expressed and the difficulties to obtain a natural substance which can have antifungal activity and stability at the same time, the researchers are continually searching new substances naturally occurring in nature with antifungal properties so that in the future and after optimization could be used as commercial proposes. Some of these formulates are being already commercialized and some of them are briefly exposed below.

Among these natural compounds, the biocides are extracted from plants and some of them are used as additives in food industry. They present different formulates according to their application mode (Wilson & Wisniewski, 1994). These natural biocides present a wide mode of action and, in general, they are composed by plant extracts, like citrus extracts which are neither toxic nor corrosive. Moreover, they are not irritate and are biodegradable with a good antimicrobial activity and fungicide properties. That is the case of CitroBio, produced in Florida (USA), in which the active ingredient is made from citric seeds. It is only contains 100% citric natural extract with a wide antimicrobial action. Other natural extract is P3-Tsunami which is considered like a product with a high effect against fungi which cause fruit and vegetable decay and also is used to control bacteria growth in cut-produces (Monterde et al., 2002).

Several plant and bacterial natural products have novel applications as plant protectants through the induction of systemic acquired resistance (SAR) processes. Commercial products that appear to induce SAR include Messenger® (EDEN Biosciences, Inc., Bothell, WA) and the bioprotectant fungicides Serenade® (AgraQuest, Davis, CA), Sonata® (AgraQuest, Davis, CA) and Milsana® (KHH BioSci, Inc., Raleigh, NC). Messenger is a harpin protein which switches on natural plant defences in response to bacterial leaf spot and fungal diseases such as *Botrytis* brunch rot and powdery mildew. Serenade is a microbial-protectant derived from *Bacillus subtilis,* with SAR activity that controls *Botrytis,* powdery and downy mildews, early blight and bacterial spot. Sonata is also a microbial-biopesticide with activity against *Botrytis,* downy and powdery mildews, rusts, *Sclerotinia* blight and rots. Milsana® is an extract from *Reynoutria sachalinensis* (giant knotweed) that induces phytoalexins able to confer resistance to powdery mildew and other diseases such as by *Botrytis*. However, elicitors with no innate antifungal activity will not appear active in most current screening high throughput screening systems. Many experimental approaches have been used to screen compounds and extracts from plants and microorganisms in order to discover new antifungal compounds.

Mints oils are well-known antifungal treatments that have been developed as natural fungicides. A mixture of mint oil and citric acid commercially available as Fungastop is a broad spectrum fungicide that reduced postharvest decay of lettuce (Martínez-Romero et al., 2008).

Although new fungicides based on natural plant extracts are continually developing, more research is necessary for optimizing applications and become a safe alternative for eliminating the chemical fungicides from agriculture. Meantime these types of plant fungicides are safe under some conditions and applied together with synthetic fungicides in order to reduce residues in an IPM strategy.

6. Acknowledgment

Thanks are due to my wife Olga and my daughter Paula for helping me to write this chapter.

7. References

Arras, G. & Piga, A. (1994). *Thymus capitatus* essential oil reducing citrus fruit decay. In: *Ait-Oubahou, A. & M. El-Otmani, M.* (Eds.), Postharvest Physiology, Pathology, and Technology for Horticultural Commodities, Recent Advances, Proc. Int. Symp., Agadir, Morocco, pp. 426-428.

Arteca, R.N. (1996). *Plant Growth Substances. Principles and Applications.* Chapman & Hall, New York, USA, 332 pp.

Baker, C.J. & Orlandi, E.W. (1995). Active oxygen in plant pathogenesis. *Annu. Rev. Phytopatol.*, 33: 299-321.

Barkai-Golan, R. (2001). *Postharvest Diseases of Fruits and Vegetables. Development and Control.* Elsevier, Amsterdam, The Netherlands, 418 pp.

Beno-Moualem, D. & Prusky, D. (2000). Early events during quiescent infection development by *Colletotrichum gloeosporioides* in unripe avocado fruits. *Phytopathology*, 90: 553-559.

Ben-Yehosua, S., Rodov, V., Kim, J.J. & Carmeli, S. (1992). Preformed and induced antifungal materials of citrus fruits in relation to the enhancement of decay resistance by heat and ultraviolet treatments. *J. Agric. Food Chem.*, 40: 1217-1221.

Bettiol, W., Mattos, L.P.V. & Morais, L.A.S. (2011). Control of postharvest green mold of orange by essential and fixed oils. *International Congress of Postharvest Pathology*, Lleida, Spain, 11-14 April. 2011.

Clough, J.M. & Godfrey, C.R.A. (1998). The strobilurin fungicides. In: *Hutsun, D. & Miyamoto, J.* (Eds.), Fungicidal Activity. Chemical and Biological Approaches to Plant Protection. John Wiley & Sons Ltd., West Sussex, England, 254 pp.

Davidson, P.M. (1997). Chemical preservatives and natural antimicrobial compounds. In: *Doyle, M.P., Beuchat, L.R. & Montville, T.J.* (Eds.). Food Microbiology-Fundamentals and Frontiers. Washington D.C., ASM Press: 520-556.

Dmitriev, A.P., Tverskoy, L.A., Kozlovsky, A.G. & Grodzinsky, D.M. (1990). Phytoalexins from onion and their role in disease resistance. *Physiological and Molecular Plant Pathology*, 37: 235-244.

Doke, N. (1983). Involvement of superoxide anion generation in the hypersensitive response of potato tuber tissues to infection with an incompatible race of Phytophthora infectans and to the hyphal wall components. *Physiol. Plant Pathol.*, 23: 345-357.

Droby, S., Prusky, D., Jacoby, B. & Goldman, A. (1986). Presence of antifungal compounds in the peel of mango fruits and their relation to latent infection of *Alternaria alternata. Physiol. Mol. Plant Pathol.*, 9: 173-183.

Droby, S.; Porat, R.E.; Weiss, B.; Shapiro, B.; Philosoph-Hadas, S. & Meir, S. (1999). Suppressing green mold decay in grapefruit with postharvest jasmonate application. *J. Amer. Soc. Hort. Sci.*, 124 (2). 184-188.

Eckert, J.W. (1978). Pathological diseases of fresh fruits and vegetables. In: *Hultin, H.O. & Milner, N.* (Eds.), Postharvest Biology and Biotechnology, Food and Nutrition Press, Wesport, 161-209.

Farmer, E.E. & Ryan, C.A. (1992). Octadecanoid presursors of jasmonic acid activate the synthesis of wound-inducible proteinase inhibitors. *Plant Cell*, 4: 129-134.

Garrod, B. & Lewis, B.G. (1979). Location of the antifungal compound falcarindiol in carrot root tossue. *Trans. Br. Mycol. Scoc.*, 72: 515-517.

Goodman, R.N. & Novacky, A.J. (1994). *The Hypersensitive Reaction in Plants to Pathogens.* A.P.S. Press, St. Paul.

Grange, M. & Ahmed, S. (1988). *Handbook of Plants with Pest Control Properties.* John Wiley & Sons, New York.

Gwinn, K.D., Ownley, B.H., Greene, S.E., Clark, M.M., Taylor, C.L., Springfield, T.N., Trently, D.J., Green, J.F., Reed, G.A. & Hamilton, S.L. (2010). Role of essential oils in control of *Rhizoctonia* damping-off in tomato with bioactive monarda herbage. *Phytopathology*, 100: 493-501.

Harding, V.K. & Heale, J.B. (1980). Isolation and identification of the antifungal compounds accumulating in the induced resistance response of carrot root slices to *Botrytis cinerea. Physiol. Plant Pathol.*, 17: 277-289.

Indrakeerthi, S.R.P. & Adikaram, N.K.B. (1996). Papaya latex, a potential postharvest fungicide. In: Proc. Australian Postharvest Hortic. Conf. 'Science and Technology for the Fresh Food Revolution', Melbourne, Australia, pp. 423-427.

Knight, S.C., Anthony, V.M., Brady, A.M., Greenland, A.J., Heaney, S.P., Murray, D.C., Powell, K.A., Schulz, M.A., Spinks, C.A., Worthington, P.A. & Youle, B. (1997). Rationale and perspectives on the development of fungicides. *Ann. Rev. Phytopathol.*, 35: 349-372.

Link, K.P. & Walker, J.C. (1933). The isolation of catechol from pigmented onion scales and its significance in relation to disease resistance in onions. *Journal of Biological Chemistry*, 100: 379-383.

Martínez J.A. & Bañón S. (2007). Differential effects of gibberellic acid on the growth of *Botrytis cinerea* isolated from various ornamental plants. *XVI International Plant Protection Congress,* Glasgow, Scotland, UK, 15-18 October 2007. Congress Proceedings, volume 2:780-781.

Martínez, J.A., Navarro A., Fernández J.A. & Bañón S. (2007). Using paclobutrazol to delay the growth of *Botrytis cinerea* isolated from *Chamelaucium uncinatum. Australasian Plant Pathology*, 36:39-45.

Martínez, J.A., Valdés, R., Gómez-Bellot, M.J. & Bañón, S. (2011). Effects of índole-3-acetic acid on *Botrytis cinerea* isolates obtained from potted plants. *63rd International Symposium on Crop Protection* (ISCP 2011), Ghent, Belgium, 24th May 2011.

Martínez, J.A. & Hernández, M.C. (2007). Utilización de aceites esenciales de enebro y tomillo para controlar el desarrollo de *Penicillium digitatum* y *P. italicum* sobre frutos de naranja. *XXI Congreso Nacional de Microbiología* (SEM 2007), Seville, Spain.

Martínez-Romero, D., Serrano, M., Bailén, G., Guillén, F., Zapata, P. J., Valverde, J. M., Castillo, S., Fuentes, M., & Valero, D. (2008). The use of a natural fungicide as an alternative to preharvest synthetic fungicide treatments to control lettuce deterioration during postharvest storage. *Postharvest Biology & Technololy*, 47:54-60.

Minamikawa, T., Akazawa, T. & Utitani, I. (1963). Analytical study of umbelliferone and scopoletin synthesis in sweet potato roots infected by *Ceratocystis fimbriata*. *Plant Physiol.*, 38: 493-497.

Monterde A.; Salvador, A.; Cuquerella, J. & Martínez-Jávega, J.M. (2002). Aplicación de productos naturales e la postcosecha de cítricos. *Levante Agrícola. Especial Postcosecha*, 361. 263-265.

Narayanasamy, P. (2006). *Postharvest Pathogens and Disease Management*. John Wiley & Sons, Inc., New Jersey, USA, 578 pp.

Neuhaus, J.M. (1999). Plant chitinases (PR-3, PR-4, PR-4, PR-8, PR-11). In: *Datta, S.K. & Muthukrishnan, S.* (Eds.), Pathogenesis-Related Proteins in Plants. CRC Press, Boca Raton, London, New York, Washington, DC, pp. 77-105.

Prusky, D. & Keen, N.T. (1993). Involvement of preformed antifungal compounds in the resistance of subtropical fruits to fungal decay. *Plant Dis.*, 72:114-119.

Prusky, D. & Keen, N.T. (1995). Inducible preformed compounds and their involvement in the resistance of plants to pathogens. In: *Reuveni, R.* (Ed.), Novel Approaches to Integrated Management, Lewis Publishers, CRC Press, Boca Raton, pp. 139-152.

Prusky, D. (1997). Mechanisms of resistance of fruits and vegetables to postharvest diseases. In: Disease Resistance in Fruit. Proceedings of an International Workshop, Chiang Mai, Thailand: 19-33.

Richardson, D.M. (2009). Assessment of the impact on crop protection by the 'cut-off criteria' in a new regulation for authosisation of plant protection products. *Communications in Agricultural and Applied Biological Sciences*, 74(1): 3-8.

Rodov, V., Ben-Yehosua, S., Kim, J.J., Shapiro, B. & Ittah, Y. (1992). Ultraviolet illumination induces scoparone production in kumquat and orange fruit and improves decay resistance. *J. Amer. Soc. Hort. Sci.*, 117: 788-792.

Rong, T. & Ting, Z. (2000). Antifungal activity of monoterpenoids against postharvest pathogens *Botrytis cinerea* and *Monilinia fructicola*. *Journal of Essential Oil Research*, 12:113-121.

Schiffmann-Nadel, M. & Littauer, F.S. (1956). Mode of infection of *Penicillium digitatum* and *P. italicum* in citrus fruits. *Ktavim*, 6: 61-67.

Schlumbaum, A., Mauch, F., Vogeli, U. & Boller, T. (1986). Plant chitinases are potent inhibitors of fungal growth. *Nature*, 324: 365-367.

Schönbeck, F. & Schroeder, C. (1972). Role of antimicrobial substances (tuliposides) in tulips attacked by *Botrytis* spp. *Physiological Plant Pathology*, 2: 91-99.

Seigler, D.S. (1998). *Plant Secondary Metabolism*. Kluwer Academic Publishers, Norwell, Massachusetts, USA, 761 pp.

Sharon A; Elad Y; Barakat R. & Tudzynski P (2004). Phytohormones in *Botrytis*-plant interactions. In: *Elad, Y., Williamson, B., Tudzynski, P. & Delen, N.* (Eds.) *Botrytis*: Biology, Pathology and Control. (Eds.) Kluwer Academic Publishers: Dordrecht, Netherlands, pp. 163-179.

Sharon, N. (1997). Functional aspects of microbial and plant lectins. *Nova Acta Leopoldina NF75*, 301:13-26.

Tomiyama, K., Sakuma, T., Ishizaka, N., Katsui, N., Takasugi, M. & Masamune, T. (1968). A new antifungal substance isolated from resistant potato tuber tissue infected by pathogens. *Phytopathology*, 58: 115-116.

Tomlin, C.D.S. (ed.) (1994). *The Pesticide Manual*. 10th Edition. British Crop Protection Council, Farnham, UK, and the Royal Society of Chemistry, Cambridge, UK.

van Baarlen, P., Legendre, L. & van Kan, J.A.L. (2004). Plant defence compounds against *Botrytis* infection. In: *Elad, Y., Williamson, B., Tudzynski, P. & Delen, N.* (Eds.) *Botrytis*: Biology, Pathology and Control. (Eds.) Kluwer Academic Publishers: Dordrecht, Netherlands, pp. 146-161.

Verpoorte, R. (2000). Plant secondary metabolism. In: *Verpoorte, R. & Alfermann, A.W.* (Eds.), Metabolic Engineering of Plant Secondary Metabolism, Kluwer Academic Publishers, Dordrecht, The Netherlands, 293 pp.

Vick, B. A. & Zimmerman, D. C. (1984). Biosynthesis of jasmonic acid by several plant species. *Plant Physiology*, 1984; 75: 458-461

Wedge, D. E. & Camper, N. D. (2000). Connections between agrochemicals and pharmaceuticals. In: *Cutler, H.G., Cutler, S. J.* (Eds.), Biologically Active Natural Products: Agrochemicals and Pharmaceuticals., ACS Symposium Series, American Chemical Society, Washington DC, 2000

Wedge, D.E. & Smith, B.J. (2006). Discovery and evaluation of natural product-based fungicides for disease control of small fruits. *Biological Control of Plant Pathogens and Diseases*, 1-14.

Williams, D.H., Stone, M.J., Hauck, P.R. & Rahman, S.K. (1989). 'Why are secondary metabolites (natural products) biosynthesised? *J. Nat. Prod.*, 52: 1189-1208.

Wilson, C.L. & Wisniewski, M.E. (1994). *Biological Control of Postharvest Diseases – Theory and Practice*, CRC Press, Boca Raton, 182 pp.

Wood Mackenzie Consultans Limited, Edinburg and London (1997). Agrochemical products. Part 1. The key agrochemical product groups. In: Agrochemical Service, Update of the Products Section, May 1997: 1-74.

2

Applications of Actinobacterial Fungicides in Agriculture and Medicine

D. Dhanasekaran[1], N. Thajuddin[1] and A. Panneerselvam[2]
[1]*Department of Microbiology, School of Life Sciences, Bharathidasan University, Tiruchirappalli, Tamilnadu,*
[2]*P.G. & Research Department of Botany & Microbiology, A.V.V.M. Sri Pushpam College, (Autonomous), Poondi, Tamil Nadu, India*

1. Introduction

Actinobacteria are found in virtually every natural substrate, such as soils and compost, freshwater basins, foodstuffs and the atmosphere. Deep seas, however, do not offer a favorable habitat. These organisms live and multiply most abundantly in various depths of soil and compost, in cold and in tropical regions. Alkaline and neutral soils are more favorable habitats than acid soils and neutral peats are more favorable than acid peats.

The application of fungicides and chemicals can control crop diseases to a certain extent, however, it is expensive and public concern for the environment has led to alternative methods of disease control to be sought, including the use of microorganisms as biological control agents. Microorganisms are abundant in the soil adjacent to plant roots (rhizosphere) and within healthy plant tissue (endophytic) and a proportion possess plant growth promotion and disease resistance properties. Actinobacteria are gram-positive, filamentous bacteria capable of secondary metabolite production such as antibiotics and antifungal compounds. A number of the biologically active antifungal compounds are obtained from the actinobacteria. A number of these isolates were capable of suppressing the fungal pathogens *Rhizoctonia solani*, *Pythium* sp. and *Gaeumannomyces graminis* var. *tritici*, both *in vitro* and in plants indicating the potential of the actinobacteria to be used as biocontrol agents.

The principal reason behind the actinobacteria having such important roles in the soil and in plant relationships comes from the ability of the actinobacteria to produce a large number of secondary metabolites, many of which possess antibacterial activity. Actinobacteria produce approximately two-thirds of the known antibiotics produced by all mircoorganisms. The genus *Streptomyces* produces nearly 80% of the actinobacterial antibiotics, with the genus *Micromonospora* producing one-tenth as many as the *Streptomyces*. In addition to the production of antibiotics the actinobacteria produce many secondary metabolites with a wide range of activities. Activities of the secondary metabolites include antifungal agents that degrade cell walls and inhibit the synthesis of mannan and β-glucan enzymes, antiparasitic agents and insecticidal agents.

Actinobacteria produce a number of plant growth regulatory compounds, some of which have been used commercially as herbicides. Not all secondary metabolites are anti-microbial. Others are enzyme inhibitors, immunomodulators and antihypertensives. The actinobacteria produce over 60% of secondary metabolites produced by microorganisms, with *Streptomyces* accounting for over 80%.

In some cases actinobacteria form a pathogenic relationship with plants. *Streptomyces scabies* is a soil-borne actinobacterium that is the principal causal agent of scab diseases, which affect a variety of underground tuberous vegetables such as potato. *S. scabies* produces thaxtomin, a family of phytotoxins, that induce the development of necrotic lesions in potato. There is a 100% correlation between pathogenicity and the ability to produce thaxtomin. Scab suppressive soils have been identified and it has been found that the lenticels on these tubers are colonised by *Streptomyces* (Schottel *et al.,* 2001). Suppressive strains of *Streptomyces* isolated from a naturally scab suppressive soil produced antibiotics that inhibited *S. scabies in vitro* (Neeno-Eckwall and Schottel, 1999).

Streptomyces species have also been implicated in the biological control of a number of other pathogens. *S. ambofaciens* inhibited *Pythium* damping-off in tomato plants and *Fusarium* wilt in cotton plants. *S. hygroscopius* var. *geldanus* was able to control *Rhizoctonia* root rot in pea plants and the inhibition was due to the production of the antibiotic geldanamycin. *Streptomyces lydicus* WYEC108 inhibited *Pythium ultimum* and *R. solani in vitro* by the production of antifungal metabolites (Yuan and Crawford, 1995). A number of other actinobacteria that are used in inhibiting the human and animal pathogens such as *Aspergillus niger, Penicillium* sp., *Mucor* sp., *Rhizopus* sp. *Candida albicans, Cryptococcus neoformans.* This chapter describes the potential applications of fungicidal substances from actinobacterial origin, screening methods, mode of action of fungicides against plant and animal fungal pathogens.

1.1 Antagonistic actinobacteria

The actinobacteria first recognized as potential destroyers of fungi and bacteria by Gasperini (1890). Tims (1932) studied an actinobacteria antagonistic to *Pythium* of sugarcane. Waksman (1937) made a detailed survey of actinobacteria possessing antagonistic effect upon the activity of other microorganisms in their studies on decomposition.

Dhanasekaran *et al.,* (2009a) screened 78 *Streptomyces* isolates for their antimicrobial activity against pathogenic fungi by agar overlay assay method. Among the 78 isolates, 18 isolates showed antifungal activity. The maximum percentage of the isolates of *Streptomyces,* which showed antifungal antagonistic activity, was found in sea shore soil (13/27 isolates, 48.14 %) followed by salt pan soil (4/9 isolates, 44.44 %), estuarine soil (3/12 isolates, 25 %) and agricultural field soil (5/30 isolates, 16.6 %). Among the 18 isolates tested, all the isolates showed extracellular antifungal activity including 8 isolates having both extra and intracellular antifungal activity (Fig.1; Plate 1). They also studied the antifungal actinobacteria in marine soil of Tamilnadu against *Candida albicans, Aspergillus niger* using agar overlay, diffusion assay method (Dhanasekaran *et al.,* (2005b) and estuarine *Streptomyces* against the *Candida albicans* (Dhanasekaran *et al.,* (2009b)

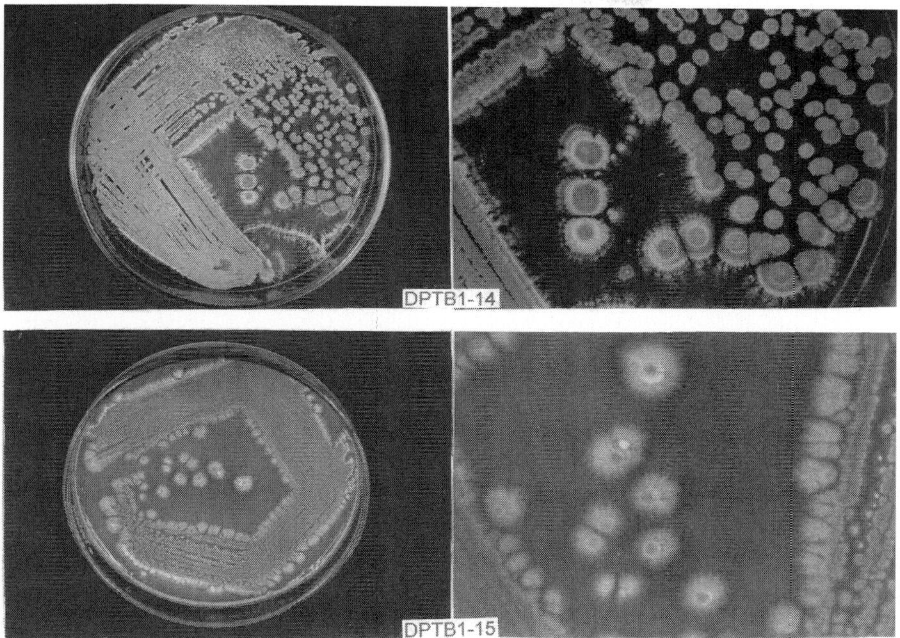

Plate view

A portion enlarged (3x)

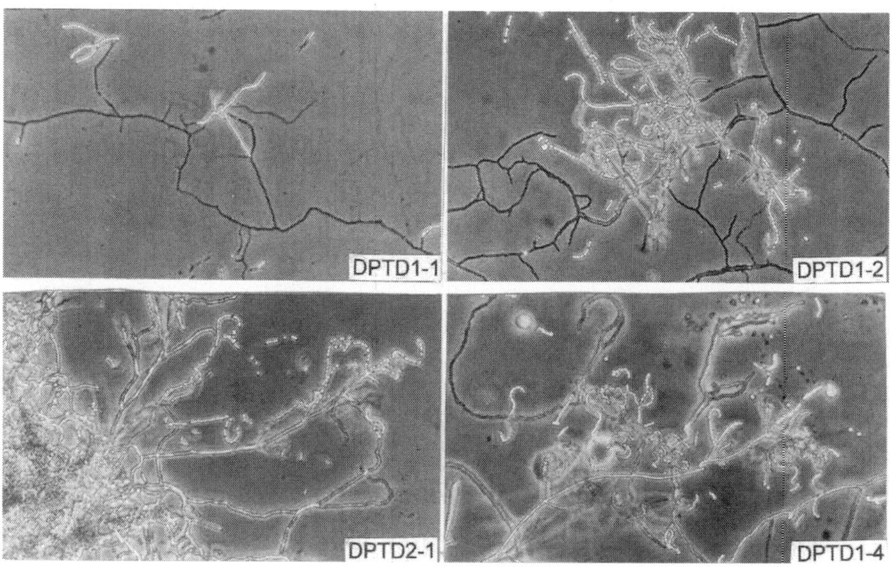

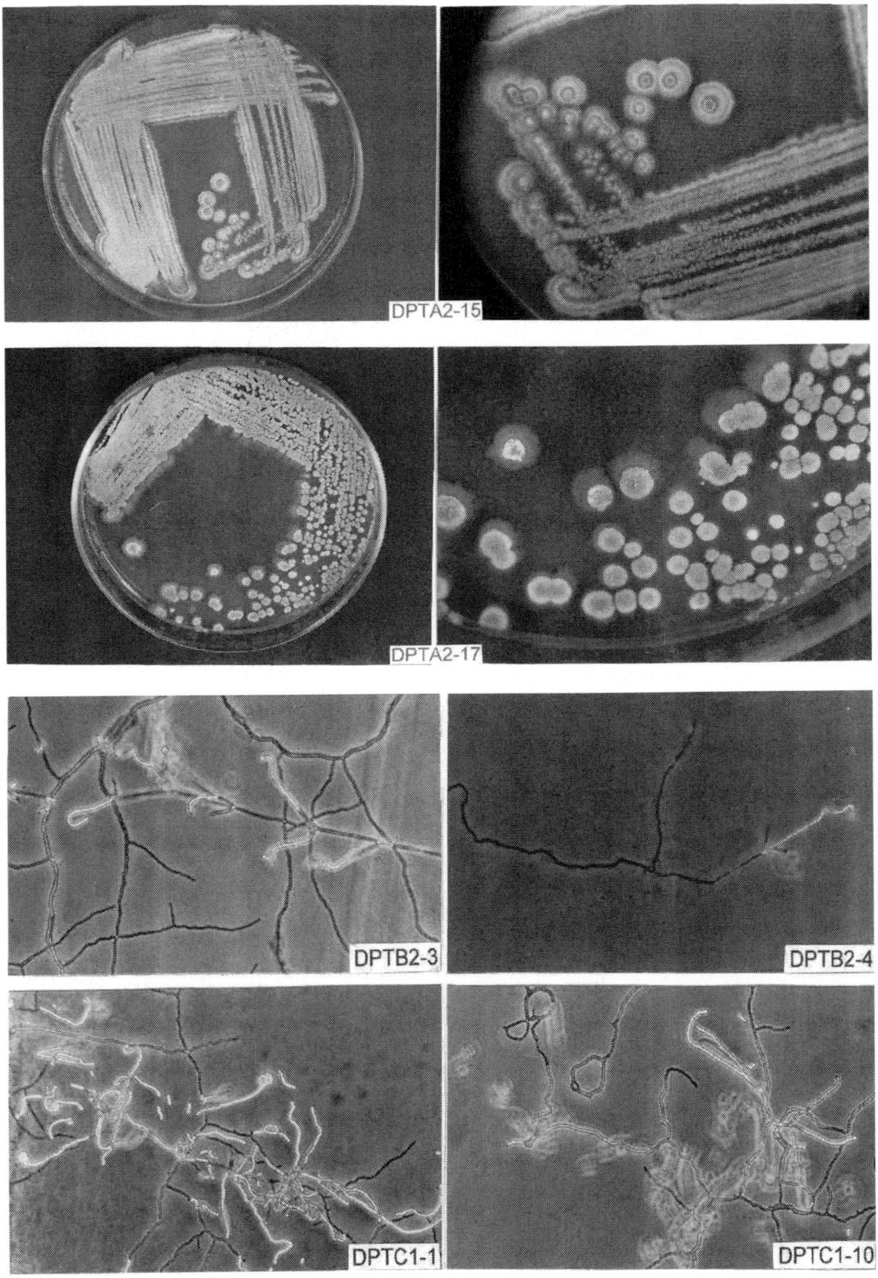

Plate 1. Cultural and microscopic view of *Streptomyces* isolates aerial mycelium with spores

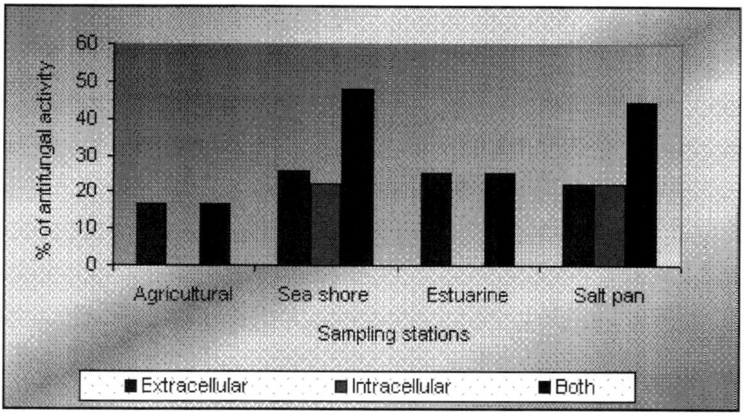

Fig. 1. Antifungal activity of Extra and Intracelluar compounds of *Streptomyces* isolates

2. *In vitro* screening methods of fungicidal substances produced by actinobacteria

2.1 Crowded plate technique

A series of test tubes containing 9 ml of sterile water was taken. From the stock culture, 1 ml suspension was transferred aseptically to the 1st tube (10^{-1}) and mixed well. Further serial dilutions were made to produce 10^{-5} suspensions. Suspension (0.1 ml) from each test tube was spread on sterile soyabean-casein digest medium (SBCD), actinobacteria isolation agar (AIA) medium and starch-casein agar medium plates aseptically in a laminar air flow cabinet. The plates were incubated at 27 ± 2°C for 84 h. The plates were observed intermittently during incubation. After 72 h, whitish pin-point colonies, characteristic of actinobacteria and with clear zone of inhibition around them were observed. The pinpoint colonies with inhibitory or clear zone of inhibition were selected and purified and used as a potent isolate for fungicidal compound production.

2.2 Agar streak method

The Fungicidal activity of the soil actinobacterial isolates were analyzed by agar streak method. Each of the isolate was streaked as a straight line on Starch casein agar (SCA) medium and incubated at 27°C for 6 days. After the 6th day, different fungal pathogens were streaked at right angle, but not touching each other, and then incubated at 28°C for 48 h. If the organism is susceptible to the antibiotic produced by actinobacteria, then it will not grow near the actinobacteria. The zone of inhibition against each test fungal pathogen was noted.

2.3 Agar disk method

The *Streptomyces* isolate was smeared on SCA medium as a single streak and incubated at 28°C for 4-6 days, from well grown streaks 6 mm agar disks of *Streptomyces* colony mass was prepared by using sterile cork borers. Disks were then aseptically transferred to PDA plates having fresh lawn cultures of *Aspergillus* isolate. Controls included using plain disks from

SCA medium. Plates were incubated at 24°C for 4-6 days and fungicidal activity was evaluated by measuring the diameter of inhibition zones (mm).

2.4 Dual culture method

Antifungal activity of actinobacterial isolates were tested by dual culture technique using PDA medium. A mycelial disc of the fungal pathogen (5mm dia.) was placed at one end of the Petri plate. The actinobacterial antagonists were streaked 1 cm away from the periphery of the Petri plate just opposite to the mycelial disc of the pathogen. Visual observation on the inhibition of pathogenic fungal growth was recorded after 96 hours of incubation in comparison with the PDA plate simultaneously inoculated with fungal pathogen only as control.

Percent of test pathogen inhibition by the actinobacterial isolate was evaluated by dual culture technique. The radial growth of mycelium in mm was measured and percent inhibition (PI) was calculated.

$$PI=\frac{C-T}{C}\times 100$$

Where, C is the growth of test pathogen (mm) in the absence of the antagonistic isolate; T is the growth of test pathogen (mm) in the presence of the antagonistic isolate.

2.5 Agar overlay method

To evaluate the fungicidal activity of the actinobacteria, phytopathogenic filamentous fungi were used as test microorganisms. The actinobacteria were spot inoculated onto SCA medium and incubated at 28°C for 14 days. After this period, the antagonism between actinobacteria and the test fungal pathogen was evaluated using the agar over lay method. For this procedure, 10 ml of Sabouraud soft agar medium was added and inoculated with 10^6 spores/ml of filamentous fungi. All plates were incubated at 28°C and incubation time of 7-10 days for fungi.

2.6 Well diffusion method

The sterilized Sabouraud's dextrose agar medium (Dextrose 4.0 g, Mycological peptone 1.0 g, Agar 2.0 g, pH 5.0, Distilled Water 100 ml,) was poured to a petridish in a uniform thickness and kept aside for solidification. Using sterilized swabs, even distribution of lawn culture was prepared using desired fungi such as *A. niger, P. notatum, C. albicans* in SDA plates. Using sterile well cutter two wells were made in plates at required distance. 20µl of different solvents treated test fungicidal compound was added in to one well and another well was loaded with corresponding control (solvent without compound). The plates were incubated at room temperature for 48 hours. After incubation, the zone of inhibition was analyzed and recorded.

2.7 Paper disc assay

2.7.1 Preparation of disc

The filter paper disc were impregnated with 5µg/µl antifungal compound + 25µl distilled water, similar procedures were used to prepare the other concentrations of the disc such as 10µg/µl + 20µl distilled water, 15µg/µl + 15µl distilled water, 20µg/µl + 10µl distilled water, 30µg/µl + 1µl distilled water.

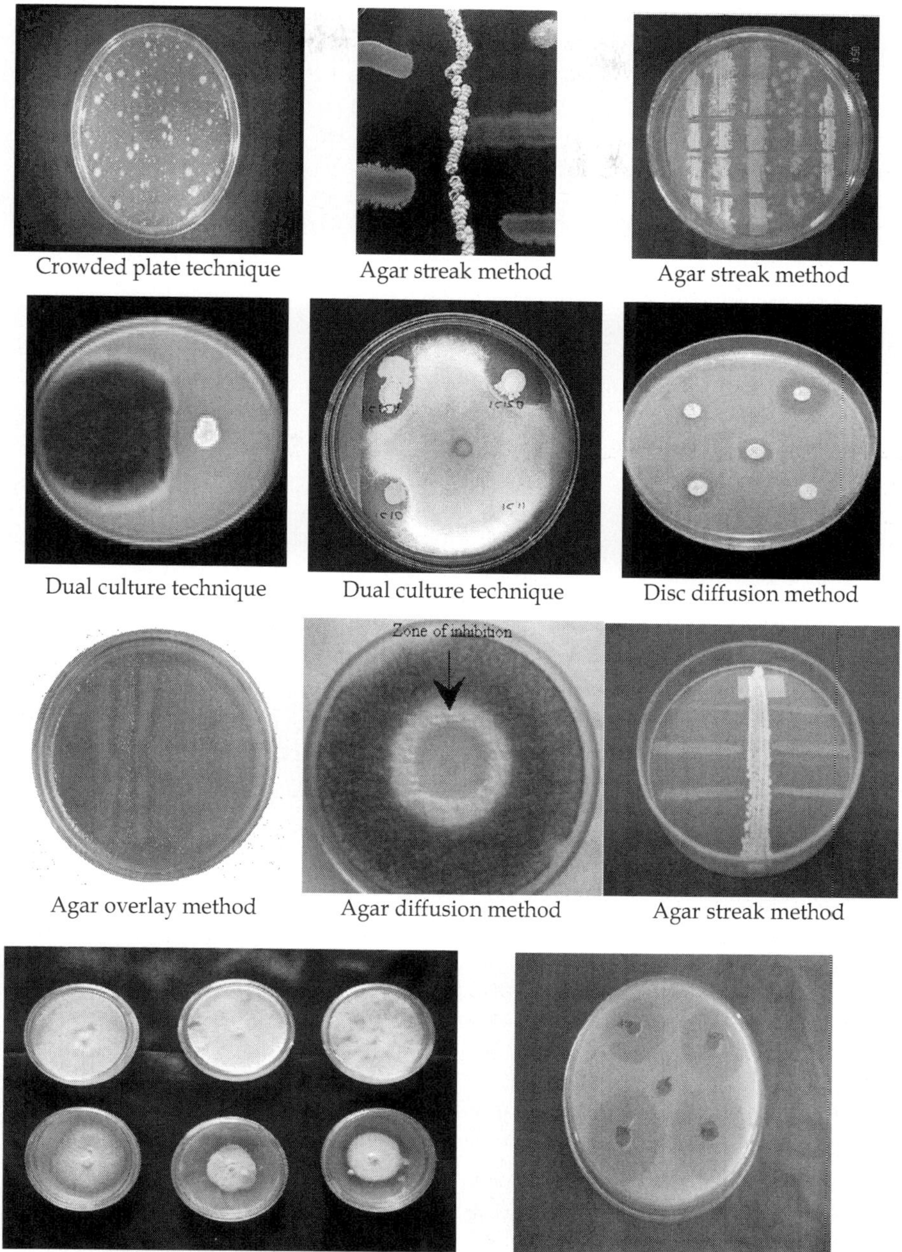

Plate 2. *In vitro* screening methods for Fungicidal substances produced by actinobacteria

2.7.2 Inoculum preparation

All the clinical pathogens were prepared in 0.85% saline corresponding to No. 0.5 McFarland turbidity standard. All cultures were incubated on a shaker at 37°C for 18 h and then diluted to 1/10 the concentration to yield a culture density of approximately 10^8CFU/ml. The pathogens used in the study such as *Candida albicans, Cryptococcus neoformans* and *Aspergillus flavus* for sensitivity assay.

2.7.3 Antifungal assay

Mueller Hinton agar (Beef extract 0.2 g, Peptone 1.75g, Starch 0.15g, Agar 2.0g, Distilled water 100 ml, pH 7.5) prepared with lawn culture using desired test organisms. The inoculated plates were kept aside for few minutes. The discs with fungicidal compound were placed over the medium. After diffusion, the plates were incubated at 28°C for 48hours for antifungal analysis. After incubation, the zone of inhibition was analyzed and recorded.

2.8 Determination of Minimum Inhibitory Concentrations (MIC)

To measure the MIC value, two-fold serial dilutions of 50, 25, 12.5, 6.25, 3.125, 1.562 and 0.781 mg ml^{-1} of the fungicidal compound was prepared in solvent and assayed by well diffusion method. The MIC was defined as the lowest concentration able to inhibit any visible fungal growth.

3. Fungicidal potential of actinobacteria against phytopathogen in crop plants

Fungal phytopathogens pose serious problems worldwide and cause a number of plants and animal diseases such as ringworm, athlete's foot, and several more serious diseases. Plant diseases caused by fungi include rusts, smuts, rots, and may cause severe damage to crops. Fungi are some of the world's largest and possibly oldest individuals.

Agrochemical treatment may result in environmental impact and pose a threat to humans and animals. As a result, there has been an increase in research on potential Biocontrol agents, aimed at finding a definitive solution or at least reducing pesticide use in the treatment of phytopathogenic diseases.

Actinobacteria have been considered as potential Biocontrol agents of plant diseases. Several investigators have described the *in vitro* and *in vivo* activities of the actinobacteria. Their modes of action includes parasitism of hyphae (El-Tarabily and Sivasithamparam, 2006), oospores or fungal sclerotia (Crawford *et al.*, 1993) competition with pathogens (Kunoh, 2002), antibiotic production (Igarashi, 2004), siderophores (Khamna *et al.*, 2009), as herbicides (Hasegawa *et al.*, 2006), and via enzymes such as cellulases, hemicellulases, chitinases, amylases, and glucanases (Yuan and Crawford, 1995).

In addition, actinobacteria may affect plant growth (Igarashi, 2004). According to Kunoh (2002), endophytic *Streptomyces* may play an important role in the development and health of plants, because it affects plant growth due to its assimilation of nutrients and production of secondary metabolites. The tomato (*Lycopersicon esculentum*) is highly susceptible to phytopathogen attack, and tomato crops are most intensively treated with agrochemicals.

Among the many fungal pathogens that attack tomatoes are *Phytophthora infestans, Alternaria solani, Sclerotinia sclerotiorum, Rhizoctonia solani, Fusarium oxysporum,* Most of these pathogens not only spread disease in tomato plants, but also affect other crops. According to Cao *et al.* (2004a), *R. solani* can develop in both farmed and unfarmed soils, spreading disease in many crops, including rice. *F. oxysporum* attacks banana plants, causing a disease known as fusarium wilt (Cao *et al.,* 2004b) and also infects wheat (Taechowisan *et al.,* 2003). *R. solanacearum* is an important soil pathogen causing bacterial wilt in more than 200 plant species, including the potato, tomato, pea, tobacco, banana and others (Tan *et al.,* 2006).

Marten *et al.* (2001) reported that RhizovitR from *Streptomyces rimosus* is used in the control of a wide range of fungi such as *Pythium* spp., *Phytophthora* spp., *Rhizoctonia solani, Alternaria brassicola,* and *Botrytis* sp. Liu *et al.* (2004a) also reported that *S. rimosus* showed a high antagonism activity against *Fusarium solani, F. oxysporium* f sp. *cucumarinum, Verticillium dahliae, R. solani, Fulvia fulva, Botrytis cinerea, A. alternata, Sclerotinia sclerotiorum* and *Bipolaris maydis.* The antifungal antibiotic, which is produced by *S. rimosus,* was purified by silica gel column chromatography. Its ultraviolet (UV) spectrum was consistent with that of polyene macrolide, which had the same absorption peaks at 291, 305, and 318 nm. Antifungal activity can be kept for 20 months at room temperature (12–30°C, pH 5.4) (Liu *et al.,* 2004b). So *S. rimosus* will be employed as a target to search for new biocontrol agents or drugs to satisfy public demands, and much interests will be generated (Table1).

Name of the actinobacteria	Fungicidal activity against	Disease control	Investigator(s) and year
Actinomycetes	*Fusarium oxysporum* *F. cubense* *Penicillium graminicolum*	Vascular wilt in Tomato	Meredith, 1946
Streptomyces antibioticus	*Helminthosporium sativum*	Seedling blight and root rot of cereals	Stevenson, 1956
Streptomyces aureus	*H. oryzae* *C. lunata*	Brown leaf spot of Rice Stem blight disease of Cassava	Chakrabarthy and Chandra, 1979
S. violaceoniger	*Macrophomia phaseolina*	Charcoal root rot of cotton, soybean, peanut, and corn.	Hussain and El-Gammal, 1980 Ivanova *et al.*, 1998
S. globus	*Alternaria solani* *A. niger* *C. pallescens* *Phytophthora* sp.	Early blight in potato and Tomato Pod, seed rot of peanut Black mould of onion Wilt and root rot	Paul and Banerjee, 1983 Nair *et al.*, 1994 Hwang *et al.*, 1996.
S. arabicus	*A. brassicae*	Blight, black leaf spot in cauliflower, cabbage, radish	Sharma *et al.*, 1985; El-Shahed, 1994
Streptomyces sp.	*Botrytis* sp. *Helminthosporium* sp. *Fusarium* sp. *Pyricularia* sp. *Sclerotinia sclerotiorum*	Blights and fruit rots in alliums, beans, carrot, celery, citrus Leaf, crown, and root diseases in bluegrass and rye - Blast in Rice Cottony rot, watery soft rot, stem rot in flowers and vegetables	Tanaka *et al.*, 1987 De and Gupta, 1991 Matsuyama, 1991 Wang and Shen, 1992
Streptomyces sp.	*Phytophthora* sp.	Blights of leaves and shoots and root and crown rots in Ericaceae (Rhododendrons, azaleas, etc.)	Omura *et al.*, 1988; Omura, 1990; Hwang *et al.*, 1994; Tang *et al.*, 2000
S.griseochromogenes	*S. scabies* Botrytis sp. *Phytophthora*	Scab on potatoes, beets, radish, rutabaga, turnip, carrot and parsnips - Late blight or potato blight.	Cheng *et al.*, 1989 Philips and Mc Closkey, 1990 Eckwall and Schottel, 1997 Xiao *et al.*, 2002
Streptomyces	*Phytophthora capsici*	Blight in *Capsicum annuum*	Kook and Kim, 1995
Marine *Streptomyces* sp. Strain AP77	*Pythium*	Damping-off and seedling diseases	Woo and Kamei, 2003

Name of the actinobacteria	Fungicidal activity against	Disease control	Investigator(s) and year
Streptomyces sp.	*F. moniliforme, ,* *F. oxysporum,* *Botrytis pumilis* *C. albicans*	Ear-rotting in maize Wilt on banana, tomato - -	Haque *et al.*, 1992 Abussaud, 1996 Malviya *et al.*, 1994 Saadoun *et al.*, 2000 a,b Singh et al., 2009
Streptomyces sp.	*A.fumigatus, A. niger, M. hiemalis,* *P. roqueforti, P. rariotii, C. albicans,* *Cryptococcus humicolus* *P. cinnamoni, Pestalotiopsis sydowiana,* *B. cinerea, Sclerotinia homoeocarpa,* *A. alternata, M. circinellodies, P.citirum,* *P. niotinanae, P. aphanidermatum,* *P. oligandrum, P. porphyrae, P. ultimo,* *Ustilogo maydei*	- - - Roots of disease-Rhododendron - Corn smut is a disease of maize	Frandberg *et al.*, 2000 Shimizu *et al.*, 2000 Woo and Kamei., 2003 Moncheva *et al.*, 2002 Augustine *et al.*, 2004 Augustine and Kapadnis, 2005 Dhanasekaran *et al.*, 2005 b
Marine actinomycetes	*R. solani, Pyricularia oryzae, H. oryzae* and *Colletotrichum falcatum*	- Red rot disease of sugarcane	Kathiresan *et al.*, 2005 Dhanasekaran *et al.*, 2005c
Streptomyces sp. GAAS7310	*Rhizopus nigricans* *Geotrichum candidum Link* *Fusarium oxysporum f.* sp.*cubense* *Cladosponum carpophilum Thum.* *Alternaria solani (E. et M.) Jones et Grout* *Cladosporium* sp. *Sclerotinia sclerotiorum (Lib.) de Bary* *Peronophthora litchi Chen* *Botrytis* sp. *Colletotrichum nigrum EL.et Halst* *Curvularia lunata (Wakker) Boed* *Phomopsis vexans (Sacc et Syd) Harter*	Rot in Peaches, cherries Sour rot in peach, tomato Panama disease in banana Scab in Plum Early blight in tomato Internal blight in papaya fruit White Rot of Sunflower Brown blight in Litchi Leaf rot in Potato Anthracnose in Chilli Leaf Spot in Millet Phomopsis blight in Brinjal	Chen *et al.*, 2005 Thakur *et al.*, 2007
Streptomyces	*Aspergillus flavus, A. niger, Candida albicans, Fusarium semitectum, Rhizoctonia solani and Botrytis cinera*	- - -	Yadav *et al.*, 2009
Streptomyces, Microbispora, Micromonospora and *Nocardia*	*Bipolaris sorokiniana* *Verticillium alboatrum* *Rhizoctonia* sp.,	Root rot, Spot blotch of barley Wilt in alfalfa and cotton -	Oliveira, *et al.*, 2010
Actinoplanes sp. HBDN08	*Cladosporium cucumerinum,* *Corynespora cassiicola,* *Rhizoctonia solani,* *Phytophthora capsici, and Phytophthora infestans.*	Scab on water melon seedlings Target spot in Tomato - Blight on pumpkin, squash, pepper	Zhang *et al.*, 2010
Streptomyces sp. 201	*Rhizoctonia solani*	Damping off-cause of death of seedlings in tomato	Thakur *et al.*, 2009

Table 1. Fungicidal activity of actinobacterial isolates against plant pathogens

4. Fungicidal potential of actinobacteria against human fungal pathogens

Some species of fungi produce mycotoxins that are very toxic to humans. For example, the fungus *Claviceps purpurea* causes the ergot poisoning. An individual infected with the mycotoxin experiences hallucination, gangrene, and blood flow restrictions in limbs. Humans usually get infected with the fungus after eating cereal grains contaminated with *C. purpurea.*

The incidence of opportunistic mycoses and the number of different fungal pathogens are increasing dramatically. During the 1980s, the frequency of nosocomial candidemia increased as much as 500% over the decade (Mitchell 1998). High mortality and increasing antifungal drug resistance are also major concerns. These trends will continue unless better preventive or treatment measures are developed. The traditional approach is to increase the screening programmes that are still being initiated in various countries for the isolation of antibiotic producing microorganisms from the environment, especially marine and terrestrial soil, which provide a rich source for these organisms, particularly the actinobacteria (Labeda & Shearer 1990). It has been estimated that approximately two-thirds of naturally occurring antibiotics have been isolated from actinobacteria (Takizawa *et al.* 1993). Of these antibiotics, the majority were isolated from the genus *Streptomyces* (Goodfellow & O'Donnell 1989).

Saadoun *et al.* (2000b) identified several *Streptomyces* isolates from soils in northern Jordan which were bioassayed for their antifungal activity against several food-associated fungi and moulds isolated from olive-mill residue. Dhanasekaran et al. (2008) reported the antifungal compound 4' phenyl-1-napthyl-phenyl acetamide from *Streptomyces* sp. DPTB16. It showed significant antifungal activity against *Candida albicans* followed by *Aspergillus niger, A. fumigatus, A. flavus* and minimum inhibitory activity was observed with *Mucor* sp. and *Penicillium* sp. Kumar and Kannabiran (2010b) reported the antifungal activity of *Streptomyces* VITSVK5 spp. against drug resistant *Aspergillus* clinical isolates from pulmonary tuberculosis patients.

Dermatophyte infections are one of the earliest known fungal infections of mankind and are very common throughout the world. There are three genera of dermatophytes, such as *Trichophyton, Microsporum* and *Epidermophyton*. Dermatophytoses are world wide in distribution with high prevalence in tropical and sub-tropical countries due to the hot and humid climate which favours their growth. As the dermatophytes have developed resistance to antimycotic drugs and due to a lack of safe and effective antifungal antibiotics, there is an urgent need for nontoxic, safe and cost effective antifungal antibiotics. Deepika *et al.* (2009) reported the actinobacteria exhibiting antidermatophytic activity against *Trichophyton rubrum* were identified among 100 isolates by cross streak method. Among them only two actinobacterial isolates DKD 6 and DKD 7 exhibiting potential antidermatophytic activity and further characterized and identified as *Streptomyces* sp (Table 2; Plate 3).

Organism	Activity against	Disease control	Investigator(s) and year
Streptomyces spp.	*Trichophyton* sp., *Fusarium* sp., *Penicillium* sp.	Dermatophytoses -Trichophytosis in humans and animals	Leben *et al.*, 1952
Streptomyces aureus	*T. mentagrophytes*	Tinea in man and animals	Chakrabarthy and Chandra, 1979
S. hygroscopicus	*T. mentagrophytes* and *C. albicans*	Tinea in man and animals Oral thrush in man and animals	Gurusiddaiah *et al.*, 1979
S. globus	*T. rubrum, T. mentagrophytes* *C. albicans*	Tinea capitis in Man and animals -	Hwang *et al.*, 1996.
S. aerocolonigens	*C. albicans*	Oropharyngeal candidiasis (thrush) and vulvovaginal candidiasis (vaginal Candidiasis)	Nishio *et al.*, 1989
S. roseiscleroticus (Sultriecin) and *S. hygroscopicus* (Yatakemycin)	*Cryptococcus neoformans* *Blastomyces dermatitides* *A. fumigatus* *F. moniliforme,* *Petriellidium boydii* *C. albicans* *T. mentagrophytes*	Cryptococcal meningitis Dermatitidis in skin bronchopulmonary aspergillosis - Visceral infections- endocarditis - Ringworm infections of man, domestic and captive animals such as horses, chinchillas, dogs, cats, calves, and monkeys, as well as many wild animals such as foxes, muskrats, squirrels	Ohkuma *et al.*, 1992 Imamura *et al.*,1993 Atalan, 1997; Zheng *et al.*, 2000 Datta *et al.*, 2001
S. violaceusniger (new-macrolide)	*C. neoformans,C. albicans, C. tropicalis,* *C. parapsiolis,C. glabrata, A. fumigatus,* *C. gloeosporides A. flavus,* *T. mentagrophytes, T. rubrum,* *M. canis, M. gypseum,, M. grisea*	- - - Tinea barbae, unguim, pedis Ectothrix Hair infection, Tinea barbae	Ubukata *et al.*, 1995 a Fulgueira *et al.*, 2004
Streptomyces sp.	*Candida albicans,* *Cryptococcus terreus* *Aspergillus terreus* *Aspergillus niger* *Penicillium funiculosum* *Trichophyton rubrum* *Alternaria alternata* *Alternaria brassicicola* *Fusarium moniliforme* *Aspergillus parasiticus* *Aspergillus ochraceus*	- Lungs pneumonia - - - - Upper respiratory tract infections and asthma Mycotic keratitis Aflatoxicosis Mycotoxicosis	Banga *et al.*, 2008 Deepika *et al.*, 2009 Valan Arasu *et al.*, 2009 Arumugam *et al.*, 2010 Kumar and Kannabiran, 2010a Duraipandiyan *et al.*,2010
Streptomyces sp. DPTB16	*Aspergillus flavus, A. niger, A. fumigatus* *Mucor* sp. *Penicillium* sp. *Candida albicans*	Bronchopulmonary aspergillosis Pulmonary Zygomycosis Pulmonary grnuloma, Keratitis	Dhanasekaran *et al.*, 2008; 2009a

Table 2. Fungicidal activity of actinobacterial isolates against human pathogens

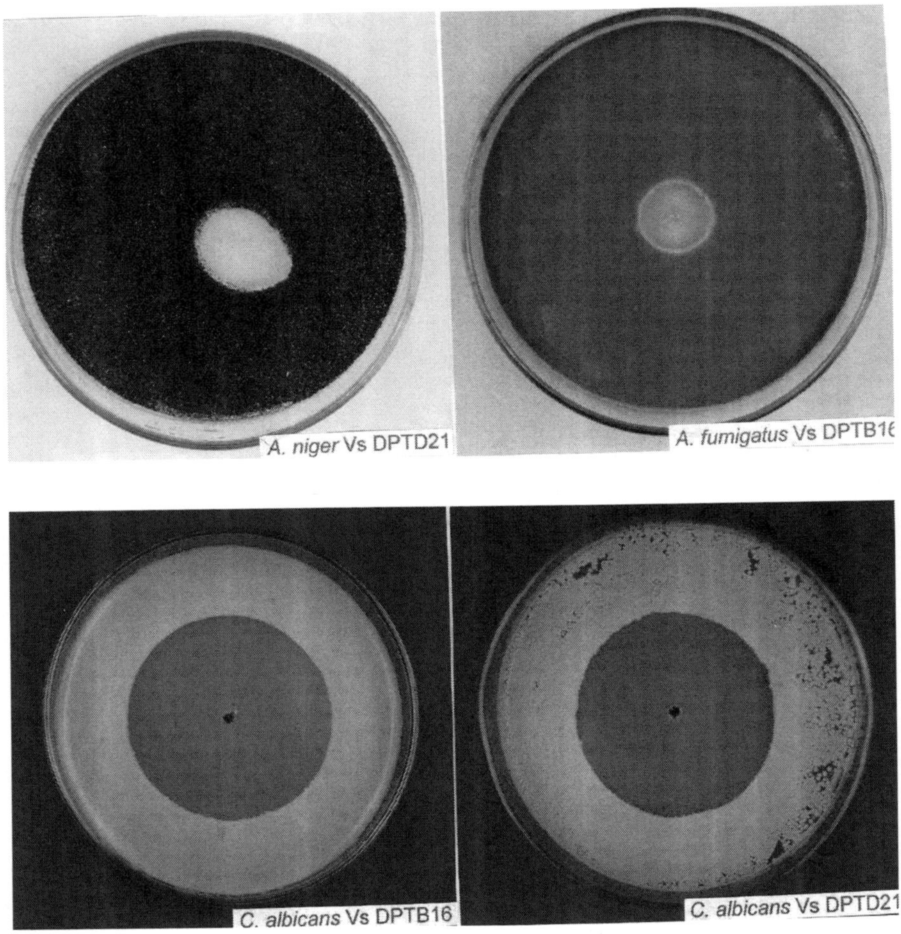

Plate 3. Antifungal activity 4' phenyl-1-napthyl-phenyl acetamide and methyl substituted β-lactum compounds of *Streptomyces* isolates DPTB16 and DPTD21

5. Characterization of actinobacterial fungicidal compounds

A novel antifungal antibiotic, FR-900848 was isolated from *Streptoverticillium ferverns* and its physico-chemical and biological properties have been reported (Yoshida and Horikoshi, 1988). Tomita *et al.* (1989) reported that Pradimicins A, B and C as new antifungal antibiotics from *Streptomyces* sp. Fushimi *et al.* (1989b) reported new phosphate ester antifungal antibiotics phoslactomycins and elucidated the structure of phoslactomycins A to F.

Yamaguchi *et al.* (1989) studied the mode of antifungal action of (S)2-amino-4-oxo-5-hydroxypentanoic acid, RI-331 derived from *Streptomyces* sp. Schwartz *et al.* (1988) studied L-671, 329, a new antifungal agent from *Streptomyces* strain.

Oki *et al.* (1989) studied cispentacin, a new antifungal antibiotic and its *in vitro* and *in vivo* antifungal activities. Novel antifungal antibiotics, Maniwamycins A and B I and II and their structure were studied and reported (Nakayama *et al.*, 1989; Takahashi *et al.*, 1989).

Konishi *et al.* (1989) studied the production, isolation, physico-chemical properties and its structure of cispentacin, a new antifungal antibiotic. A novel hepatoprotective γ-lactone, MH-031 was discovered and their physico-chemical properties and structure have been reported (Itoh *et al.*, 1991). Nishio *et al.* (1989) reported Karnamicin, a complex of new antifungal antibiotic from *Streptomyces* sp. and its taxonomy, fermentation, physico-chemical and biological properties.

Sawada *et al.* (1990) reported new antifungal antibiotics, pradimicins D and E glycine analogs of pradimicins A and C. Water-soluble pradimicin derivatives synthesis and antifungal evaluation of N, N-dimethyl pradimicins derived from *Actinomadura hibisca* was studied and reported (Oki *et al.*, 1990a,b).

Kakushima *et al.* (1990) studied the effect of stereochemistry at the C-17 position on the antifungal pradimicin A. Stephan *et al.* (1996) was observed that Kanchanamycins, new polyol macrolide antibiotics produced by *S. olivaceus*. The structures of the Kanchanamycins were determined by eletrospray MS and modern 2D NMR techniques. A Manumycin type antibiotic (SW-B) was isolated from a solid agar culture of *S. flavus* strain A-11. The structure was determined by MS and by 1 and 2D NMR spectroscopy (Kook *et al.*, 1996).

Harindran *et al.* (1999) isolated a new antifungal antibiotic, HA-1-92 from the biomass of *Streptomyces* CDRIL-312. The antibiotic is presumed to be an oxehexaene macrolide and showed promising antifungal activity against yeasts and filamentous fungi. The structure elucidation and antifungal activity of plants an anthracycline antibiotic, daunomycin, isolated from *Actinomadura roseola* against *Phytophthora* blight in pepper have been reported (Kim *et al.*, 2000). A new tetraene polyene macrolide antibiotic was isolated from the culture broth of *S. arenae* var. *ukrainiana* and its structure was determined on the basis of spectral data such as UV, IR, ^{1}H, ^{13}C NMR and Mass spectroscopy (Gupte *et al.*, 2000). The isolation and structure elucidation of a new antifungal and antibacterial antibiotic produced by *Streptomyces* sp. have been reported (Bordoloi *et al.*, 2001).

Hwang *et al.*, (2001) isolated the antifungal substances SH-1 and SH-2 from *Streptomyces humidus* strains SE 55 cultures by various purification procedures and identified as phenyl acetic acid and sodium phenyl acetate respectively based on the nuclear magnetic resonance, electron ionization mass spectral analysis and inductively coupled plasma mass spectral data SH-1 and SH-2. The two compounds were as effective as the fungicide metalaxyl in inhibiting spore germination and hyphal growth. Raytapadar and Paul (2001) found a broad-spectrum antifungal *Streptomyces* isolate IDA-28 from Indian soil, which was characterized and identified as *Streptomyces aburaviensis* var. *ablastmyceticus* (MTCC 2469).

Frandberg *et al.* (2000) observed antifungal compounds on solid substrate that inhibit radial growth of fungi among Ascomycetes, Basidomycetes, Deuteromycetes, Oomycetes and Zygomycetes and strongly affected hyphal branching and morphology of the fungus such as *Aspergillus niger*, *Mucor hiemalis*, *Penicillium roqueforti* and *Paecilomyces variotii*.

Ellis (2002) reported that Amphotericin B is a polyene macrolide antibiotic derived from the actinomycete, *Streptomyces nodosus*. Amphotericin B has a relatively broad spectrum of action and is useful in treating cases of Candidiasis, extracutaneous sporotrichosis, Mucormycosis and some cases of hyalohyphomycosis and Phaeohyphomycosis.

Igarashi *et al.* (2003) screened for novel antifungal compound, Yatakemycin from the *Streptomyces* species TP - AO 356. Yatakemycin were obtained by solvent extraction of the fermentation broth and preparative HPLC. NMR elucidated the structure of Yatakemycin and CID - MS/MS experiments as a novel antibiotic belonging to a family of CC - 1065 and duocarmycins known to be DNA alkylating agents. Yatakemycin inhibited the growth of pathogenic fungi such as *Aspergillus fumigatus* and *Candida albicans* with the MIC values of 0.01 - 0.03 µg/ml more potent than amphotericin B (MIC 0.1 - 0.5 µg/ml). It also showed potent cytotoxicity against cancer cell lines with the IC_{50} of 0.01 - 0.3 µg/ml.

Datta *et al.* (2001) studied the Ju-2 a novel phosphorous-containing antifungal antibiotic from *Streptomyces kanamyceticus* M8. Ellaiah *et al.* (2005) studied the chracteristics of oligosaccharide antibiotic by ^{1}H NMR and ^{13}C NMR spectra and elucidated the structural formula as $C_{14}H_{86}O_{17}$. Separation, purification and structural elucidation of Irumamycin and 17-hydroxy-venturicidin were established by IR, ESI-MS, ^{1}H and ^{13}C NMR data (Fourati *et al.*, 2005). The numerous fungicidal compounds from actinobacterial genera are summarized in the Table. 3

6. *In silico* molecular mechanism of action of fungicidal compounds

Computational biology has now become an indispensable part in research to understand the biological process in a better way at a short period. *In silico* molecular docking is an application area in bioinformatics, which studies the interaction between the molecules by fitting the molecules in a 3D space. The interactions between protein-protein, protein - DNA, protein - small molecule (drug) and even within carbohydrates and lipid molecules can be studied using docking tools.

Docking tools are automated, available both in online and offline, most of the offline tools are commercial tools. Each tool has its own advantage. Some of the online tools are patch dock (bioinfo3d.cs.tau.ac.il/PatchDock/), Zdock server (zdock.bu.edu/), Dock blaster (blaster.docking.org/). Few non commercial docking tools are Hex, Autodock, Dock, MS-dock and few commercial tools are Flexidock, GOLD, HADDOCK.

Molecular docking studies can also be performed for fungicides. As fungicides are the chemical compounds used to kill or inhibit fungi or fungal spores. The fungicides act primarily by inhibiting any of the process: electron transport chain, nucleic acid synthesis, mitosis and cell division, protein synthesis, lipid and membrane synthesis, sterol biosynthesis. The enzyme involved in any of the above mentioned process can be considered as a target receptor and the fungicide as ligand.

In case of a fungicide that targets sterol biosynthesis (DM inhibitors), CYP 51 enzyme (14 -α demethylase) involved in the ergosterol biosynthesis can be chosen as a target (Yang, *et al.*, 2009). The docking studies require the 3D structure of the target enzyme. If the 3D structure of the target enzyme is not already available, then homology modeling can be performed.

The docking tools calculate the binding energy between the fungicide and the enzyme target. The binding energy is denoted as E value. The E value for the docked complex should be more negative. The more negative the E value, more stable the docking complex formed.

S.No.	Compound	Source	References
1.	Rapamycin	*Streptomyces hygroscopicus*	Lomovskaya *et al.*, (1997)
2.	FK520 Ascomycin	*Streptomyces hygroscopicus* var.*ascomyceticus*	Wu *et al.*,(2000)
3.	Jinggangmycin	*Streptomyces hygroscopicus*	Jian *et al.*,(2006)
4.	Amphotericin B	*Streptomyces nodosus*	Caffrey *et al.*,(2001)
5.	CE-108	*Streptomyces diastaticus*	Perez-Zuniqua *et al.*, (2004)
6.	Rimocidin	*Streptomyces diastaticus* var. 108	Seco *et al.*, (2004)
7.	Shurimycins A and B	*Streptomyces hygroscopicus*	Kamazawa *et al.*, (1994)
8.	Nikkomycins	*Streptomyces ansochromogenus*	Luo *et al.*,(1998)
9.	Aminoglycoside antibiotics, Istamycins	Streptomyces tenjimariensis	Hotta *et al.*, (1980)
10.	Scopafungin	*Streptomyces hygroscopicus* var. *enhygrus* var. nova UC-2397	Samain *et al.*,(1982)
11.	Polyether antibiotic-Ferensimycins	*Streptomyces myxogenes*	Kusakabe *et al.*,(1982)
12.	Auramycins A and B	*Streptomyces galilaeus* OBB-111	Hoshino *et al.*, (1982)
13.	Aminoglycoside antibiotic-Boholmycin	*Streptomyces hygroscopicus* H617-25	Saitoh *et al.*, (1988)
14.	Maniwamycins	*Streptomyces prasinopilosus*	Nakayama *et al.*, (1989)
15.	Phoslactomycins	*Streptomyces nigrescens*	Fushimi *et al.*, (1989a)
16.	Griseochelin	*Streptomyces griseus*	Grafe *et al.*, (1984)
17.	Resormycin	*Streptomyces platensis*	Iquarashi *et al.*, (1997)
18.	Ileumycin	*Streptomyces lavendulae*	Kawakami *et al.*, (1978)
19.	Kalafungin	*Streptomyces tanashiensis* strain Kala UC5063	Johnson and Dietz (1968)
20.	Lomofungin	*Streptomyces lomodensis*	Johnson and Dietz (1969)
21.	Axenomycins	*Streptomyces lisandri* nov. sp.	Bruna *et al.*,(1973)
22.	Candiplanecin	*Ampullariella reguralis* subsp.*mannitophila* subsp. nov.	Itoh *et al.*,(1981)
23.	Milbemycins	*Streptomyces hygroscopicus* sub sp. *aureola rimosus*	Takahashi *et al.*,(1993)
24.	Kanchanamycins	*Streptomyces olivaceus*	Stephan *et al.*,(1996)
25.	Pimaricin	*Streptomyces natalensis*	Recio *et al.*,(2004)
26.	Blasticidin S	*Streptomyces griseochromogenes*	Zhanq *et al.*,(1998)
27.	Leptomycin	*Streptomyces lividans*	Hu *et al.*,(2005)
28.	RS-22 A, B and C	*Streptomyces violaceusniger*	Ubukata *et al.*, 1995a,b
29.	Methyl substituted β-lactum compound	*Streptomyces*DPTD21	Dhanasekaran, 2005a
30.	4' phenyl-1-napthyl-phenyl acetamide	*Streptomyces*DPTB16	Dhanasekaran *et al.*, 2008

Table 3. Fungicidal secondary metabolites produced by actinobacteria

Computational advancement also provides way to modify the functional group of the fungicide, leading to the creation of analogues of the fungicide. These docking approaches provide insight into the structure based drug designing. Structure based drug designing assist in the creation of novel fungicides that can be used to treat already existing drug resistant fungal pathogens.

The interaction between the fungicide and the enzyme target will always by hydrogen bond formation between electropositive and electronegative atom. For an enzyme target to get disrupted, the H bond formation of the fungicide should be within the active site of the target.

The knowledge about the active site of the target enzyme can be obtained using online active site predicting tools. Few such tools include: CASTp - Computed Atlas of Surface Topography of proteins (http://sts.bioengr.uic.edu/castp/calculation.php), Q-site finder (http://www.modelling.leeds.ac.uk/qsitefinder/), and Pocket finder (http://www.modelling.leeds.ac.uk/pocketfinder/).

The antifungal compound isolated from the marine *Streptomyces* sp. DPTB16 was characterized as 4-Phenyl-1-Napthyl Phenyl Acetamide and its 1D structure was also elucidated via spectral analysis (Fig.2). The structure of the compound was submitted to Pubchem compound database with accession number CID: 49786168 (Fig.3,4,5).

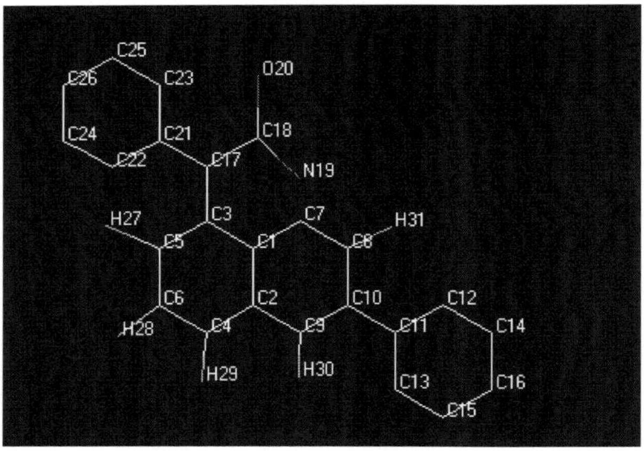

Fig. 2. The elemental representation of 4 - Phenyl 1-Napthyl Phenyl Acetamide

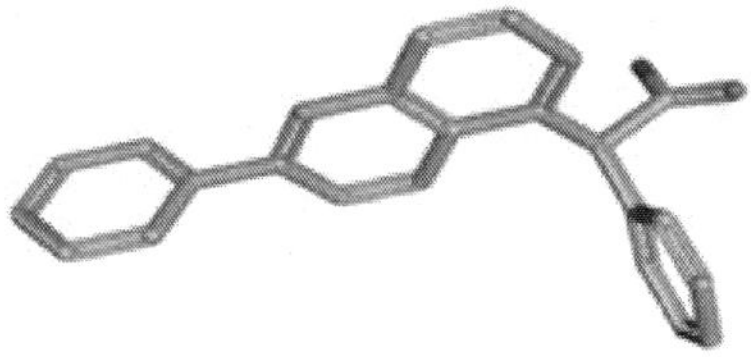

Fig. 3. The 3D structure of 4P1NPA in Pubchem compound database (CID: 49786168)

In silico tools are fruitful to gain insight into the mode of action of fungicides and help the research to move a step ahead leading to the rational drug discovery.

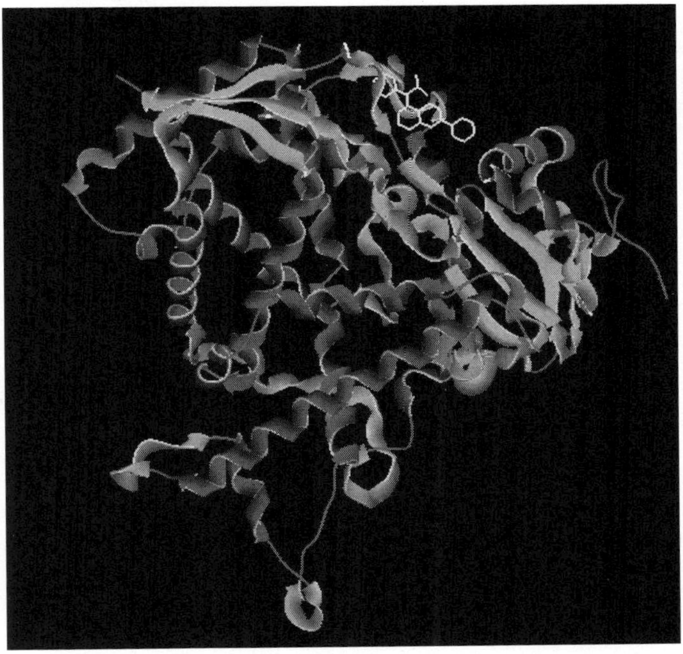

Fig. 4. Cytochrome 51 docking complex with a fungicidal compound(4P1NPA)

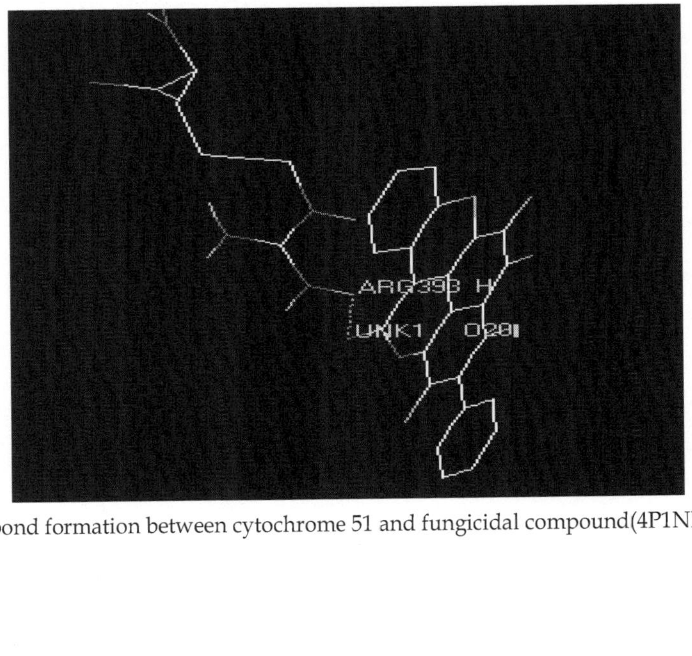

Fig. 5. H bond formation between cytochrome 51 and fungicidal compound(4P1NPA)

7. Conclusion

Actinobacteria isolated from marine habitat have potential for novel fungicidal compounds. The significance of finding actinobacteria in marine soil sample lies in the intrinsic economic importance in biotechnological perspectives. The chapters reinforce the view that the unexplored actinobacteria for bioprospecting novel fungicidal compound in the development of Biocontrol agents and formulations of drugs. Further investigation should address the relationship between structure relationship activity of fungicidal compounds, rapid methods for large scale production, purification and application in managing fungal infection in agriculture crops, human and animals.

8. References

Abussaud, 1996. Characteristics of *Streptomyces* strains isolated from soils in two landfill areas in North Jordan. Acta Microbiol. Immunol. Hungarica, 43 (1): 47-53.

Arumugam, M, Mitra A, Jaisankar P,Dasgupta S, Sen T,Gachhui R, Mukhopadhyay U K and J. Mukherjee. 2010. Isolation of an unusual metabolite 2-allyloxyphenol from a marine actinobacterium, its biological activities and applications, Appl Microbiol Biotechnol 86:109–117

Atalan, E., 1997. Analysis of streptomycetes taxa by pyrolysis mass spectrometry. Turkish J. Biol., 21(4): 487-493.

Augustine, S. K., Bhavsar, S. P., Baserisallhi and Kapadnis, B.P., 2004. Isolation, characterization and optimization of antifungal activity of an actinomycetes of soil origin. *Indian J. Exp. Biol.*, 42: 928-932.

Augustine, S.K. and Kapadnis, B.P., 2005. Bioactive compounds from actinomycetes with a potential to inhibit pathogenic fungi. *J. Microb. World*, 7(2): 328-331.

Banga, J, Vpraveen,V, Singh,V, Tripathi, C. K. M. and V Bihari, 2008. Studies on medium optimization for the production of antifungal and antibacterial antibiotics from a bioactive soil actinomycetes. Med Chem Res 17:425–436.

Bordoloi, G.N., Kumarim, B., Guha, A., Bordoloi, M., Yadav, R.N., Roy, M.K. and Bora, T.C., 2001. Isolation and structure elucidation of a new antifungal and antibacterial antibiotic produced by *Streptomyces* sp. 201. Biosci. Biotechnol. Biochem., 65(8): 1856-1858.

Bruna, CD, Ricciardi ML and Sanfi lippo A, 1973. Axenomycins, new cestocidal antibiotics. Antimicrob Agents Chemother 3:708–710.

Caffrey P, Lynch S, Flood E, Finnan S and Oliynyk M., 2001. Amphotericin biosynthesis in *Streptomyces nodosus*: deductions from analysis of polyketide synthase and late genes. Chem Biol 8:71–723.

Cao, L., Qiu, Z., Daı´, X., Tan, H., Lin, Y., Zhou, S., 2004b. Isolation of endophytic actinobacteria from roots and leaves of banana (*Musa acuminata*) plants and their activities against *Fusarium oxysporum* f. sp. cubense. World J. Microbiol. Biotechnol. 20, 501-504.

Cao, L., Qiu, Z., You, J., Tan, H., Zhou, S., 2004a. Isolation and characterization of endophytic *Streptomyces* strains from surface-sterilized tomato (*Lycopersicon esculentum*) roots. Lett. App. Microbiol. 39, 425-430.

Chakrabarty, S. and Chandra, A.L., 1979. Antifungal activity of A-7, a new tetraene antibiotic. Indian J. Exp. Biol., 17 (3): 313-315.

Chen, G, Lin, B, Lin, Y, Xie, F Lu. W and WF Fong. 2005. A New Fungicide Produced by a *Streptomyces* sp. GAAS7310. J. Antibiot. 58(8): 519–522,

Cheng, X.C., Kihara, T., Ying, S., Uramoto, M., Osada, H., Kusakabe, H., Wang, B.N., Kobayashi, Y., Ko, K., Yamaguchi, I., Shen, Y.C. and Isono, K., 1989. A new antibiotic, tautomycetin. J. Antibiot. 42(1): 141-144.

Crawford, D.L., Lynch, J.M., Whipps, J.M., Ousley, M.A., 1993. Isolation and characterization of actinomycete antagonists of a fungal root pathogen. Appl. Environ. Microbiol. 59, 3899e3905.

Datta, I., Banerjee, M., Mukherjee, S.K. and Majumdar, S. K., 2001. JU-2, a novel phosphorous-containing antifungal antibiotic from *Streptomyces kanamyceticus* M8. Indian J. Exp. Biol. 39(6): 604-606.

De, K. and Gupta, M.K., 1991. Antifungal activity of some soil actinomycetes. *Indian J. Microbiol.*, 31(1): 53-54.

Deepika,T. L., Kannabiran, K and D. Dhanasekaran, 2009. Diversity of antidermatophytic *Streptomyces* in the coastal region of Chennai, Tamil Nadu, India, Journal of Pharmacy Research, 2(1),22-26.

Dhanasekaran D. Selvamani, S., Panneerselvam A and N. Thajuddin, 2009b. Isolation and characterization of actinomycetes in Vellar Estuary, Annagkoil, Tamil Nadu. African Journal of Biotechnology, 8 (17): 4159-4162.

Dhanasekaran, D. 2005a. Biochemical and molecular characterization and antimicrobial compounds of *Streptomyces* spp. from Cuddalore District, Tamilnadu, India. Ph. D Thesis, Bharathidasan University, Tiruchirappalli-24. pp 48.

Dhanasekaran, D., Paneerselvam, A. and Thajuddin, N, 2008. An antifungal compound: 4' phenyl -1-napthyl -phenylacetamide from *Streptomyces* sp. DBTB16. Facta Universitatis *Series*: Medicine and Biology, 15(1), 7 – 12.

Dhanasekaran, D., Panneerselvam, A. and Thajuddin, N., 2005b. Antifungal actinomycetes in marine soils of Tamil Nadu. *Geobios*, 32: 37-40.

Dhanasekaran, D., Sivamani, P., Panneerselvam, A., Thajuddin, N., Rajakumar, G. and Selvamani, S., 2005c. Biological control of Tomato seedling damping off with *Stretomyces* sp. *Pl. Pathol., J.*, 4(2): 91-95.

Dhanasekaran,D, Panneerselvam, A and N. Thajuddin, 2009a. Distribution and ecobiology of antagonistic *Streptomyces* from agriculture and coastal soil of Tamil Nadu, India, Journal of Culture collections,6,10-20

Duraipandiyan V. Sasi A.H. Islam V.I.H. M. Valanarasu and S. Ignacimuthu. 2010. Antimicrobial properties of actinomycetes from the soil of Himalaya. J Mycol Med 20: 15—20

Eckwall, E.C. and Schottel, J.L., 1997. Isolation and characterization of an antibiotic produced by the scab disease suppressive *Streptomyces diastatochromogenes* strain. J. Indus. Microbiol. Biotechnol., 19(3): 220-225.

Ellaiah, P., Adinarayana, G., Saisha, V and Vasu P. 2005. An oligoglycosidic antibiotic from a newly isolated *Streptomyces albovinaceus,* Indian J. of Microbiol., 45: 33-36

Ellis, D. 2002. Amphotericin B spectrum and resistance. Journal of Antimicrob chemother. 49 Suppl 1:7-10.

El-Shahed, K.Y.I., 1994. Production of antifungal antibiotics Polyoxins by *Streptomyces cacoi* var *asoensis* NRC-19. Egyptian J. Microbiol., 29(3): 315-328.

El-Tarabily, K.A., Sivasithamparam, K., 2006. Non-streptomycete actinobacteriaas biological agents of soil-borne fungal plant pathogens and asplant growth promoters. Soil Biol. Biochem. 38, 1505-1520.

Fourati, B.F.L., Fatso, S., Ben Ameur, M.R., Mellouli, L. and Laatsch, H., 2005. Purification and structure elucidation of antifungal and antibacterial activities of newly isolated *Streptomyces* sp. strains US80. *Res. Microbiol.*, 156(3): 341-347.

Frandberg, E., Peterson, C., Lundgren, L. N. and Schnurer, J., 2000. *Streptomyces halstedii* K122 produces the antifungal compounds bafilomycin B1 and C1. Canadian J. Microbiol., 46: 753-758.

Fulgueira, C. L., Amigot, S. L. and Madni, C., 2004. Growth inhibition of toxigenic fungi by a proteinaceous compound from *Streptomyces* sp. c/33-6. Curr. Microbiol., 48(2): 135-139.

Fushimi, S., Furihata. K. and Seto, H., 1989b. Studies on new phosphate ester antifungal antibiotics phoslactomycins II. Structure elucidation of phoslactomycins A to F. J. Antibiot., XLII(7): 1026-1036.

Fushimi, S., Nishikawa, S., Shimazu, A. and Seto, H., 1989a. Studies on new phosphate ester antifungal antibiotics phoslactomycins I. Taxonomy, fermentation, purification and biological activities. J. Antibiot., XLII(7): 1019-1025.

Gasperni, G., 1890. Researches morphologiques et biologiques sur un microorganisme de atmosphere, le *Streptothrix foersterii* Cohn. Ann. Microgr., 2: 449-474.

Goodfellow, M. and O'Donnell, A. G., 1989. Search and discovery of industrial significant actinomycetes. In: Microbial Products: New Approaches (S. BaumberG, I. Hunter and M. Rhodes,Editors), pp. 343–383. Cambridge: Cambridge University Press.

Grafe, U., Schade, W. and Roth, M., 1984. Griseochelin, a novel carboxylic acid antibiotic from *Streptomyces griseus*. J. Antibiot. 37(7): 836-846.

Gupte, T. E., Chatterjee, N. R., Nanda, R. K. and Naik, S.R., 2000. Isolation, physio-chemical properties and structure elucidation studies on HA-2-91, a new tetraene polyene antifungal antibiotic produced by *Streptomyces arenae* var *ukrainiana*. Indian J. Chem. 39 (12): 936-940.

Gurusiddaiah, S., Winward, L.D., Burger, D. and Graham, S.O., 1979. Pantomycin, a new antimicrobial antibiotic. Mycologia, 71(1): 103-118.

Haque, S. K. F., Sen, S. K. and Pal, S.C., 1992. Screening and identification of antibiotic producing strains of *Streptomyces*. Hindustan Antibiot. Bull. 34(3-4): 77-83.

Harindran, J., Gupte, T.E. and Naik, S.R., 1999. HA-1-92, a new antifungal antibiotic produced by *Streptomyces* CDRIL-312: Fermentation, isolation, purification and biological activity. World J. Microbiol. Biotechnol., 15: 425-430.

Hasegawa, S., Meguro, A., Shimizu, M., Nishimura, T., Kunoh, H., 2006. Endophytic actinobacteria and their interactions with host plants. Actinomycetologia 20, 72-81.

Hoshino, T., Tazoe, M., Nomura, S. and Fujiwara, A., 1982. New anthracycline antibiotics, auramycins and sulfurmycins II. Isolation and characterization of 10 minor components (C-G). J. Antibiot., 35(10): 1271-1279.

Hotta, K., Saito, N. and Okami, Y., 1980. Studies on new aminoglycoside antibiotics, istamycins, from an actinomycetes isolated from a marine environment I. The use of plasmid profiles in screening antibiotic-producing streptomycetes. J. Antibiot., 33(12): 1502-1509.

Hu, Z, Reid R and Gramajo H., 2005. The leptomycin gene cluster and its heterologus expression in *Streptomyces lividans*. J Antibiot (Tokyo) 58:625–633.

Hussain, A.M. and El-Gammal, A., 1980. An antibiotic produced by *Streptomyces violaceoniger*. Egyptian J. Bot.,, 23(3): 187-190.

Hwang, B. K., Lee, J. Y., Kim, B. S. and Moon, S.S., 1996. Isolation, structure elucidation and antifungal activity of a manumycin type antibiotic from *Streptomyces flaveus*. J. Agri. Food. Chem., 44(11): 3653-3657.

Hwang, B.K., Lim, S.W., Kim, B.S., Lee, J.Y. and Moon, S.S., 2001. Isolation and *in vivo* and *in vitro* antifungal acitivity of phenyl acetic acid and sodium phenyl acetate from *Streptomyces humidus*. Appl. Environ. Microbiol., 67(8): 3739-3745.

Hwang, K. K., Ahn, S.J., and Moon, S. S., 1994. Production, purification and antifungal activity of the antibiotic nucleoside, tubercidin produced by *Streptomyces violaceoniger*. Canadian J. Bot., 72(4): 480-485.

Igarashi, Y., 2004. Screening of novel bioactive compounds from plant-associated actinobacteria. Actinomycetologia 18, 63-66.

Igarashi, Y., Futamata, K., Fujita, T., Sekine, A., Senda, H., Naoki, H., Furumai, T. 2003. Yatakemycin, a novel antifungal antibiotic produced by *Streptomyces sp.* TP - A0356. Journal of Antibiotics (Tokyo). 56(2) : 107-13.

Imamura, N., Nishijima, M., Adachi, K. and Sano, H., 1993. Novel antimycin antibiotics, urauchimycins A and B produced by marine actinomycete. J. Antibiot., 46(2): 241-246.

Iquarashi, M, Kinoshita N, Ikeda T, Kameda M, Hamada M and Takeuchi T., 1997. Resormycin, a novel herbicidal and antifungal antibiotic produced by a strain of *Streptomyces platensis*. I. Taxonomy, production, isolation and biological properties. J Antibiot (Tokyo) 50:1020-1025

Itoh, Y, Torikata A, Katayama C, Haneishi T and Arai M., 1981. Candiplanecin, a new antibiotic from *Ampullariella regularis* subsp. *mannitophila* subsp. nov. II. Isolation, physico-chemical characterization and biological activities. J Antibiot (Tokyo) 34:934–937

Itoh, Y., Shimura, H., Ito, M., Watanabe, N., Yamagishi, M., Tamai, M. and Hanada, K., 1991. A novel hepatoprotective beta-lactone, MH-031 I. Discovery, isolation, physico-chemical properties and structural elucidation. J. Antibiot., 44(8): 832-838.

Ivanova, E.P., Nocolau, D.V., Yumoto, N., Taguchi, T., Okamoto, K., Tatsu, Y. and Yoshikawa, S., 1998. Impact of the conditions of cultivation and adsorption on antimicroibal acitivity of marine bacteria. Mar. Biol., 130: 545-551.

Jian X, Pang X, Yu Y, Zhou X and Deng Z., 2006. Identification of genes necessary for jinggangmycin biosynthesis from *Streptomyces hygroscopicus* 10-22. Antonie van Leeuwenhoek 90:29–39.

Johnson LE and Dietz A., 1968. Kalafungin, a new antibiotic produced by *Streptomyces tanashiensis* strain kala. Appl Microbiol 16:1815-1821.

Johnson LE and Dietz A., 1969. Lomofungin, a new antibiotic produced by *Streptomyces lomondensis* sp. Appl Microbiol 17:755–759.

Kakushima, M., Nishio, M., Numata, K.I., Konishi, M. and Oki, T., 1990. Effect of stereochemistry at the C-17 position on the antifungal activity of pradimicin A. J. Antibiot., XLIII(8): 1028-1030.

Kamazawa S, Asami Y, Awane K, Ohtani H, Fukuchi C, Mikawa T and Hayase T., 1994. Structural studies of new macrolide antibiotics, shurimycins A and B. J Antibiot (Tokyo) 47:688–96.

Kathiresan, K., Balagurunathan, R. and Masilamani Selvam, M., 2005. Fungicidal activity of marine actinomycetes against phytopathogenic fungi. Indian J. Biotechnol., 4: 271-276.

Kawakami Y, Matsuwaka S, Otani T, Kondo H and Nakamura S., 1978. Ileumycin, a new antibiotic against *Glomerella cingulata*. J Antibiot (Tokyo) 31:112–116.

Khamna, S., Yokota, A., Lumyong, S., 2009. Actinobacteria isolated from medicinal plant rhizosphere soils: diversity and screening of antifungal compounds, indole-3-acetic acid and siderophore production. World J. Microbiol. Biotechnol. 25, 649-655.

Kim, B.S., Moon, S.S., Hwang, B.K., 2000. Structure elucidation and antifungal activity of an anthracycline antibiotic, daunomycin, isolated from *Actinomadura roseola*. J. Agri. Food Chem. 48(5): 1875-1881.

Konishi, M., Nishio, M., Saitoh, K., Miyaki, T., Oki, T. and Kawaguchi, H., 1989. Cispentacin, a new antifungal antibiotic I. Production, isolation, Physico-chemical properties and structure. J. Antibiot., XLII(12): 1749-1755.

Kook, H. B. and Kim, B. S., 1995. *In vivo* efficacy and *in vitro* activity of tubercidin, an antibiotic nucleoside, for control of *Phytophthora capsici* blight in *Capsicum annuum*. Pest. Sci. 44(3): 255-260.

Kook, H. B., Lee, J.Y., Kim, B.S. and Moon, S.S., 1996. Isolation, structure elucidation and antifungal activity of a manumycin type antibiotic from Streptomyces flaveus. J. Agri. Food Chem., 44 (11): 3653-3657.

Kumar, S and K. Kannabiran. 2010a. Diversity and Optimization of process parameters for the growth of *Streptomyces* VITSVK9 spp. isolated from Bay of Bengal, India. J Nat Env Sci, 1(2):56-65

Kumar, S and K. Kannabiran. 2010b. Antifungal activity of *Streptomyces* VITSVK5 spp. against drug resistant *Aspergillus* clinical isolates from pulmonary tuberculosis patients. Journal of Medical Mycology,20(2): 101-107

Kunoh, H., 2002. Endophytic actinobacteria: attractive biocontrol agents.J. Gen. Plant Pathol. 68, 249-252.

Kusakabe, Y., Mizuno, T., Kawabata, S., Tanji, S. and Seino, A., 1982. Ferensimycins A and B, two polyether antibiotics taxonomy, fermentation, isolation, characterization and structural studies. J. Antibiot., 35(9): 1119-1129.

Labeda, D.P. & Shearer, M.C. 1990 Isolation of actinomycetes for biotechnological applications. In Isolation of Biotechnological Organisms from Nature, ed. Labeda, D.P. pp. 1-19, McGraw-Hill Publising Company. ISBN 0-07-035701-3.

Leben, C., Stessel, G.J. and Keitt. G.W., 1952. Helixin, an antibiotic active against fungi and bacteria. Mycologia, 44: 159-169.

Liu *et al.*, 2004a Q. Liu, Y.H. Wu and J.C. Yu, Screening for antagonistic actionmyces isolates from greenhouse soil in northeast china, Soil 36:. 573–575.

Liu *et al.*, 2004b Q. Liu, Y.H. Wu and J.C. Yu, Purification of active components SN06 in fermentation of *Streptomyces rimosus* MY02, Acta Phytopath. Sinica 31: 353–358.

Lomovskaya N, Fonstein L, Ruan X, Stassi D, Katz L and Hutchinson CR., 1997. Gene disruption and replacement in the rapamycin producing *Streptomyces hygroscopicus* strain ATCC 29253. Microbiology 143:875–883.

Luo A, Gao C, Song Y, Tan H and Liu Z.,1998. Biological responses of a *Streptomyces* strain producing nikkomycin to space fl ight space. Med Eng (Beijing) 11:411–414.

Malviya, H.K., Golwalkar, M.P., Narayanan, M.S. and Tandon, G.G., 1994. Preliminary studies on *Streptomyces* sp. CS –14 showing broad spectrum antibiotic activity. Hindustan Antibiot. Bull., 36(1-2): 21-29.

Marten, P., Bruckner, S., Minkwitz, A., Luth, P., Bergm G., 2001. RhizovitR: Impact and formulation of a new bacterial product. In: Koch, E., Leinonen P. (Eds.), Formulation of Microbial Inoculants: Proceedings of a meeting held in Braunschweig, Germany. COST Action 830/Microbial inoculants for agriculture and environment, Germany, pp. 78–82.

Matsuyama, N., 1991. Purification and characterization of antifungal substance AC-1 produced by a *Streptomyces* sp. AB-88 M. Ann. Phytophathol. Soc. Japan, 57(4): 591-594.

Meredith, C.H., 1946. Soil actinomycetes applied to banana plants in the field. Phytopathology., 36: 983-987.

Mitchell, T.G. 1998 Medical mycology research and training: needs and opportunities. ASM News 64, 17-23.

Moncheva, P., Tishkov, S., Dimitrova, N., Chipeva, V., Nikolova, S. A. and Bogatzevska, N., 2002. Characteristics of soil actinomycetes from Antartica. J. Culture Collections, 3: 3-14.

Nair, M.G., Amitabh, C., Thorogod, D.L. and Chandra, A., 1994. Gopalamicin, an antifungal macrodiolide produced by soil actinomycetes. J. Agric. Food Chem., 42(10): 2308-2310.

Nakayama M, Takahashi Y, Itoh H, Kamiya K, Shiratsuchi M and Otani G. 1989. Novel antifungal antibiotics maniwamycins A and B. I. Taxonomy of the producing organism, fermentation, isolation, physico-chemical properties and biological properties. J Antibiot (Tokyo). 42(11):1535-40.

Nakayama, M., Takahashi, Y., Itoh, H., Kamiya, K., Shiratsuchi, M. and Otani, G., 1989. Novel antifungal antibiotics Maniwamycins A and B I. Taxonomy of the producing organism, fermentation, isolation, physico-chemical properties and biological properties. J. Antibiot., XLII(11): 1535-1540.

Neeno-Eckwall, E.C., and Schottel, J.L., 1999. Occurance of antibiotic resistance in the biological control of potato scab disease. Biological Control 16, 199-208.

Nishio, M., Tomatsu, K., Konishi, M., Tomita, K., Oki, T. and Kawaguchi, H., 1989. Karnamicin, a complex of new antifungal antibiotics I. Taxonomy, fermentation, isolation and physico-chemical and biological properties. J. Antibiot.., XLII(6): 852-868.

Ohkuma, H., Naruse, N., Nishiyama, Y., Tsuno, T., Hoshino, Y., Sawada, Y., Konishi, M. and Oki, T., 1992. Sultriecin, a new antifungal and antitumor antibiotic from *Streptomyces roseiscleroticus*. Production, isolation, structure and biological activity. J. Antibiot., 45(8): 1239-1249.

Oki, T, Hirano, M., Tomatsu, K., Numata, K.I. and Kamei, H., 1989. Cispentacin, a new antifungal antibiotic II. *In vitro* and *in vivo* antifungal activities. *J.* Antibiot., XLII(12): 1756-1762.

Oki, T., Kakushima, M., Nishio, M., Kamei, H., Hirano, M., Sawada, Y. and Konishi, M., 1990a. Water-soluble pradimicin derivatives synthesis and antifungal evaluation of N, N-dimethyl pradimicins. J. Antibiot., XLIII(10): 1230-1235.

Oki, T., Tenmyo, O., Hirano, M., Tomatsu, K. and Kamei, Hl, 1990b. Pradimicins A, B and C: New antifungal antibiotics II. In vitro and in vivo biological activities. J. Antibiot., XLIII(7): 763-770.

Oliveira, M. F. Silva M.G, S.T. Van Der Sand. 2010. Anti-phytopathogen potential of endophytic actinobacteria isolated from tomato plants (*Lycopersicon esculentum*) in southern Brazil, and characterization of *Streptomyces* sp. R18(6), a potential biocontrol agent. Research in Microbiology 161: 565-572.

Omura, S., 1990. Phthoxazolin, a specific inhibitor of cellulose biosynthesis, produced by a strain of *Streptomyces* sp. J. Antibiot., 43(8): 1034-1036.

Omura, S., Tanaka, L., Hisatome, K., Miura, S., Takahashi, Y., Nagakawa, A., Imai, H. and Woodruff, H.B., 1988. Phthoramycin, a new antibiotic active against plant pathogen, *Phytophthora* spp. J. Antibiot., 41(12): 1910-1912.

Paul, A.K. and Banerjee, A.K., 1983. A new antifungal antibiotic produced by *Streptomyces galbus*. Folia Microbiol., 28(5): 386-396.

Perez-Zuniqua FJ, Seco EM, Cuesta T, Dequenhardt F, Rohr J, Vallin C, Iznaqa Y, Perez ME, Gonzalez L and Malpartida F., 2004. CE-108, a new macrolide tetraene antibiotic. J Antibiot (Tokyo) 57:197–204.

Philips, D.R. and Mc Closkey, J.A., 1990. Isolation and characterization of phosmidosine. A new antifungal nucliotide antibiotic. J. Antibiot. 44(4): 375-381.

Raytapadar, S. and Paul, A.K., 2001. Production of an antifungal antibiotic by *Streptomyces aburaviensis* 1DA-28. Microbiol. Res., 155(4): 315-323.

Recio E, Colinas A, Rumbero A, Aparicio JF and Martin JF., 2004. PI factor, a novel type quorum-sensing inducer elicits pimaricin production in *Streptomyces natalensis*. J Biol Chem 279:41586–41593

Saadoun, I. and Al-Momani, F., 2000a. Activity of North Jordan streptomycete isolates against *Candida albicans*. World J. Microbiol. Biotechnol. 16: 139-142.

Saadoun, I., Hameed, K., Al-Momani, F., Malkawi, H., Meydam, M. & Mohammad, M.J. 2000b. Characterization and analysis of antifungal activity of soil streptomycetes isolated from North Jordan. Egyptian Journal of Microbiology 35, 16: 139-142

Saitoh, K., Tsunakawa, M., Tomita, K., Miyaki, T., Konishi, M. and Kawaguchi, H., 1988. Boholmycin, a new aminoglycoside antibiotic I. Production, isolation and properties. J. Antibiot., 41(7): 855-861.

Samain D, Cook JC and Rinehart KL., 1982. Structure of scopafungin, a potent nonpolyene antifungal antibiotic. J Am Chem Soc 104:4129–4141.

Sawada, Y., Hatori, M., Yamamoto, H., Nishio, M., Miyaki, T. and Oki, T., 1990. New antifungal antibiotics pradimicins FA-1 and Fa-2: D-serine analogs of pradimicins A and C. J. Antibiot., XLIII(10): 1223-1229.

Schottel, JL., Shimizu, K, and Kinkel, LL. 2001. Relationships of *in vitro* pathogen inhibition and soil colonization to potato scab biocontrol by antagonistic *Streptomyces* spp. Biol Control 20:102–112.

Schwartz, R.E., Giacobbe, R.A., Bland, J. and Monaghan, R., 1988. L-671, 329, a new antifungal agent I. Fermentation and isolation. J. Antibiot., XLII(2): 163-167. Schottel, J.L., Shimizu, K., and Kinkel, L.L. 2001. Relationships of *in vitro* pathogen inhibition and soil colonization to potato scab Biocontrol antagonistic *Streptomyces* spp. Biological Control 20, 102-112.

Seco EM, Zuniga FJP, Rolon MS and Malpartida F., 2004. Starter unit choice determines the production of two tetraene macrolides, rimocidin and CE-108, in *Streptomyces diastaticus* var. 108. Chem Biol 11:357–366.

Sharma, A.K., Gupta, J.S. and Singh, S.P., 1985. Effect of temperature on the antifungal activity of *Streptomyces arabicus* against *Alternaria brassicae.* Geobios, 12(3-4): 168-169.

Shimizu, M., Nakagawa, Y., Sato, Y., Furumai, T., Igarashi, Y., Onaka, H., Yoshida, R. and Kunch, H., 2000. Studies on endophytic actinomycetes (1) *Streptomyces* sp. Isolated from *Rhododendron* and its antimicrobial activity. J.Gen. Pl. Pathol., 66(4): 360-366.

Singh L.S.,Mazumder,S and T.C. Bora. 2009. Optimisation of process parameters for growth and bioactive metabolite produced by a salt-tolerant and alkaliphilic actinomycete, *Streptomyces tanashiensis* strain A2D Journal de Mycologie Médicale 19, 225-233.

Stephan H, Kempter C, Metzger JW, Jung G, Potterat O, Pfefferle C and Fiedler HP., 1996. Kanchanamycins, a new polyol macrolide antibiotics produced by *Streptomyces olivaceus* Tu 4018. II. Structure elucidation. J Antibiot (Tokyo) 49:765–769.

Stevenson, I.L., 1956. Antibiotic activity of actinomycetes in soil as demonstrated by direct observation techniques. J. Gen. Microbiol., 15 : 372-380.

Taechowisan, T., Peberdy, J.F., Lumyong, S., 2003. Isolation of endophyticactinobacteria from selected plants and their antifungal activity. World J.Microbiol. Biotechnol. 19, 381-385.

Takahashi, A.S., Nakayama, M., Watanabe, I., Deushi, T., Ishiwata, H., Shiratsuchi, M. and Otani, G., 1989. Novel antifungal antibiotics maniwamycins A and B II. Structure determination. J. Antibiot. XLII(11): 1541-1546.

Takahashi, S., Miyaoka, H., Tanaka, K., Enokita, R. and Okazaki, T., 1993. Milbemycins alpha11, alpha12, alpha13, alpha14 and alpha15: a new family of milbemycins from *Streptomyces hygroscopicus* sub sp. *Aureolac rimosus.* J. Antibiot. 46(9): 1364-1371.

Takizawa, M., Colwell, R.R. and Hill, T.R. 1993 Isolation and diversity of actinomycetes in the Chesapeake Bay. Applied and Environmental Microbiology 59: 997-1002.

Tan, H.M., Cao, L.X., He, Z.F., Su, G.J., Lin, B., Zhou, S.N., 2006. Isolation of endophytic actinobacteria from different cultivars of tomato and their activities against *Ralstonia solanaceraum* in vitro. World J. Microbiol. Biotechnol. 22: 1275-1280.

Tanaka, M., Fulkushima, T., Takahashi, Y., Iwai, Y. and Omura, S., 1987. Globopeptin, a new antifungal peptide antibotic. J. Antibiot., 40(2): 242-244.

Tang, Y.Q., Sattler, I., Thiericke, R., Grabley, S. and Feng, X.Z., 2000. Feigrisolides A, B, C and D, new lactones with antibacterial activities from *Streptomyces griseus,* as well as medium cytotoxic and antiviral activities. J. Antibiot., 53(9): 934-943.

Thakur D., Bora, T.C., Bordoloi, G.N. and S. Mazumdar. 2009.Influence of nutrition and culturing conditions for optimum growth and antimicrobial metabolite production by *Streptomyces* sp. 201. Journal de Mycologie Médicale 19, 161—167.

Thakur D., Yadav, A., Gogoi, B.K and T.C. Bora. 2007. Isolation and screening of Streptomyces in soil of protected forest areas from the states of Assam and Tripura, India, for antimicrobial metabolites Journal de Mycologie Médicale 17, 242—249.

Tims, F.C., 1932. An actinomycete antagonistic to a *Pythium* root parasite of sugarcane. Phytopathology, 22: 27.

Tomita, K., Nishio, M., Sattoh, K., Yamamoto, H., Hoshino, Y, Ohkuma, H., Konishi, M., Miyaki, T. and Oki, T., 1989. Pradimicins A, B and C: new antifungal antibiotics I. Taxonomy, production, isolation and physico-chemical properties. J Antibiot., XLIII(7): 755-762.

Ubukata M, Shiraishi N, Kobinata K, Kudo T, Yamaguchi I, Osada H, Shen YC and Isono K., 1995. RS-22 A, B and C: new macrolide antibiotics from *Streptomyces violaceusniger*. I. Taxonomy, fermentation, isolation and biological activities. J Antibiot (Tokyo) 48:289–92.

Ubukata, M., Morita, T.I. and Osada, H., 1995a. RS-22A, B and C: new macrolide antibiotics from *Streptomyces violaceusniger* II. . Physio-chemical properties and structure elucidation. J. Antibiot., 48(4): 293-299.

Valan Arasu M, Duraipandiyan V, Agastian P, Ignacimuthu S. 2009. *In vitro* antimicrobial activity of *Streptomyces* spp. ERI-3 isolated from Western ghats rock soil (India). J Mycol Med ,19:22—8.

Waksman, S.A., 1937. Associative and antagonistic effects of microorganisms I. Historical review of antagonistic relationships. Soil Sci., 43: 51-68.

Wang, Z.H. and Shen, C.Y., 1992. On the effect of tautomycin - a new antibiotic to the fungus *Sclerotinia sclerotiorum*. Acta Phytopathol., 22(1): 59-64.

Woo, J.H. and Kamei, Y., 2003. Antifungal mechanism of an anti-pythium protein (SAP) from the marine bacterium *Streptomyces* sp. strain AP77 is specific for *P. porphyrae*, a causative agent of red rot disease in *Porhyra* spp. Appl. Microbiol. Biotechnol., 62(4): 407-413.

Wu K, Chunq L, Revill WP, Katz L and Reeves CD., 2000. The FK520 gene cluster of *Streptomyces hygroscopicus* var. ascomyceticus (ATCC 14891) contains genes for biosynthesis of unusual polyketide extender units. Gene 251:81–90.

Xiao, K., Kinkel, L.L. and Samac, D.A., 2002. Biological control of *Phytophthora* a root rots on *alfalfa* and soybean with *Streptomyces*. Biologi. Cont., 23: 285-295.

Yadav A.K.,Kumar, R. Saikia, R. Bora, T.C. and D.K. Arora. 2009. Novel copper resistant and antimicrobial *Streptomyces* isolated from Bay of Bengal, India. Journal de Mycologie Médicale (2009) 19, 234-240.

Yamaguchi, M., Yamaki, H., Shinoda, T., Tago, Y., Suzuki, H., Nishmura, T. and Yamaguchi, H., 1989. The mode of antifungal action of (S)2-amino-4-oxo-5-hydroxypentanoic acid, RI-331. J. Antibiot., XLIII(4): 411-416.

Yang, J., Zhang, Q., Liao,M., Xiao,M, Xiao W, Yang, S and J.Wana. 2009. Expression and homology modelling of sterol 14α-demethylase of *Magnaporthe grisea* and its interaction with azoles, Pest Manag Sci, 65: 260–265

Yoshida, M. and Horikoshi, K., 1988. Preparation and use of FR900848 from *Streptoverticillium ferverns*. Europian Pat. Appl., 286: 330.

Yuan, W.M., Crawford, D., 1995. Characterization of *Streptomyces lydicus* WYE108 as potential biocontrol agent against fungal root and seed rots. Appl. Environ. Microbiol. 61, 3119-3128.

Zhang, J, Wang, XJ, Yan YJ, Jiang, L, Wang, JD, Li BJ, and WS. Xiang. 2010. Isolation and identification of 5-hydroxyl-5-methyl-2-hexenoic acid from *Actinoplanes* sp. HBDN08 with antifungal activity, Bioresource Technology 101: 8383–8388

Zhanq Q, Gould SJ and Zabriskie TM., 1998. A new cytosine glycoside from *Streptomyces griseochromogenes* produced by the use of in vivo of enzyme inhibitors. J Nat Prod 61:648–651.

Zheng, Z., Zeng, W., Huang, Y., Yang, Z., Li, J., Cai, H. and Su, W., 2000. Detection of antitumor and antimicrobial activities in marine organism associated actinomycetes isolated from the Taiwan Strait, China. FEMS Microbiol. Lett., 188(1): 87-91.

3

Naturally Occurring Antifungal Agents and Their Modes of Action

Isao Kubo, Kuniyoshi Shimizu and Ken-ichi Fujita
Department of Environmental Science, Policy and Management,
University of California, Berkeley, California
USA

1. Introduction

Yeast fermentations are involved in the manufacturing of foods such as bread, beer, wines, vinegar, and surface ripened cheese. Most yeasts of industrial importance are of the genus *Saccharomyces* and mostly of the species *S. cerevisiae.* These ascospore-forming yeasts are readily bred for desired characteristics. However, yeasts are undesirable when they cause spoilage to sauerkraut, fruit juices, syrups, molasses, honey, jellies, meats, wine, beer, and other foods (Frazier and Westhoff, 1988). Finishing process of the fermentation is usually either through filtration or pasteurization. However, the use of the latter is limited to certain foods since it is a heat treatment and hence denaturalizes proteins, and the former is also limited to clear liquids. Neither process can be applicable to some foods such as sauerkraut and "miso" (soy bean pastes). *Zygosaccharomyces bailii,* is a food spoilage yeast species. It is known for its capacity to survive in stress environments and, in particular, in acid media with ethanol, such as in wine. In addition, spoilage of mayonnaise and salad dressing by this osmophilic yeast is well described. Therefore, safe and effective antifungal agents are still needed.

In our continuing search for naturally occurring antimicrobial agents, a bicyclic sesquiterpene dialdehyde, polygodial (**1**) (see Figure 1 for structures), was isolated from various plants (Kubo, 1995). This sesquiterpene dialdehyde exhibited potent antifungal activity particularly against yeasts such as *Saccharomyces cerevisiae* and *Candida albicans* (Taniguchi et al., 1988), although it possessed little activity against bacteria (Kubo et al., 2005). Because of the potent antifungal activity, polygodial can be used as a leading compound to search for new antifungal drugs. This involves the study of their structure and antifungal activity relationships (SAR). However, the study of SAR required the synthesis of a series of analogues differing in the hydrophobic bicyclic portion, and because of this, polygodial may not be practical to use as a leading compound.

Subsequently, 2*E*-alkenals and alkanals were characterized from various edible plants such as the coriander *Coriander sativum* L. (Umbelliferae) (Kubo et al., 2004), the olive *Olea europaea* L. (Oleaceae) (Kubo et al., 1995a; Bisignano et al., 2001) and the cashew *Anacardium occidentale* (Anacardiaceae) (Muroi et al., 1993), and these aldehyde compounds exhibited broad antimicrobial activity (Table 1) (Kubo et al., 1995b). The maximum antimicrobial activity of 2*E*-alkenals is dependent on the balance of the hydrophobic alkyl (tail) chain length from the hydrophilic aldehyde group (head) (Kubo et al., 1995b and 2003a). The hydrophobicity of molecules is often associated with biological action (Hansch and Dunn, 1972). However, the rationale for this observation, especially the role of the hydrophobic portion, is still poorly understood and widely debated. Although the antifungal action of polygodial may differ from those of the aliphatic aldehydes to some extent, 2*E*-alkenals with different chain lengths are a superior model for SAR study because these molecules possess the same hydrophilic portion, the enal group, which explains the role of the hydrophobic alkyl portion. In addition, a series of 2*E*-alkenals and their related analogues are common in many plants (Kim et al., 1995; Kubo and Kubo, 1995; Kubo et al., 1996 and 1999; Kubo and Fujita, 2001; Trombetta et al., 2002) and readily available. Therefore, a homologous series of aliphatic 2*E*-alkenals and the corresponding alkanals, from C_5 to C_{13} were studied to gain new insights into their antifungal action on a molecular basis using *S. cerevisiae* ATCC 7754 as a model organism (Kubo et al., 2001a).

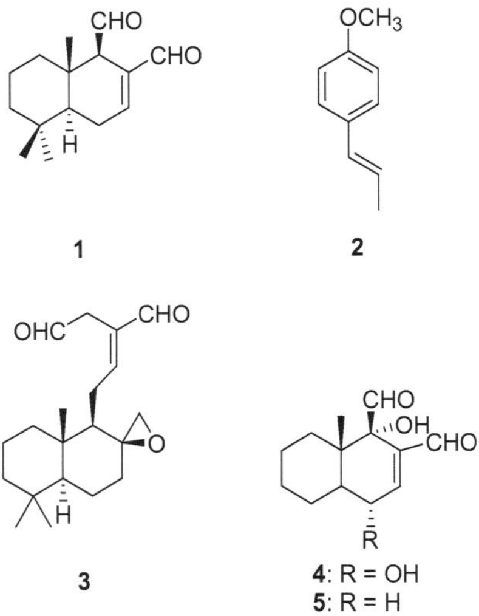

Fig. 1. α,β-Unsaturated aldehydes and related compounds

2. 2*E*-alkenals

The antimicrobial activity of a homologous series of 2*E*-alkenals characterized from plants has previously been reported (Kubo and Kubo, 1995; Kubo et al., 1995a; Bisignano et al., 2001) and is generally similar to being described for the corresponding alkanols (Kubo et al., 1995b). Their MIC and MFC values against *S. cerevisiae* are listed in Table 2. In general, the differences between the MIC and MFC values are not more than 2-fold, suggesting no residual fungistatic activity. As the carbon chain length increases the activity is increased, and the activity disappears after the chain length reaches the maximum activity. This so-called cutoff is a known phenomenon. For example, 2*E*-dodecenal (C_{12}) was very effective against *S. cerevisiae* with a MIC of 12.5 μg/mL, while 2*E*-tridecenal (C_{13}) no longer showed any activity up to 800 μg/mL. Interestingly, 2*E*-dodecenal exhibited the most potent MIC against *S. cerevisiae* but did not exhibit the most potent MFC. More precisely, 2*E*-dodecenal is fungistatic against *S. cerevisiae* but not fungicidal. The most potent fungicide in the 2*E*-alkenal series was 2*E*-undecenal (C_{11}) with a MFC of 6.25 μg/mL, followed by 2*E*-decenal (C_{10}) with a MFC of 12.5 μg/mL.

Microorganisms Tested	MIC (*MBC or MFC*)		
	2*E*-hexenal	2*E*-nonenal	2*E*-undecenal
Bacillus subtilis	400 (*>800*)	50 (*>800*)	25 (*>800*)
Brevibacterium ammoniagenes	400 (*400*)	100 (*100*)	25 (*25*)
Micrococcus luteus	400 (*400*)	50 (*50*)	12.5 (*25*)
Streptococcus mutans	800 (*800*)	100 (*100*)	50 (*50*)
Staphylococcus aureus	400 (*400*)	50 (*50*)	12.5 (*12.5*)
S. aureus (MRSA)	400 (*400*)	50 (*50*)	12.5 (*12.5*)
Propionibacterium acnes	200 (*200*)	25 (*25*)	6.25 (*6.25*)
Escherichia coli	400 (*400*)	200 (*200*)	>800 (–)
Pseudomonas aeruginosa	400 (*400*)	>800 (–)	>800 (–)
Enterobacter aerogenes	400 (*400*)	200 (*200*)	>800 (–)
Proteus vulgaris	200 (*200*)	25 (*25*)	12.5 (*12.5*)
Salmonella choleraesuis	200 (*200*)	25 (*25*)	12.5 (*12.5*)
Saccharomyces cerevisiae	200 (*200*)	25 (*25*)	25 (*25*)
Zygosaccharomyces bailii	400 (*400*)	50 (*50*)	25 (*25*)
Candida albicans	100 (*100*)	25 (*50*)	25 (*25*)
Pityrosporum ovale	50 (*50*)	25 (*25*)	12.5 (*12.5*)
Penicillium chrysogenum	50 (–)	50 (–)	6.25 (*6.25*)
Trichophyton mentagrophytes	50 (–)	12.5 (–)	1.56 (*3.13*)
Aspergillus niger	200 (–)	200 (–)	100 (*200*)

Numbers in *Italic* type in parenthesis are *MBC* or *MFC*. –, Not tested.

Table 1. Antimicrobial activity (μg/mL) of 2*E*-hexenal, 2*E*-hexenal and 2*E*-undecenal.

Aldehydes Tested	2*E*-Alkenal		Alkanal	
	MIC	MFC	MIC	MFC
C_5	100	200	–	–
C_6	100	200	1600	1600
C_7	100	200	400	400
C_8	100	100	200	200
C_9	25	25	100	100
C_{10}	12.5	12.5	25	50
C_{11}	6.25	6.25	25	50
C_{12}	12.5*	100	200*	>800
C_{13}	>800	>800	>800	>800
C_{14}	>400	–	–	–

The cells of *S. cerevisiae* were grown in ME broth at 30 °C without shaking.
*, The values are variable. –, Not tested.

Table 2. Antifungal activity (μg/mL) of aldehydes against *S. cerevisiae*.

The fungicidal activity of 2*E*-undecenal against *S. cerevisiae* was confirmed by the time kill curve experiment. Cultures of 2*E*-undecenal, with a cell density of 5.8 X 10^5 CFU/mL, were exposed to two different concentrations of 2*E*-undecenal. The number of viable cells was determined following different periods of incubation with 2*E*-undecenal. The result verifies that the MIC and MFC of 2*E*-undecenal are the same. It shows that ½MIC slowed growth, but that the final cell count was not significantly different from the control. Notably, lethality occurred remarkably quickly, within the first 1 h after adding 2*E*-undecenal. This rapid lethality very likely indicates that antifungal activity of 2*E*-undecenal against *S. cerevisiae* is associated with the disruption of the membrane (Fujita and Kubo, 2002).

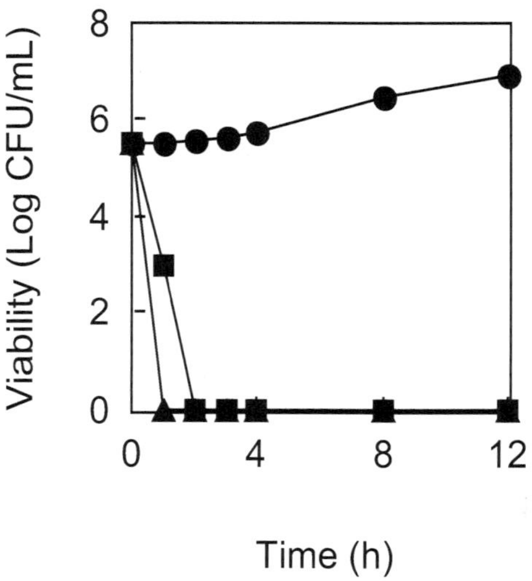

Fig. 2. Time kill curve of 2*E*-undecenal against *S. cerevisiae*. A 16-h culture was inoculated into ME broth containing 0 µg/mL (•), 6.25 µg/mL (■), and 12.5 µg/mL (▲) of 2*E*-undecenal.

Further support for the membrane action was also obtained in experiments that showed the rapid decline in the number of viable cells after the addition of 2*E*-undecenal both at the stationary growth-phase and in the presence of cell growth inhibitors, as shown in Figure 3. Namely, 2*E*-undecenal rapidly killed *S. cerevisiae* cells in which cell division was inhibited by cycloheximide. This antibiotic is known to inhibit protein synthesis in eukaryotes, thereby restricting cell division. The fungicidal effect of 2*E*-undecenal appears independent of the necessary functions accompanying the reproduction of yeast cells, which are macromolecule biosyntheses of DNA, RNA, protein and cell wall components. Hence, the antifungal mechanism of 2*E*-undecenal is associated in part with membrane functions or derangement of the membrane.

In our preliminary test, octanal showed the similar antifungal activity against *S. cerevisiae*, so that the above-mentioned antifungal activity should not be specific to 2*E*-alkenals because the conjugated double bond is unlikely essential to elicit the activity. This prompted us to test antifungal activity of the same series of alkanals against *S. cerevisiae* for comparison. The results are listed in Table 2. The activity of alkanals is slightly less than those of the corresponding 2*E*-alkenals. Similar to 2*E*-alkenal series, dodecanal (C_{12}) was effective with a MIC of 200 µg/mL, but did not exhibit any fungicidal activity up to 800 µg/mL. Thus, *S. cerevisiae* cells appeared to adapt to dodecanal stress, eventually recovering and growing normally. In connection with this, undecanal (C_{11}) and decanal (C_{10}) were the most potent with MFCs of 50 µg/mL. Although the current study was emphasized 2*E*-alkenals because of their more structural similarity with polygodial, the data obtained with alkanals are basically the same as those obtained with 2*E*-alkenals. In the case of short (<C_9) chain 2*E*-alkenals, the activity did not increase with each additional CH_2 group in the alkyl chain, indicating their mode of antifungal action may somewhat differ from that of alkanals.

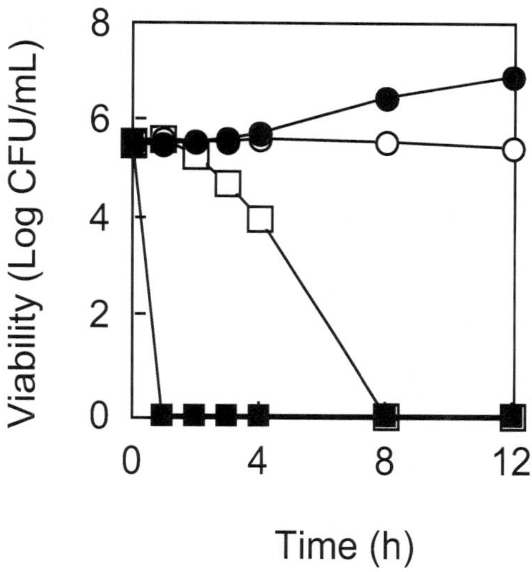

After 5.8 x 10^5 cells were incubated in ME broth for 16 h, compounds were added as follows; 50 µg/mL cycloheximide (O), 12.5 µg/mL 2*E*-undecenal (■), no compound (•). After further 2-h incubation, 2*E*-undecenal was added in cycloheximide treated cells (□). Viability was estimated by the number of colonies formed on YPD plate after incubation at 30 °C for 48 h.

Fig. 3. Fungicidal effect of 2*E*-undecenal in cycloheximide treated cells.

The fungicidal activity of undecanal against *S. cerevisiae* was confirmed by the time kill curve experiment as shown in Figure 4. Cultures of *S. cerevisiae*, with a cell density of 5.8 X 10^5 CFU/mL, were exposed to two different concentrations of undecanal. The number of viable cells was determined following different periods of incubation with undecanal. Figure 4 verifies that the MIC and MFC of undecanal are the same. It shows that ½MIC slowed growth, but that the final cell count was not significantly different from the control. Notably, lethality occurred remarkably quickly, within the first 1 h after adding undecanal, indicating that undecanal possesses a membrane disruptive effect, in a similar manner described for 2*E*-undecenal.

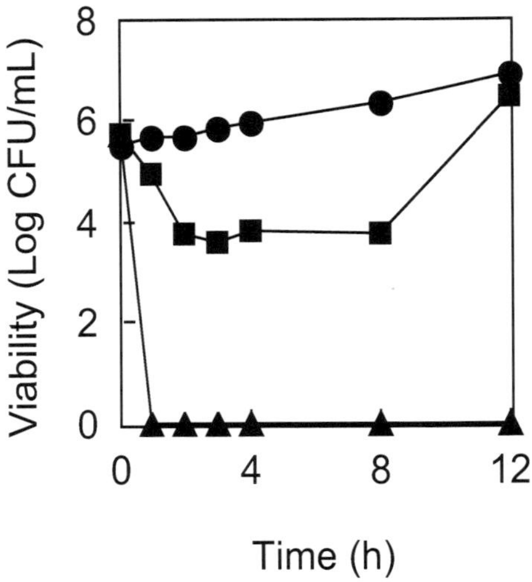

A 16-h culture was inoculated into ME broth containing 0 μg/mL (●), 25 μg/mL (■), and 50 μg/mL (▲) of undecanal. Viability was estimated by the number of colonies formed on YPD plate after incubation at 30 °C for 48 h.

Fig. 4. Time kill curve of undecanal against *S. cerevisiae.*

It is known that *S. cerevisiae* produces the acidification of the external medium during growth on glucose. This external acidification is closely associated with the metabolism of the sugar and its magnitude depends on the buffering capacity of the growth medium (Busa and Nuccitelli, 1984). The H^+-ATPase (P-type) is important not only in the regulation of internal pH but also the energy-dependent uptake of various metabolites (Coote et al., 1994). 2*E*-Alkenals inhibit the external acidification by inhibiting the H^+-ATPase as shown in Figure 5. Their antifungal activity is also partly due to the inhibition of this H^+-ATPase. Interestingly, the potency of H^+-ATPase inhibition in each 2*E*-alkenal differs and the cutoff phenomenon does not occur. It is an interesting question how these 2*E*-alkenals inhibit H^+-ATPase. The 2*E*-alkenals with the chain length less than C_8 and longer than C_{12} exhibited weaker fungicidal activity. This inhibition pattern is not specific to only 2*E*-alkenals but also that of alkanals. It seems that medium-chain (C_9-C_{11}) 2*E*-alkenals have a better balance between the hydrophilic and hydrophobic portions of the molecules to act as surfactants. It should be remembered here that 2*E*-dodecenal exhibited fungistatic activity with a MIC of 12.5 μg/mL against *S. cerevisiae* but did not show any fungicidal activity up to 100 μg/mL.

In the aforementioned acidification inhibitory activity, the effect of the fungicidal 2*E*-undecenal was gradually enhanced, whereas cells treated with fungistatic 2*E*-dodecenal gradually recovered with time, as shown in Figure 6. Yeast cells appeared to adapt to 2*E*-dodecenal stress, eventually recovering and growing normally, similar to that of weak-acid stress (Holyoak et al., 1996). Among the alkanals tested, dodecanal was the most effective against *S. cerevisiae* with a MIC of 200 μg/mL but not fungicidal. This can be explained by the same manner described for 2*E*-dodecenal.

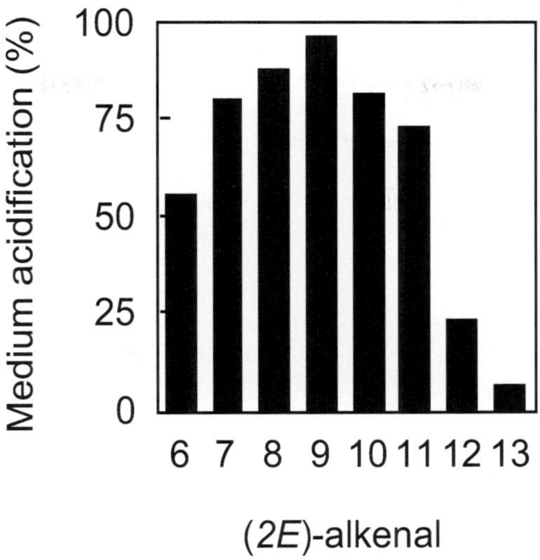

Fig. 5. Inhibition of medium acidification by 2*E*-alkenals (400 μg/mL) for short time incubation. The acidification was assayed for 10 min. The inhibition ratio (%) was calculated as follows; $(1-[H^+]_{inhibitor}/[H^+]_{inhibitor\ free}) \times 100$.

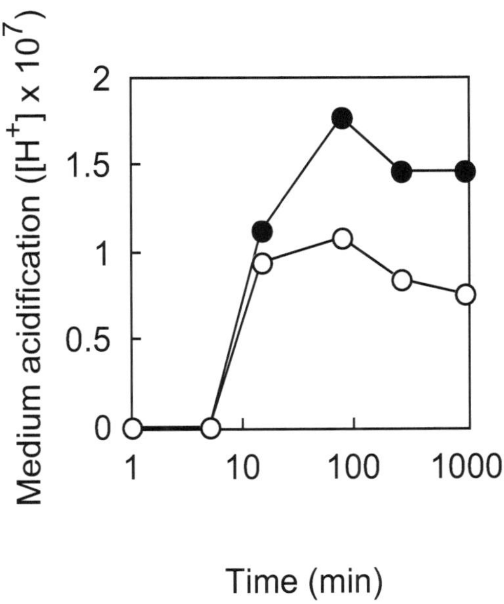

Fig. 6. Effects of incubation time on the inhibition of 2*E*-undecenal (○) and 2*E*-dodecenal (●) to the medium acidification by the plasma membrane H^+-ATPase of *S. cerevisiae*. Alkenals were tested at the concentration of 5 mM.

The data obtained so far indicates that the medium-chain 2*E*-alkenals act as nonionic surfactants at the lipid-protein interface, in a similar manner reported for alkanols (Kubo et al., 1995b). For example, the absence of a functioning state of the H^+-ATPase could be due to its relative sensitivity to functional disruption by 2*E*-alkenals. It is suggested that the intrinsic proteins of the membranes are held in position by hydrogen bonding, as well as by hydrophobic and electrostatic forces, and that hydrogen bonding also mediates the penetration of membranes by proteins. The binding of nonionic surfactants such as the aliphatic aldehydes can only involve relatively weak headgroup interactions, such as hydrogen bonding, so that the predominant interactions will be hydrophobic, involving the alkyl tails. As proposed above, hydrogen bonds are formed or broken by the aliphatic aldehydes, and redirected. Thereby the conformation of membrane protein may be changed. For example, the H^+-ATPase in particular could lose its proper conformation, which would lead to cell death. The H^+-ATPase is the most abundant plasma membrane protein, constituting over 20% of the total membrane protein in *S. cerevisiae*, but the above mentioned fungicidal mechanism of the aliphatic aldehydes seems nonspecific. This can be explained as the amphipathic medium-chain aldehydes are nonionic surfactants and disrupt the hydrogen bonding in the lipid-protein interface in *S. cerevisiae*. As surfactants, the binding site of the aliphatic aldehydes should not be specific and their broad antimicrobial spectrum supports this postulate.

Further supporting generalized surfactant action at the plasma membrane is that the 2*E*-alkenals do not appear to inhibit the major energy production pathway. *S. cerevisiae* is a facultative anaerobic organism that is able to survive without a functional respiratory chain by relying on the fermentation of sugars to supply its energy demand, which is the state yeast prefer when sugars are present in significant amounts. 2*E*-Alkenals are inhibitory to the yeast while in this fermentative state. 2*E*-Alkenals also inhibit the growth of *S. cerevisiae* growing on non-fermentable carbon sources such as ethanol-, lactate-, acetate- and glycerol-containing media. Since no suppression of fungicidal activity seen as would be expected by removal of the potential target, it is unlikely that 2*E*-alkenals' lethal action in yeast is caused by inhibiting components of the respiration or fermentation pathway.

In addition, further support for the surfactant concept was obtained in an additional experiment that indicates antifungal 2*E*-undecenal rapidly adsorbed onto the surface of *S. cerevisiae* cells but 2*E*-hexenal did slightly, as shown in Figure 7. It appears that *S. cerevisiae* showed different affinities to 2*E*-alkenal having different alkyl chain length. The hydrophilic enal moiety was adsorbed by an intermolecular hydrogen bond by attaching itself to hydrophilic portion of the membrane surface. The adsorbing sites may not be specific but need to be clarified.

Given the surfactant-like properties of medium-chain 2*E*-alkenals, it is possible to suggest that 2*E*-alkenals act at the lipid-protein interface of integral proteins, such as ion channels and/or transport proteins, denaturing their functional conformation in a similar manner found for alkanols (Kubo et al., 1995b and 2003a). The common nature among these 2*E*-alkenals should be considered in that the electron negativity on the aldehyde oxygen atom forms an intermolecular hydrogen bond with a nucleophilic group in the membrane, thereby creating disorder in the fluid bilayer of the membrane. The fluidity of the cell membrane can be disturbed maximally by hydrophobic compounds of particular hydrophilic enal group. Thus, the amphipathic medium-chain 2*E*-alkenals disrupt the hydrogen bonding in the lipid-protein interface in *S. cerevisiae*. The data obtained are consistent with an effect on the bulk membrane rather than a direct interaction of the specific target proteins, and 2*E*-alkenals' non-specificity of antimicrobial activity supports this assumption. The possibility of the antimicrobial activity of the medium-chain 2*E*-alkenals is due to their nonionic surfactant property, but this may not be the case for short-chain (<C_9) 2*E*-alkenals. The short chain 2*E*-alkenals enter the cell by passive diffusion across the plasma membrane and/or through porin channels (Schulz, 1996). Once inside the cell, their α,β-unsaturated aldehyde (enal) moiety is chemically highly reactive and hence, they may readily react with biologically important nucleophilic groups such as sulfhydryl, amino, or hydroxyl (Schauenstein et al., 1977). Sulfhydryl groups in proteins and lower molecular weight compounds such as glutathione are known to play an important role in the living cell. Microorganisms protect themselves against hydrogen peroxide in various ways (Brul and Coote, 1999), and some of the most ubiquitous systems include glutathione. 2*E*-Alkenals causes depletion of cytoplasmic and mitochondrial glutathione, which functions in eliminating reactive oxygen species, similar to found for polygodial (Machida et al., 1999). This 2*E*-alkenal mediated depletion of intercellular glutathione can be explained by a direct interaction between the enal moiety and the sulfhydryl group of glutathione by a Michael-type addition. This may reveal the reason why 2*E*-alkenals exhibit in general more potent and broader antimicrobial activity than those of the corresponding alkanals and alkanols. In the case against *S. cerevisiae*, 2*E*-hexenal exhibited the fungicidal activity against this yeast with an MFC of 200 μg/mL, whereas hexanol did not show any activity up to 1600 μg/mL.

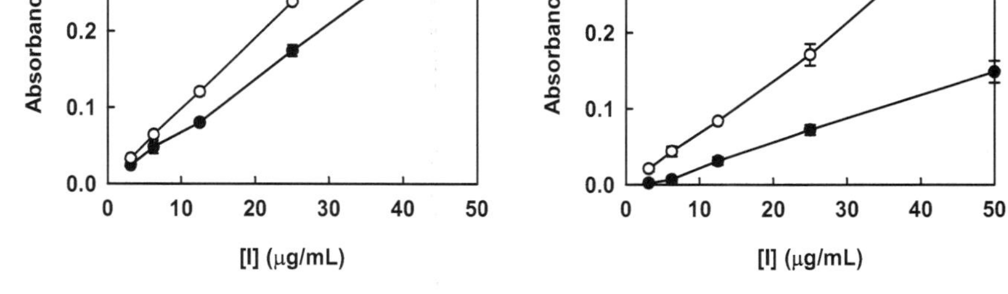

Fig. 7. Binding of 2*E*-hexenal (*a*) and 2*E*-undecenal (*b*) to *S. cerevisiae* cells. After each 2*E*-alkenal was mixed with (•) or (O) without yeast cells (10^8 cells/mL), the suspension was vortexed for 5 sec. Absorbance of the supernatant obtained by centrifugation for 5 min was measured.

Moreover, the leakage of carboxyfluorescein (CF) in liposomes of phosphatidylcholine (PC) following exposure to 2*E*-alkenals was previously reported (Trombetta et al., 2002). Interestingly, 2*E*-alkenals caused rapid CF leakage from PC liposomes and the effectiveness order correlated well with the alkyl chain length. Thus, 2*E*-nonenal was more effective in inducing CF leakage from PC liposomes than that of 2*E*-hexenal. This previous report also supports the surfactant concept. Short chain 2*E*-alkenals are involved more in biochemical processes.

The process by which antimicrobial agents reach the action sites in living microorganisms is usually neglected in the cell-free experiment, but this must be taken into account in the current study. The inner and outer surfaces of the membrane are hydrophilic while the interior is hydrophobic, so the increased lipophilicity of 2*E*-alkenals should affect their movement further into the membrane lipid bilayer portions. It should be logical to assume that most of the lipophilic 2*E*-alkenal molecules being dissolved in the medium are partially incorporated into the lipid bilayers (Franks and Lieb, 1986) in which they may react with biologically important substances. The amount of 2*E*-alkenals entering into the cytosol or lipid bilayer is dependent on the length of the alkyl chain. Hence, the length of the alkyl chain is associated with eliciting activity to a large extent (Kubo and Kubo, 1995; Kubo et al., 1995a).

The current SAR study of 2*E*-alkenals was initiated largely as a model to understand the modes of the potent antifungal action of polygodial. The data described so far demonstrates the similarity between polygodial and 2*E*-alkenals in many aspects, but there are also significant differences. For example, polygodial loses its potent antifungal activity in YPD medium but not aliphatic aldehydes as shown in Table 3. This observation is consistent with the previous report that a primary amino group reacts with the dialdehyde moiety of polygodial and inactivates it (Cimino et al., 1987; Fujita and Kubo, 2005), but not 2*E*-alkenals. YPD contains very high levels of components with amine groups. The result indicates that polygodial forms a pyrrole derivative with the compounds possessing a primary amino group. Therefore, the binding site of polygodial may be, at least in part, a primary amino group in living systems.

Aldehydes Tested	ME	RPMI1640	YPD
Polygodial	1.56	1.56	>100
2*E*-Undecenal	6.25	6.25	12.5
Undecanal	25	25	50

Table 3. Antifungal (MIC) activity (μg/mL) of aldehydes against *S. cerevisiae* in different media.

On the other hand, neither 2*E*-alkenals nor alkanals can form a pyrrole derivative. Notably, isolated mitochondrial ATPase (F-type) is strongly inhibited by polygodial (Lunde and Kubo, 2000), while it is weakly inhibited by the 2*E*-alkenals. In connection with this, 2*E*-alkenals and alkanals were found to inhibit the succinate-supported respiration of intact mitochondria isolated from rat liver, similar to those found for alkanols (Hammond and Kubo, 2000). However, results already discussed above show that these slight mitochondrial inhibitory activities are not primary responsible for cellular inhibition. The antifungal mechanism of polygodial seems to be associated in part with its specific dialdehyde structural features and differs from aliphatic 2*E*-alkenals.

The volume of the hydrophobic portions also seems to be related to the activity since antifungal activity of aliphatic α,β-unsaturated aldehydes are weaker than that of bicyclic sesquiterpene, polygodial. For example, the best fungicidal activity of the 2*E*-alkenal series against *S. cerevisiae* is 2*E*-undecenal with a MFC of 6.25 μg/mL, which is 2-fold less potent than that of polygodial. In the case of alkanals, the most potent undecanal is 16-fold less effective. There are two ways to increase the activity. First, the activity can be enhanced by combining with synergists. For example, the MFC of 2*E*-undecenal against *S. cerevisiae* was enhanced 16-fold when it was combined with ½MFC of anethole (**2**) (Kubo and Kubo, 1995). This combination strategy may be superior to enhance and broaden the total biological activity and, more importantly, it may hinder the development of resistant mechanisms in microorganisms. It should be noted that fungistatic compounds did not provide the stable enhancing activity in combination with other antifungal compounds. In fact, the combination data of the above mentioned fungistatic 2*E*-dodecenal varied. Second, the activity may be enhanced by increasing the volume of the hydrophobic portion through synthetic modification. For example, the volume of polygodial is unlikely the maximum since a more bulky labdane diterpene dialdehyde, aframodial (**3**), exhibited even more potent activity as listed in Table 4. It seems that the activity increased with increasing the volume of the hydrophobic portions. On the other hand, mukaadial (**4**) did not exhibit any activity up to 200 μg/mL but warburganal (**5**) still exhibited the activity but lesser extent than polygodial (Kubo and Himejima, 1992). It is therefore apparent that the activity decreased for each additional hydroxyl group to polygodial. However, the rationale for these still remains unclear.

Aldehydes Tested	MIC	MFC
Polygodial (**1**)	1.56	3.13
Aframodial (**3**)	0.78	1.56
Mukaadial (**4**)	>200	–
Warburganal (**5**)	3.13	6.25

The cells of *S. cerevisiae* were grown in ME broth at 30 °C without shaking. –, Not tested.

Table 4. Antifungal activity (μg/mL) of polygodial and its related compounds against *S. cerevisiae*

Safety is a primary consideration for antifungal agents, and hence, the aldehydes characterized as antifungal agents from edible plants should be superior compared to non-natural antifungal agents. In addition, aldehydes have another superior property as antifungal agents compared to sorbic acid, a common commercial antifungal agent. As a weak acid antifungal agent, the activity of sorbic acid is pH dependent and increases as the pH of the substrate decreases (Sofos et al., 1983), as shown in Table 5. At higher pH values (>5), sorbic acid did not show any antifungal activity up to 1600 μg/mL due to a higher degree of dissociated molecules. In contrast, the aldehydes are not affected by pH. This would appear to be of greater overall value than other pH-sensitive antimicrobials, since many foods have near neutral pH values.

pH	2*E*-Undecenal	Sorbic acid	Undecanal
3	3.13	400	25
5	6.25	1600	50
7	6.25	>1600	50
9	6.25	>1600	50

Table 5. pH Effect of fungicidal (MFC) activity (μg/mL) of 2*E*-undecenal, sorbic acid and undecanal against *S. cerevisiae*.

3. Alkanols

As long as the antifungal activity is concerned, medium chain length 2*E*-alkenals may be potent enough to use as antifungal agents. However, it still needs to be considered, especially from practical points of view. For example, since α,β-unsaturated aldehyde is a chemically highly reactive group and readily reacts with biologically important nucleophilic groups, such as sulfhydryl, amino, or hydroxyl groups (Schauenstein, 1977), some practical application may limit the use of 2*E*-alkenals. As aforementioned the surfactant concept (biophysical processes) is a major contributor to their antifungal activity, so it is worth testing alkanols because they are usually considered as surfactants. In addition, alkanols are chemically stable compounds and unlikely react with any biologically important substances in the cytosol or lipid bilayer. Their maximum antimicrobial activity is dependent upon the hydrophobic alkyl (tail) chain length from the hydrophilic hydroxyl group (head). It should be noted that antimicrobial agents, which primarily act as surfactants, may have the potential, since they may target the extracytoplasmic region, and thus do not need to enter the cell, thereby avoiding most cellular pump-based resistance mechanisms. Alkanols are considered to be stable, colorless, inexpensive, biodegradable (Swisher, 1970), and essentially nontoxic to humans (Opdyke, 1973). More importantly, alcohols are among the most versatile of all organic compounds, and free and esterified alcohols are known to occur widely in nature.

A series of aliphatic primary alkanols from C_6 to C_{13} against *S. cerevisiae* were tested for their antifungal activity against *S. cerevisiae* using a 2-fold serial broth dilution method. The results are listed in Table 6, indicating basically similar to those found for 2*E*-alkenals. In agreement with many other studies of the homologous series of alkanols, the antifungal activity of the alkanols increased with number of carbons in the chain until dodecanol and undecanol, which had the best MIC and MFC, in this experiment. Noticeably, the activity disappeared after the chain length reached the best MIC and MFC, known as the so-called cutoff phenomenon. For example, dodecanol (C_{12}) was the most effective with an MIC of 12.5 μg/mL, while tridecanol (C_{13}) showed no activity up to 1600 μg/mL. Dodecanol is the most effective fungistatic but did not show any fungicidal activity up to 1600 μg/mL. The MIC of dodecanol slowed growth for the first 24 h, but the growth recovered shortly after and became no longer different from the control. The cutoff point may migrate by the slight difference in growth conditions such as inoculum size of yeast cells or medium composition. Namely it seems to be important that the cutoff point exists but not crucial where is the cutoff point. Alkanols can form hydrogen bonds with water and as a result, simple alkanols are fairly soluble in water. However, as the hydrocarbon content increases, especially to more than six carbons, there is a general decline in solubility. As the hydrocarbon chain becomes longer, its hydrophobic properties come to dominate the properties of the molecule so that the medium-chain (C_9-C_{11}) alkanols are amphipathic molecules.

Alkanols Tested	μg/mL		log *P*	Remarks
	MIC	MFC		
C_6	1600	>1600	1.938	
C_7	800	>1600	2.469	
C_8	400	>1600	3.001	
C_9	50	100	3.532	↓
C_{10}	25	50	4.063	
C_{11}	12.5	25	4.595	Surfactant
				↓
C_{12}	12.5**	>1600	5.126	Partially soluble in phospholipid
C_{13}	>1600	-	5.657	

-, Not tested. **, The value is variable.

Table 6. Antifungal activity and log *P* of alkanols against *S. cerevisiae*.

Among the alkanols tested, undecanol (C_{11}) was the most potent against *S. cerevisiae* with a MFC of 25 µg/mL (0.15 mM). No differences in MIC and MFC were noted, suggesting that undecanol's activity was fungicidal. This fungicidal effect was confirmed by a time kill curve method as shown in Figure 8. Cultures of *S. cerevisiae*, with a cell density of 6 X 10^5 CFU/mL, were exposed to two different concentrations of undecanol. The number of viable cells was determined following different periods of incubation with undecanol. The results show that ½MIC slowed growth but the final cell count was not significantly different from the control. At the MFC lethality occurred quickly, within the first 8 hours, indicating a membrane disruptive effect. Similar results were obtained with hexanol but fungicidal activity was not seen until 24 h, indicating that short-chain alkanols act in somewhat different ways.

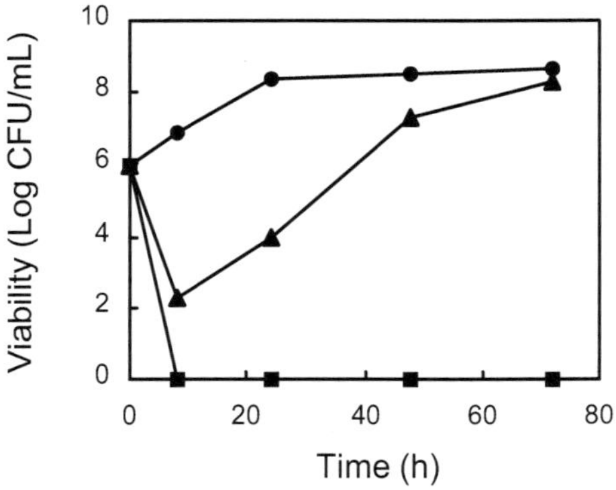

Fig. 8. Time kill curve of undecanol against *S. cerevisiae*. A 16-h culture was inoculated into ME broth containing 0 µg/mL (▲), 12.5 µg/mL (■), and 25 µg/mL (●) of undecanol.

It is known that *S. cerevisiae* produces the acidification of the external medium during growth on glucose. This external acidification is closely associated with the metabolism of the sugar and its magnitude depends on the buffering capacity of the growth medium. The H^+-ATPase is important not only in the regulation of internal pH but also the energy-dependent uptake of various metabolites. Interestingly, alkanols were found to inhibit this acidification process by inhibiting the H^+-ATPase (Kubo et al., 2003b). As a result, the antifungal activity of alkanols is, at least in part, due to their inhibition of the H^+-ATPase shown in Figure 9. Interestingly, the inhibitory value of each alkanol differs and the cutoff phenomenon occurred between C_{12} and C_{13}. The alkanols of the chain length less than C_8 and longer than C_{12} exhibited much weaker inhibition activity. This inhibition pattern is not specific to only alkanols but also that of alkanals and fatty acids. The longer chain ($>C_{12}$) alkanols are soluble in the membrane phospholipid, and is thought to be incorporated into the hydrophobic domain of the membrane. In contrast, the shorter chain ($<C_9$) alkanols enter the cell by passive diffusion across the plasma membrane. It seems that only amphipathic medium-chain (C_9-C_{11}) alkanols act as surface-active compounds (surfactants). It should be remembered that dodecanol exhibited fungistatic activity with MIC of 25 µg/mL but did not show any fungicidal activity up to 1600 µg/mL. This alkanol inhibited the external acidification when tested after 5 min but not after 4 h. More specifically, the acidification inhibitory activity of fungicidal undecanol was

gradually enhanced, whereas cells treated with fungistatic dodecanol gradually recovered with time, as shown in Figure 10. Yeast cells appeared to adapt to dodecanol stress, eventually recovering and growing normally, similar to that of other stress.

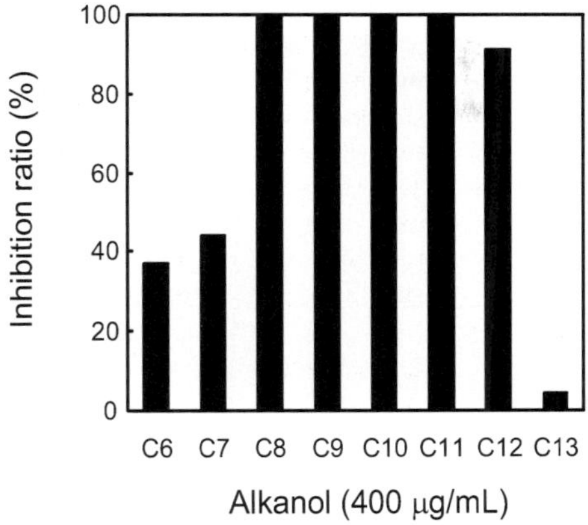

Fig. 9. Inhibition of medium acidification by alkanols (400 μg/mL) for short time incubation. The acidification was assayed for 10 min. The inhibition ratio (%) was calculated as follows; $(1-[H^+]_{inhibitor}/[H^+]_{inhibitor\ free}) \times 100$.

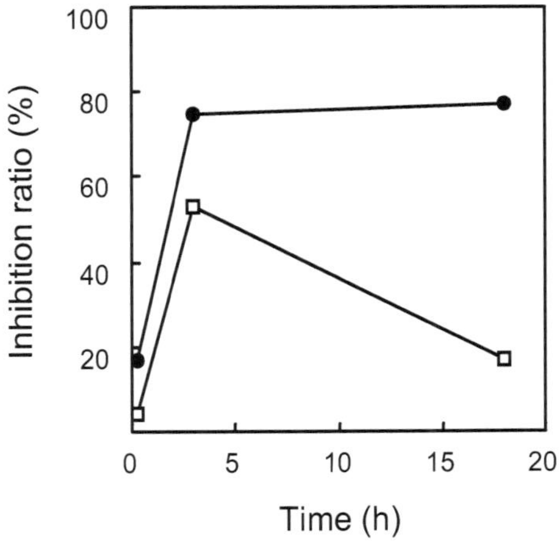

Fig. 10. Inhibition of medium acidification by undecanol (•) and dodecanol (□) (400 μg/mL) during long time incubation. The inhibition ratio (%) was calculated as follows; $(1-[H^+]_{inhibitor}/[H^+]_{inhibitor\ free}) \times 100$.

Based on the data obtained, it seems logical to assume that alkanols act at the lipid-protein interface of H^+-ATPase as nonionic surfactants. The absence of a functioning state of the H^+-ATPase could be due to its relative sensitivity to denaturation by alkanols. The binding of alkanols as nonionic surfactants can only involve relatively weak head group interactions such as hydrogen bonding. It is suggested that the intrinsic proteins of the membranes are held in position by hydrogen bonding, as well as by hydrophobic and electrostatic forces. As proposed above, hydrogen bonds are formed or broken by alkanols and then redirected. As a result, the conformation of the membrane protein may change. In particular, the H^+-ATPase could lose its proper conformation. In addition to H^+-ATPase, alkanols may destroy the native membrane-associated functions of integral proteins, such as ion channels and transport proteins. This can be supported, for instance, by alkanols inhibit the uptake of glucose and other nutrients by *S. cerevisiae* in a noncompetitive way. It appears that alkanols, as well as the corresponding alkanals and 2*E*-alkenals, are nonionic surfactants. Because of the lack of specificity, alkanols act in an unspecific manner on the lipid-protein interaction. This can be explained as the amphipathic medium-chain alkanols are considered as more appropriately balanced nonionic surfactants and more strongly disrupt the lipid-protein interface in *S. cerevisiae*. The shorter chain ($<C_9$) alkanols enter the cell by passive diffusion across the plasma membrane. In contrast, the longer chain ($>C_{12}$) alkanols are soluble in the membrane phospholipid and is thought to be incorporated into the hydrophobic domain of the membrane without perturbing the lipid. The partitioning of radiolabelled long-chain alcohols into biological membranes and lipid bilayers can support this. The cutoff in antifungal activity observed could be due to a corresponding cutoff in the absorption of long chain alcohols into lipid-bilayer portions of membranes. It should be noted that the carbon number for cutoff slightly varies by experimental conditions, tridecanol shows no antifungal activity against *S. cerevisiae* under any conditions. The precise explanation for the role of alkyl chain length - that must be related to antifungal activity - still remains obscure.

The nonspecific antimicrobial mechanism of alkanols is apparently due to their nonionic surface-active properties. The common nature among these alkanols should be considered in that the electron negativity on the oxygen atom forms an intermolecular hydrogen bond with a nucleophilic group in the membrane, thereby creating disorder in the fluid bilayer of the membrane. The fluidity of the cell membrane can be disturbed maximally by hydrophobic compounds of particular hydrophilic hydroxyl group. They could enter the molecular structure of the membrane with the polar hydroxyl group oriented into the aqueous phase by hydrogen bonding and nonpolar carbon chain aligned into the lipid phase by dispersion forces. Eventually, when the dispersion force becomes greater than the hydrogen bonding force, the balance is destroyed and the activity disappears. Concerning this, the hydrophobic bonding energy between an average fatty acid ester and a completely hydrophobic peptide is approximately 12 kcal/mol. Addition of a hydrogen bond between a peptide and a fatty ester's carbonyl adds another 3-6 kcal/mol. Furthermore, alkanols first approach the binding site with the electron negativity of the hydroxyl oxygen atom. This hydrogen bond acceptor will affect the hydrogen bonds that regulate the permeability of the lipid bilayer. For example, in the lipid bilayer, the hydroxyl group of ergosterol resides near the membrane-water interface and is likely to bind to the carbonyl group of phospholipids. Alkanols may function by disrupting and disorganizing these hydrogen bonds. Ergosterol is a major component of the plasma membrane of *S. cerevisiae* and owes its membrane-closing properties to its rigid longitudinal orientation in the membrane. Since ergosterol has a profound effect on membrane structure and function, cell function will be impaired if the hydrogen bond is broken. The similar hydrogen bond-breaking concept was proposed to explain the anesthesia cutoff phenomenon.

Further support for the surfactant postulate was also obtained in experiments that showed a rapid decline in the number of viable cells after the addition of undecanol at the exponential growth-phase as shown in Figure 11. In addition, the effect of undecanol was tested during holding viable cell number in the presence of cycloheximide. This drug is known to restrict cell division by inhibiting protein synthesis; its effect against *S. cerevisiae* cells is fungistatic. Undecanol rapidly killed *S. cerevisiae* cells in which cell division was inhibited by cycloheximide. This observation excludes several modes of action for undecanol such as inhibition of DNA, RNA, protein, or cell wall component synthesis in vivo. The result observed indicates that the antifungal mechanism of undecanol is due primarily to its surfactant property, although it is not possible to confirm that membrane damage is the only cause of the lethal effect. It can be concluded that the medium-chain alkanols target, in part, the extracytoplasmic region as surfactants. This conclusion is highly desirable since they do not need to enter the cell, thus avoiding most resistance mechanisms based on cellular pump.

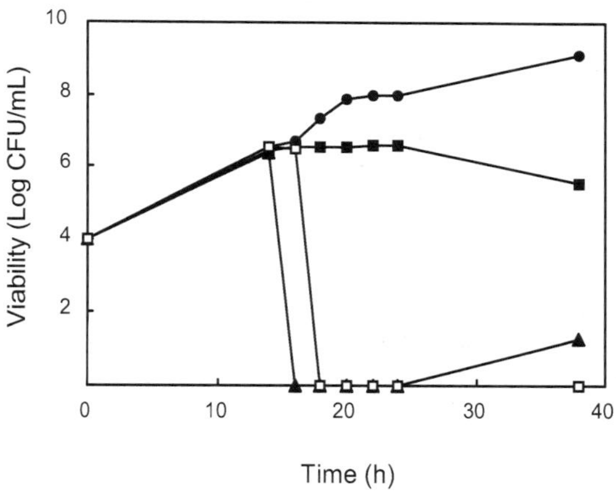

After 10^4 cells were incubated in ME broth for 14 h, compounds were added as follows; 50 μg/mL cycloheximide (■), 25 μg/mL undecanol (▲), no compound (●). After further 2-h incubation, undecanol was added in cycloheximide treated cells (□).

Fig. 11. Fungicidal effect of undecanol in cycloheximide treated cells.

The time-kill curve study showed that undecanol was fungicidal against *S. cerevisiae* at any growth stage. Figure 12 shows the effect of undecanol at various growth stages. Undecanol at MFC rapidly reduced the number of viable cells when added to the culture at the exponentially growing culture (10^6 CFU/mL) within the first 4 h. Thus, no viable cells were detected within 2 h after adding undecanol. However, undecanol was not fungicidal when added to the stationary growing culture (10^8 CFU/mL). This alcohol rapidly reduced the number of viable cells, but slowed thereafter and complete lethality did not occur. As the number of viable cells increases, the amount of drug's molecules needs to be increased to retain the fungicidal activity. The results indicate that undecanol disrupts the lipid-protein interface nonspecifically as a surfactant rather than in a direct interaction with specific target proteins such as cell-surface receptors or signal transduction proteins.

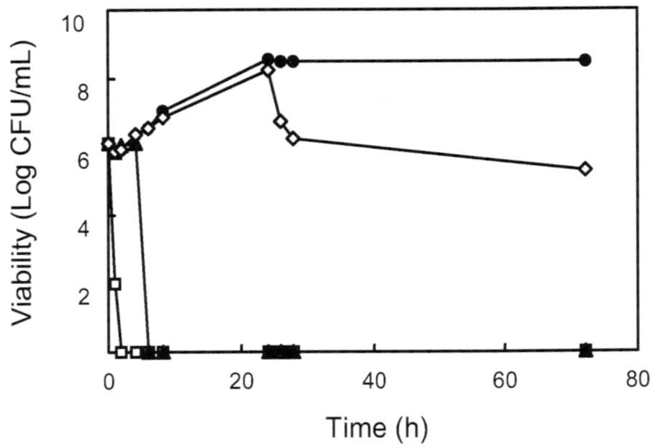

Fig. 12. Effect of undecanol (25 μg/mL) at the various growth stages of *S. cerevisiae*. A 16-h culture was inoculated into ME broth. Undecanol was added at 0 h (□), 4 h (▲), 24 h (◇), and control (•).

The same series of alkanols were recently found to inhibit the succinate-supported respiration of intact mitochondria isolated from rat liver. The potency increased with increasing chain length up to undecanol. Given each alkanol's nearly identical effect on State 3 and uncoupled respiration, action is not directly on ATP synthetase, but earlier in the respiratory process. Hexanol and decanol were also assayed against freeze-thawed (broken) mitochondria to distinguish effects on the mitochondrial substrate carrier from that on the electron transport chain. Both alcohols were only weak inhibitors of respiration in broken mitochondria, suggesting that inhibition originates from interference with the dicarboxylate carrier, which must transport succinate across the mitochondrial membranes. Alkanols may inhibit this transporter in the inner membrane as nonionic surfactants. The concentrations found to inhibit mitochondrial respiration are slightly lower than those causing fungicidal activity against *S. cerevisiae*, prompting the question whether decreased mitochondrial functions might in some way contribute to *S. cerevisiae* death. It is uncertain, however, if the respiratory inhibition mechanism is the primary mode of antifungal action of alkanols against *S. cerevisiae* because we do not know if alkanols can reach the mitochondria in vivo. The results with mitochondria also support the alkanols' nonionic surfactant concept because enzyme systems related to transport of solutes and electron transfer are located in the inner membrane of the cell envelope.

In addition, *S. cerevisiae* is a facultative anaerobic organism that is able to survive without a functional respiratory chain, by falling back on the fermentation of sugars to supply its energy demand. This latter mechanism is used preferentially, since when a combination of fermentable and non-fermentable carbon sources is available, respiration is greatly reduced and fermentation accounts for the major fraction of sugar catabolism. *S. cerevisiae* is thus able to rapidly adjust its metabolism to its environment and in particular to the availability of carbon sources. Alkanols also inhibit the growth of *S. cerevisiae* growing on non-fermentable carbon sources such as ethanol-, lactate-, acetate- and glycerol-containing media. As surfactants, alkanols exhibit fungicidal activity when *S. cerevisiae* is growing on both fermentable and non-fermentable condition.

Safety is a primary consideration for *S. cerevisiae* control agents, especially concerning their use in food products that may be utilized in unregulated quantities regularly. The phytochemicals characterized as antifungal agents against *S. cerevisiae* from edible plants should be superior to non-natural preservatives. Incidentally, alcohols are among the most versatile of all organic compounds, free and esterified alcohols occur widely in nature. In addition, alcohols have another superior property as antifungal agents compared to commonly use antifungal agents such as sorbic acid and benzoic acid. As a weak acid antifungal agent, the activity of sorbic acid is pH dependent and increases as the pH of the substrate decreases as shown in Table 7. At higher pH values (>5), sorbic acid did not show any antifungal activity up to 1600 μg/mL due to a higher degree of dissociated molecules. In contrast, undecanol and geraniol are not influenced by pH values.

pH	Undecanol	Sorbic acid	Geraniol
3	12.5	800	800
5	25	1600	800
7	12.5	>1600	800
9	12.5	>1600	800

Table 7. pH Effect of fungicidal (MFC) activity (μg/mL) of undecanol, sorbic acid, geraniol against *S. cerevisiae*.

4. Antifungal mechanisms

The fluidity of the lipid bilayer is partly regulated by hydrogen bonding. For example, the hydroxyl group of ergosterol resides near the membrane-water interface in the lipid bilayer and is likely to be bonded with the carbonyl group of phospholipids (Brockerhoff, 1974; Chauhan et al., 1984). As nonionic surfactants, the aliphatic aldehydes first approach the binding site with the electron negativity of the aldehyde oxygen atom and may function by disrupting and disorganizing the hydrogen bonds such as the above mentioned. Ergosterol is a major component of the plasma membrane of *S. cerevisiae* and owes its modulation of membrane fluidity to its rigid longitudinal orientation in the membrane. Since ergosterol has profound influences on membrane structure and function, if the hydrogen bond is broken, cell function will be impaired. If the aldehydes target the extracytoplasmic region, it is highly desirable since they do not need to enter the cell, thus avoiding most cellular pump-based resistance mechanisms. The similar hydrogen bond-breaking concept was proposed to explain the anesthesia cutoff phenomenon (Chiou et al., 1990).

In previous reports, the antifungal activity of the same series of alkanols against *S. cerevisiae* was described (Kubo et al., 1995b). Similar to 2*E*-alkenals, the short chain alkanols enter the cell by passive diffusion across the plasma membrane and/or through porin channels

(Schulz, 1996), and the long chain alkanols enter in part into the lipid bilayers (Franks and Lieb, 1986). The amount of alkanols entering into the cytosol or lipid bilayer is dependent on the length of the alkyl chain. Nonetheless, alkanols are chemically stable compounds and may not react with any biologically important substances in the cytosol or lipid bilayer. Hence, the primary antifungal action of alkanols comes largely from their ability to function as nonionic surfactants (physical disruption of the membrane). In the case of 2*E*-alkenals, their α,β-unsaturated aldehyde group should not be overlooked because this group is chemically highly reactive and readily reacts with biologically important nucleophilic groups, such as sulfhydryl, amino, or hydroxyl (Schauenstein et al., 1977). For example, the yeast plasma-membrane H^+-ATPase was reported to contain nine cysteines. 2*E*-Alkenals may bind directly to the plasma membrane H^+-ATPase probably with sulfhydryl groups of the three cysteines in the presumed transmembrane segments (C148, C312, C867). However, Petrov and Slayman (1995) reported that no single cysteine is required for activity based on their site-directed mutagenesis study. This previous result does not exclude the possibility to assume that 2*E*-alkenals first break the hydrogen bond as nonionic surfactants and then react with the freed sulfhydryl group of the H^+-ATPase as well as other plasma membrane proteins. This can be supported by the previous report that covalent modification of the conserved C148 in the transmembrane segment 2 may be important for inhibition of H^+-ATPase activity and cell growth (Monk et al., 1995). However, the observation that alkanals and 2*E*-alkenals exhibit similar antifungal activity against *S. cerevisiae* as shown in Table 2 and also inhibit glucose-induced acidification, may not support the above assumption because the conjugated double bond is not essential to elicit the activity. The possibility of this concerted function of 2*E*-alkenals is unlikely but cannot be excluded.

The leakage of carboxyfluorescein (CF) in liposomes of phosphatidylcholine (PC) following exposure to 2*E*-alkenals was previously reported (Trombetta et al., 2002), similar to those described for alkyl gallates (Fujita and Kubo, 2002). Interestingly, 2*E*-alkenals tested caused rapid CF leakage from PC liposomes and the effectiveness order correlated well with the alkyl chain length. Thus, 2*E*-nonenal was more effective in inducing CF leakage from PC liposomes than that of 2*E*-hexenal (Trombetta et al., 2002). This also supports the surfactant concept.

In general terms, aldehydes may enter the cell by passive diffusion across the membrane. Once inside the cells, the following reactions are known for the 2*E*-alkenals. Reactions with the sulfhydryl group, for which primary addition to the α,β-olefinic group occurs exclusively. Reactions with amino groups, where the formations of Schiff bases and 1,4-addition products is possible. It is generally true, however, that the reaction with the sulfhydryl groups takes place much faster, in fact by several orders of magnitude. These reactions may lead to the deactivation of enzymes - in particular of sulfhydryl enzymes. The main problem of general significance remains the certain and unequivocal experimental proof that 2*E*-alkenals recognized as highly reactive compounds is in fact essential bioregulators of metabolism. For example, sulfhydryl groups in proteins and lower molecular weight compounds such as glutathione are known to play an important role in the living cell. Bacteria protect themselves against hydrogen peroxide in various ways (Brul and Coote, 1999), and some of the most ubiquitous systems include glutathione. 2*E*-Alkenals causes depletion of cytoplasmic and mitochondrial glutathione, which functions in eliminating reactive oxygen species, similar to found for polygodial (Machida et al., 1999).

This 2*E*-alkenal mediated depletion of intercellular glutathione can be explained by a direct interaction between the enal moiety and the sulfhydryl group of glutathione by a Michael-type addition. This may reveal the reason why 2*E*-alkenals exhibit in general more potent and broader antimicrobial activity than those of the corresponding alkanals and alkanols. Thus, 2*E*-hexenal exhibited the fungicidal activity against *S. cerevisiae* with an MFC of 200 μg/mL, whereas hexanal showed the same activity at only 1600 μg/mL.

5. Antifungal assay

The maximum extent and rate of activity is known to vary with the seed culture mediums, the physiological age of the culture, and the type of culture medium. The initial assay was achieved using two-fold serial dilution method against *Saccharomyces cerevisiae* ATCC 7754. This yeast was purchased from American Type Culture Collection (Manassas, VA).

5.1 Medium

Saccharomyces cerevisiae was maintained at -80 °C in yeast nutrient broth (YNB; Difco Laboratories, Detroit, MI) containing 25% glycerol and subcultured at 30 °C in Sabouraud's dextrose agar (SDA) medium (Bactopeptone 1%, dextrose 4%, Bacto-agar 1.8%). A fresh culture was preincubated with shaking for 16 h at 30 °C in 2.5% malt extract (ME) broth (BBL) medium.

5.2 Acidification measurement

The glucose-induced medium acidification of *S. cerevisiae* was measured with a modified procedure (Haworth et al., 1993). The test strain was cultured with shaking in YPD (Glucose 2%, Bactopeptone 2%, Yeast extract 1%) broth overnight at 30 °C and washed twice with cold distilled water. The cells were diluted to 5 X 10^7 colony forming units (CFU) per mL with cold distilled water and kept on ice. The reaction mixture contained 2.7 mL of cells and 30 μL of the inhibitor in DMSO, and was preincubated at 30 °C for 5 min. A 20% glucose solution of 0.3 mL was added (final 2%) to induce acidification. After 10 min incubation, the pH of external medium was checked (Orion 8175 Ross semimicro electrode).

5.3 Antifungal assay

The test compounds were first dissolved in DMF and the concentration of DMF in each medium was always 1%. The highest concentration tested was 1600 μg/mL, unless otherwise specified. The maximum extent and rate of activity is known to vary with the seed culture mediums, the physiological age of the culture, and the type of culture medium. For example, the minimum inhibitory concentration (MIC) of anethole significantly varied with the inoculum size. All antifungal susceptibility tests in this study were performed under a standard condition using fresh inoculum from a 5 h shaking culture in ME medium, final inoculum size of 10^5 CFU/mL, and 48 h stationary incubation in ME medium, unless otherwise specified.

Broth macrodilution minimum inhibitory concentrations (MICs) were determined as previously described (Kubo and Himejima, 1992). Briefly, serial 2-fold dilutions of the test compounds were made in DMF and 30 μL of 100 X conc. solution was added to 3 mL of ME media. These were inoculated with 30 μL of seed culture to give the final inoculum of 10^5 CFU/mL. The assay tubes were incubated without shaking at 30 °C for 48 h. The MIC is the lowest concentration of test compound that demonstrated no visible growth. The minimum fungicidal concentrations (MFCs) were examined as follows. After the MIC had been determined, a 30 μL of aliquot was taken from each clear tube and added into 3 mL of drug free fresh medium. After 48 h incubation, the MFC was determined as the lowest concentration of the test compounds in which no recovery of microorganism was observed.

Time kill studies were performed to examine the effects of combinations of compounds in more detail. The culture tubes were prepared as described above and incubated at 30 °C for 16 h. A 30 μL aliquot of the culture was inoculated into 3 mL of ME broth containing appropriate concentrations of the test compounds. The initial population size for *S. cerevisiae* was 5.8 X 10^5 CFU/mL. Samples were taken at selected times during 48 h of exposure, and serial dilutions were made in sterile saline before the samples were plated onto YPD agar plates. The plates were incubated at 30 °C for 48 h before the number of CFU was determined.

5.4 Adsorption test

The test strain was cultured with shaking in YPD broth overnight at 30 °C and washed twice with 50 mM MOPS buffer (pH 6.0). After each 2*E*-alkenal was mixed with or without *S. cerevisiae* cells (10^8 cells/mL) in the above buffer at 30 °C, the suspension was vortexed for 5 seconds. Absorbance of the supernatants obtained by centrifugation for 5 min was measured at 255 nm.

6. Summary

The antifungal activity of alkanols comes from their ability to act as surfactants and the maximum activity can be designed by selecting a alkyl chain length to give the appropriate partition coefficient (log P) as a standard. Similarly, 2*E*-alkenals also act as surfactants, but their α,β-unsaturated aldehyde moiety needs to be taken into account. 2*E*-Alkenals may not act by a single defined process but have multiple functions (Figure 13). The surfactant concept, disrupting and disorganizing the lipid bilayer-protein interface nonspecifically, can be extended to answer many other problems related to membrane-bound enzymes and receptors, and the fluidity of the membrane lipids. For example, the anesthesia cutoff phenomenon among alkanols is well known and a long standing problem. Anesthesia involves many membrane-bound proteins such as synaptosomal ATPases and acetylcholine receptor (Edelfors and Ravn-Jonsen, 1990; Elliott and Haydon, 1989). The same surfactant concept, disrupting and disorganizing the lipid bilayer-protein interface, seems applicable to explain the anesthesia cutoff phenomena of alkanols. The knowledge obtained may provide insights into fungicidal action of aldehydes and alkanols on a molecular basis, and a more rational and scientific approach to use or design efficient and safe antifungal agents. Based on the data obtained, the hydrophilic head portion can be replaced by any hydrophilic groups as long as the "head and tail" structure is balanced. Hence, various additional biological activities can be introduced mainly by selecting appropriate head portions. For example, each series of alkyl gallates (Kubo et al., 2001b; Fujita and Kubo, 2001 and 2002) and alkyl protocatechuates (Nihei et al., 2003) were synthesized as antioxidation antifungal agents.

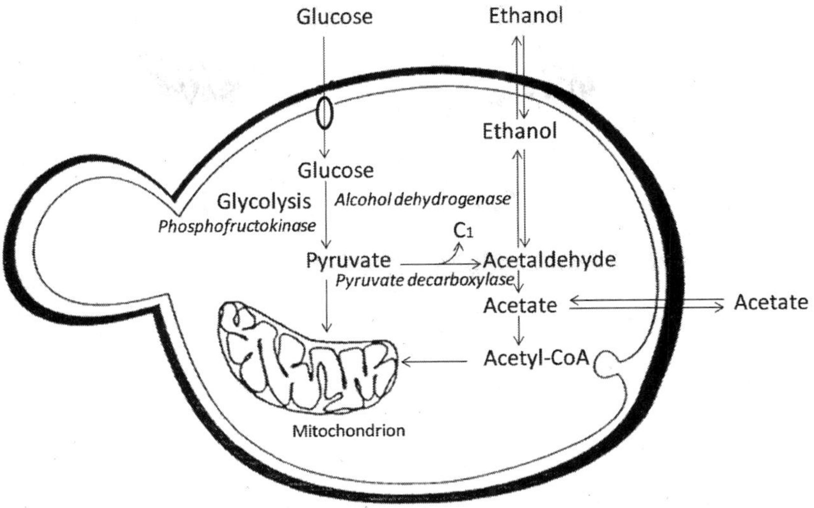

Fig. 13. Antifungal action involves multifunction. The amphipathic medium-chain aldehydes and alcohols are nonionic surfactants and disrupt the hydrogen bonding in the lipid-protein interface of integral proteins, such as ion channels and/or transport proteins, denaturing their functional conformation. 2*E*-Alkenals react with biologically important nucleophilic groups such as sulfhydryl, amino, or hydroxyl. For example, sulfhydryl groups in proteins and lower molecular weight compounds such as glutathione are known to play an important role in the living cell and 2*E*-alkenal mediated depletion of intercellular glutathione can be explained by a direct interaction between the enal moiety and the sulfhydryl group of glutathione by a Michael-type addition. Aldehydes are known to inhibit alcohol dehydrogenase competitively but not phosphofructokinase or pyruvate decarboxylase.

7. Acknowledgements

The authors are indebted to Dr. S. H. Lee, Dr. M. Himejima and Dr. C. S. Lunde for performing antimicrobial assay at earlier stage of the work, and Dr. H. Haraguchi and Dr. D. G. Hammond for performing the respiratory inhibition assay. K. F. thanks for financial support during his study at UC Berkeley.

8. References

Bisignano, G.; Lagana, M. G.; Trombetta, D.; Arena, S.; Nostro, A.; Uccella, N.; Mazzanti, G.; Saija, A. (2001) In vitro antibacterial activity of some aliphatic aldehydes from *Olea europea* L. *FEMS Microbiol. Lett.*, *198*, 9-13.

Brockerhoff, H. (2001) Model of interaction of polar lipids, cholesterol, and proteins in biological membranes. *Lipids*, *9*, 645-650.

Brul, S.; Coote, P. (1999) Preservative agents in foods, mode of action and microbial resistance mechanisms. *Int. J. Food Microbiol.*, *50*, 1-17.

Busa, W. B.; Nuccitelli, R.(1984) Metabolic regulation via intracellular pH. *Am. J. Physiol., 246*, 409-438.

Chauhan, V. P. S.; Ramsammy, L. S.; Brockerhoff, H. (1984) Molecular interactions in the hydrogen belts of membranes glucose-6-phosphate, lysophosphatidylcholine, and cholesterol. *Biochim. Biophys. Acta, 772*, 239-243.

Chiou, J. S.; Ma, S. M.; Kamaya, H.; Ueda, I. (1990) Anesthesis cutoff phenomenon: Interfacial hydrogen bonding. *Science, 248*, 583-585.

Cimino, G.; Sodano, G.; Spinella, A. (1987) Correlation of the reactivity of 1,4-dialdehydes with methylamine in biomimetic conditions to their hot taste - Covalent binding to primary amines as a molecular mechanism in hot taste receptors. *Tetrahedron, 43*, 5401-5410.

Coote, P. J.; Jones, M. V.; Seymour, I. J.; Rowe, D. L.; Ferdinabdo, D. P.; McArthur, A. J.; Cole, M. B. (1994) Activity of the plasma membrane H^+-ATPase is a key physiological determinant of thermotolerance in *Saccharomyces cerevisiae. Microbiology, 140*, 1881-1890.

Edelfors, S.; Ravn-Jonsen, A. (1990) The effects of alcohols *in vitro* on the nervous cell membrane measured by changes in the (Ca^{2+}/Mg^{2+}) ATPase activity and fluidity of synaptosomal membrane. *Pharmacol. Toxicol., 67*, 56-60.

Elliott, J. R.; Haydon, D. A. (1989) The actions of neutral anesthetics on ion conductances of nerve membranes. *Biochim. Biophys. Acta, 988*, 257-286.

Franks, N. P.; Lieb, W. R. (1986) Partitioning of long-chain alcohols into lipid bilayers: Implications for mechanisms of general anesthesia. *Proc. Natl. Acad. Sci. USA, 83*, 5116-5120.

Frazier, W. C.; Westhoff, D. C. (1988) *Food Microbiology*. McGraw-Hill, New York, pp 39-58.

Fujita, K.; Kubo, I. (2001) Plasma membrane injury by nonyl gallate in *Saccharomyces cerevisiae. J. Appl. Microbiol., 92*, 1035-1042.

Fujita, K.; Kubo, I. (2002) Antifungal activity of octyl gallate. *Int. J. Food Microbiol., 79*, 193-201.

Fujita, K.; Kubo, I. (2005) Multifunctional action of antifungal polygodial against *Saccharomyces cerevisiae*: involvement of pyrrole formation on cell surface in antifungal action. *Bioorg. Med. Chem.*, 13, 6742-6747.

Hammond, D. G.; Kubo, I. (2000) Alkanols inhibit respiration of intact mitochondria and display cutoff similar to that measured *in vivo. J. Pharm. Exp. Therap., 293*, 822-828.

Hansch, C.; Dunn, W. J. III. (1972) Linear relationships between lipophilic character and biological activity of drugs. *J. Pharm. Sci., 61*, 1-19.

Haworth, R. S., Cragoe, E. J., Jr, Fliegel, L. (1993) Amiloride and 5-(N-ethyl-N-isopropyl) amiloride inhibit medium acidification and glucose metabolism by the fission yeast Schizosaccharomyces pombe. *Biochim. Biophys. Acta., 1145*, 266–272

Holyoak, C. D.; Stratford, M.; McMullin, Z.; Cole, M. B.; Crimmins, K.; Brown, A. J. P.; Coote, P. J. (1996) Activity of the plasma membrane H^+-ATPase and optimal glycolytic flux are required for rapid adaptation and growth of *Saccharomyces cerevisiae* in the presence of the weak-acid preservative sorbic acid. *Appl. Environ. Microbiol., 62*, 3158-3164.

Kim, J. M.; Marshall, M. R.; Cornell, J. A.; Preston III, J. F.; Wei, C. I. (1995) Antibacterial activity of carvacrol, citral, and geraniol against *Salmenella typhimurium* in culture medium and on fish cubes. *J. Food Sci., 60*, 1364-1368.

Kubo, A.; Kubo, I. (1995) Antimicrobial agents from *Tanacetum balsamita*. *J. Nat. Prod.*, 58, 1565–1569.

Kubo, A.; Lunde, C. S.; Kubo, I. (1996) Indole and (*E*)-2-hexenal, phytochemical potentiators of polymyxins against *Pseudomonas aeruginosa* and *Escherichia coli*. *Antimicrob. Agents Chemother.*, *40*, 1438-1441.

Kubo, A.; Lunde; C. S.; Kubo, I. (1995a) Antimicrobial activity of the olive oil flavor compounds. *J. Agric. Food Chem.*, *43*, 1629-1633.

Kubo, I. (1995) Antifungal sesquiterpene dialdehydes from the *Warburgia* plants and their synergists. In *Studies in Natural Products Chemistry;* Vol. 17, Structure and Chemistry Part D; Rahman, A. U. Ed.; Elsevier: Amsterdam, pp. 233-249.

Kubo, I.; Fujita, K. (2001) Naturally occurring anti-Salmonella agents. *J. Agric. Food Chem.*, *49*, *5750-5754*.

Kubo, I.; Fujita, K.; Kubo, A.; Nihei, K.; Lunde, C. S. (2003a) Modes of antifungal action of (2*E*)-alkenals against *Saccharomyces cerevisiae*. *J. Agric. Food Chem.*, *51*, 3951-3957.

Kubo, I.; Fujita, K.; Lee, S. H. (2001a) Antifungal mechanism of polygodial. *J. Agric. Food Chem.*, *49*, 1607-1611.

Kubo, I.; Fujita, K.; Nihei, K.; Kubo, A.; Ogura, T. (2004) Antibacterial activity of coriander volatile compounds against *Salmonella choleraesuis*. *J. Agric. Food Chem.*, *52*, 3329-3335.

Kubo, I.; Fujita, T.; Kubo, A.; Fujita, K. (2003b) Modes of antifungal action of alkanols against *Saccharomyces cerevisiae*. *Bioorg. Med. Chem.*, 11, 1117-1122.

Kubo, I.; Himejima, M. (1992) Potentiation of antifungal activity of sesquiterpene dialdehydes against *Candida albicans* and two other fungi. *Experientia*, *48*, 1162-1164.

Kubo, I.; Muroi, H.; Himejima, M.; Kubo, A. (1995b) Structural functions of antimicrobial long-chain alcohols and phenols. *Bioorg. Med. Chem.*, *3*, 873-880.

Kubo, I.; Xiao, P.; Fujita, K. (2001b) Antifungal activity of octyl gallate: Structural criteria and mode of action. *Bioorg. Med. Chem. Lett.*, *11*, 347-350.

Kubo, I.; Xiao, P.; Nihei, K.; Fujita, K.; Yamagiwa, Y.; Kamikawa, T. (2002) Molecular design of anti-fungal agents. *J. Agric. Food Chem.*, *50*, 3992-3998.

Kubo, J.; Lee, J. R.; Kubo, I. (1999) Anti-*Helicobacter pylori* agents from the cashew apple. *J. Agric. Food Chem.*, *47*, 533-537.

Lunde; C. S.; Kubo, I. (2000) Effect of polygodial on the mitochondrial ATPase of *Saccharomyces cerevisiae*. *Antimicrob. Agents Chemother.*, *44*, 1943-1953.

Machida, K.; Tanaka, T.; Taniguchi, M. (1999) Depletion of glutathione as a cause of the promotive effects of polygodial, a sesquiterpene on the production of reactive oxygen species in Saccharomyces cerevisiae. *J. Biosci. Bioeng.*, *88(5)*, 526-30.

Monk, B. C.; Mason, A. B.; Abramochkin, G.; Haber, J. E.; Seto-Young, D.; Perlin, D. S. (1995) The yeast plasma membrane proton pumping ATP-ase is a viable antifungal target. I. Effects of the cysteine-modifying reagent omeprazole. *Biochem. Biophys. Acta*, *1238*, 81-90.

Muroi, H.; Kubo, A.; Kubo, I. (1993) Antimicrobial activity of cashew apple flavor compounds. *J. Agric. Food Chem.*, *41*, 1106-1109.

Nihei, K.; Nihei, A.; Kubo, I. (2003) Rational design of antimicrobial agents: Antifungal activity of alk(en)yl dihydroxybenzoates and dihydroxyphenyl alkanoates. *Bioorg. Med. Chem. Lett.*, *13*, 3993-3996.

Opdyke, D. L. J. (1973) Monographs on fragrance new materials. Food Cosmet. Toxicol. 11, 1011-1181.

Petrov, V. V.; Slayman, C. W. (1995) Site-directed mutagenesis of the yeast pma1 H^+-ATPase. *J. Biol. Chem.*, *270*, 28535-28540.

Schauenstein, E.; Esterbauer, H.; Zollner, H. (1977) *Aldehydes in Biological Systems*, Pion: London, pp. 172-200.

Schulz, G.E. (1996) Porins: general to specific, native to engineered passive pores. *Curr. Opin. Struct. Biol.*, *6*, 485-490.

Sofos, J. N.; Busta, F. F. Sorbates. (1983) In *Antimicrobials in Food*; Branen, A. L.; Davidson, P. M. Eds.; Dekker: New York, pp. 141-175.

Swisher, R. D. (1970) In: Surfactant Biodegradation; Swisher, R. D., Ed., Marcel Dekker, New York.

Taniguchi, M.; Yano, Y.; Tada, E.; Ikenishi, K.; Oi, S.; Haraguchi, H.; Hashimoto, K.; Kubo. I. (1988) Mode of action of polygodial, an antifungal sesquiterpene dialdehyde. *Agric. Biol. Chem.*, *52*, 1409-1414.

Trombetta, D.; Saija, A.; Bisignano, G.; Arena, S.; Caruso, S.; Mazzanti, G.; Uccella, N.; Castelli, F. (2002) Study on the mechanisms of the antibacterial action of some plant α,β-unsaturated aldehydes. *Lett. Appl. Microbiol.*, *35*, 285-290.

4

Antifungal Properties of Bioactive Compounds from Plants

F. Castillo[1], D. Hernández[1], G. Gallegos[1], R. Rodríguez[2] and C. N. Aguilar[2]
[1]*Universidad Autónoma Agraria Antonio Narro*
[2]*Universidad Autónoma de Coahuila*
México

1. Introduction

Currently, the consequences derived from application of fungicides in traditional agricultural production systems for control of crop diseases have impacted negatively this activity. Fungicides application, where the indiscriminate use and application frequency high has led to problems and constraints in the control of these diseases by loss in efficiency, increased resistance to active ingredients, ecological damage and a serious negative impact on the human health. For this reason, it is had carryed out research to develop new products, methods and strategies for diseases control. The investigation and development of bio-based products is of great interest to subtract the negative effects generated by traditional agricultural production systems. The use and application of bioactive phytochemicals with antifungal properties represent an attractive and efficient alternative to inhibit the growth of several fungal pathogens.

These bioactive compounds are naturally produced in the plants how secondary metabolites, the principal groups with antifungal activity were terpenes, tannins, flavonoids, essential oil, alkaloids, lecithin and polypeptides. These groups of compounds are important for the physiology of plants contributing properties confer resistance against microorganisms, other organisms and help preserve the integrity of the plant with continuous exposure to environmental stressors, such as ultraviolet radiation, high temperatures or dehydration.

2. Bioactive antifungal activity groups

2.1 General

Plants have developed natural defense mechanisms to protect themselves long before the man played an active role in protecting them. It is known that plants synthesize a variety of groups of bioactive compounds in plant tissues as secondary metabolites that have antifungal activity to stop or inhibit the development of mycelia growth, inhibition of germination or reduce sporulation of fungal pathogens, each these groups presented variable mechanisms of action, for example, the toxicity of polyphenols in microorganisms is attributed to enzyme inhibition by oxidation of compounds. For essential oils is postulated that cause disruption of the membrane by the action of lipophilic compounds, the use or employment as formulations of these compounds is in the form of extracts. The process of extraction of secondary metabolites from plant extracts is variable, can be obtained as aqueous extracts or powders using different solvents used for many different compounds, depending on their polarity. It is considered that these compounds obtained from plants are biodegradable and safe for use as an alternative for disease control in a traditional production system (Sepulveda *et al.*, 2003; Hernandez *et al.*, 2007; Wilson *et al.*, 1997; Bautista *et al.*, 2002; Abou-Jawdah *et al.*, 2002; Cowan, 1999).

These substances known as secondary metabolites, *secondary products*, or *natural products*, have no generally recognized, direct roles in the processes of photosynthesis, respiration, solute transport, translocation, protein synthesis, nutrient assimilation, differentiation or metabolism processes as the formation of carbohydrates, proteins and lipids. That is, particular secondary metabolites are often found in only one plant species or related group of species, whereas primary metabolites are found throughout the plant kingdom. In function to classify to chemically groups the secondary metabolites can be divided into three groups: terpenes, phenolics and nitrogen- containing compounds. This classification is due by the interrelationship with primary metabolism Figure 1.

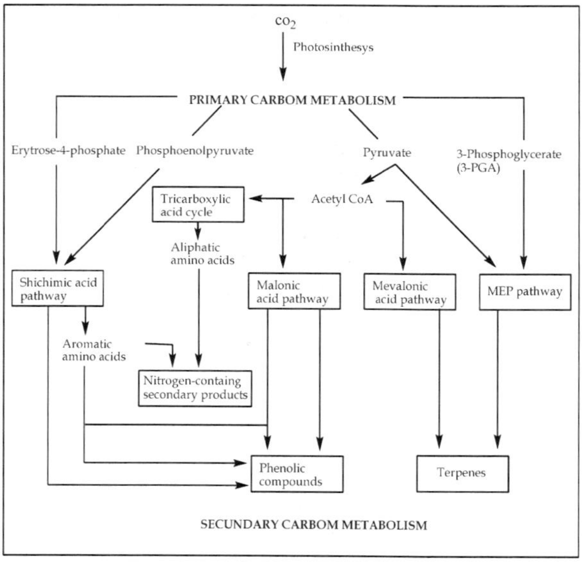

Fig. 1. A simplified view of the major pathways of secondary metabolites biosynthesis and their interrelationship with primary metabolism (Taiz & Zeiger, 2002)

2.2 Polyphenols

Plant phenolics are a chemically heterogeneous group of nearly 10,000 individual compounds: Some are soluble only in organic solvents, some are water-soluble carboxylic acids and glycosides and others are large, insoluble polymers. Present a structure of various groups replaced by hydroxyl functions benzene and its derivatives are simple phenolic compounds called phenylpropanoids (Figure 2). Allowing them to be highly soluble organic substances in water and are present in extracts of leaves, bark, wood, fruits and galls of certain ferns, gymnosperms and angiosperms (Swain, 1979). These polyphenols are important for the physiology of plants to contribute to resistance to microorganisms, insects and herbivorous animals that can affect (Haslam, 1996), help to preserve the integrity of the plant with continuous exposure to environmental stressors, including radiation ultraviolet, relatively high temperatures and dehydration (Lira *et al.*, 2007). These polyphenol antioxidants are therefore active in biological systems and probably the capacity or biological value explains its abundance in plant tissues (Meckes *et al.*, 2004).

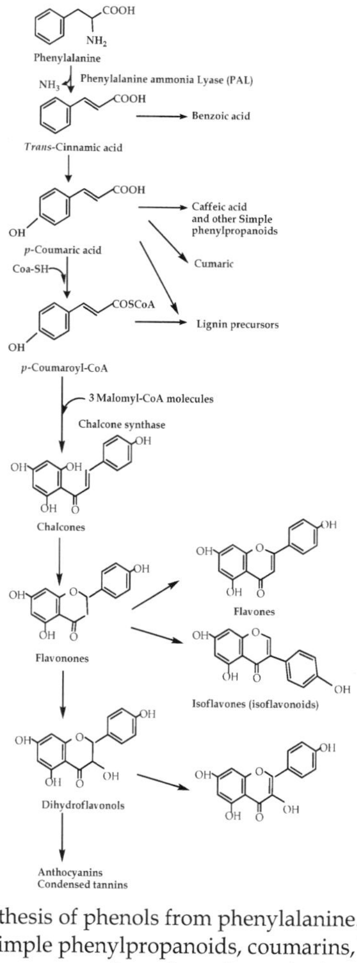

Fig. 2. Outline of the biosynthesis of phenols from phenylalanine. The formation of many plant phenolics, including simple phenylpropanoids, coumarins, benzoic acid derivatives, lignans, anthocyanins, isoflavones, condensed tannins and other flavonoides, begins with phenylalanine (Taiz & Zeiger, 2002)

2.2.1 Hydrolysable tannins (HT)

Are organic compounds, amorphous, taste astringent, weakly acidic, most soluble in water, only a few in organic solvents are yellow, red, or brown and are located in the cytoplasm and cell vacuole of plant tissues. Esters of glucose are partially or fully attached to different polyols such as ellagic acid, say, m-digallic, hexahydroxydiphenic acid or its derivatives (Figure 3). Obtained by hydrolysis with acids, bases and hydrolytic enzymes to break the glycosidic bond to liberate the sugar and phenolic compounds in it. (Gonzalez *et al.*, 2009).

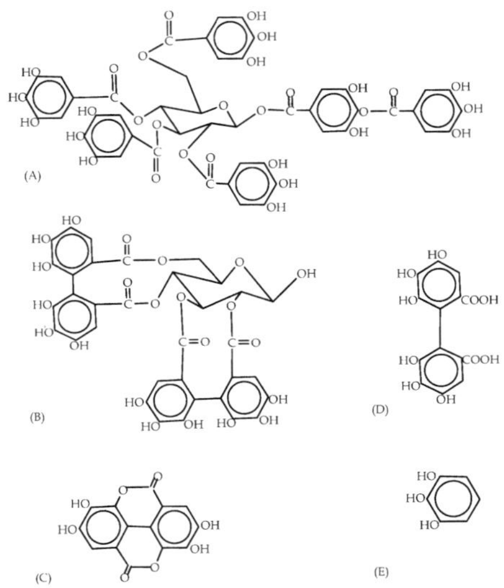

Fig. 3. Hydrolysable tannins and some of its derivatives: A) gallotannins, B) ellagitannins. C) ellagic acid, D) hydroxyphenolic acid, E) gallic acid

The hydrolysable tannins are divided into the following subgroups: The gallotannins, which by enzymatic hydrolysis give more sugar and gallic acid of phenolic compounds that comprise it (Figure 4) and ellagitannins, which give ellagic acid enzymatic hydrolysis more sugar or a derivative as hexahydrophenic acid (Figure 4).

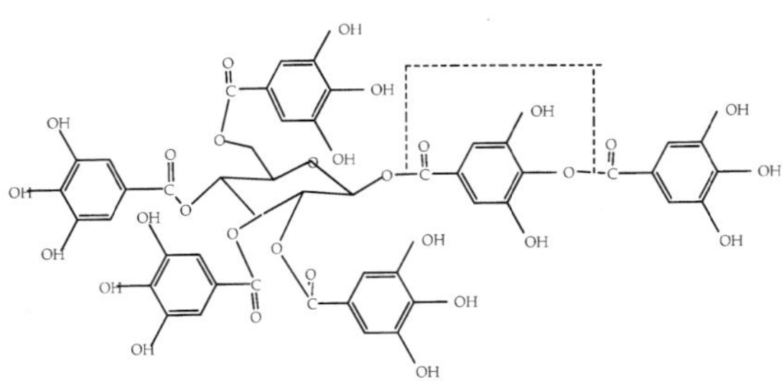

Fig. 4. Chemical structure of a gallotannins

2.2.2 Condensed tannins (CT)

Also called proanthocyanidins (PAS), are derived from the oxidation reaction that produces anthocyanidins (ACS) red in acid-alcohol solution (Figure 5). Are polymers of flavan 3-ol (catechin) and 3-4 flavan diol (leucoanthocyanidins) and have no sugar residues and their carbohydrate content is low or negligible. Are polymers of high molecular weight (1000 to 3000 Daltons), which gives them a relative immobility. Its complexity and easy to form bonds with proteins make them difficult to study. Condensed tannins include flavonoids, which in turn are subdivided into anthocyanidins and leucoanthocyanidins and catechin (Makkar *et al.*, 2007; Taiz & Zeiger, 2002).

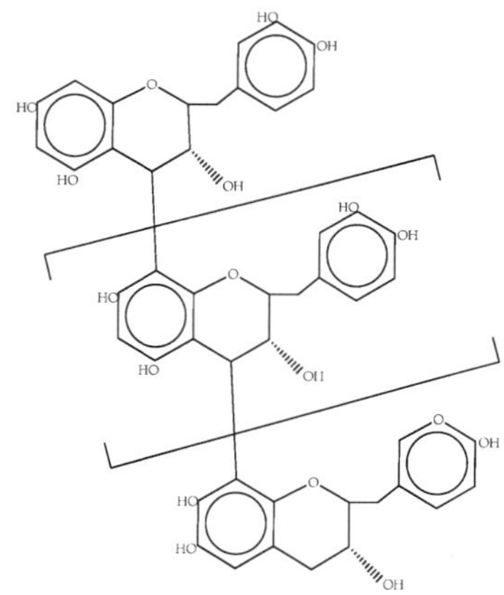

Fig. 5. Condensed tannins or proanthocyanidins

The substituents in the groups R1, R2 and R3, can have an effect on the reactivity of tannin (Figure 6). The group R2 is an OH radical can sometimes be esterified gallic acid (known as Gallo-catechin). For example an increase in the ratio prodelphynidins/procyanidins enhance the ability of condensed tannins to complex proteins.

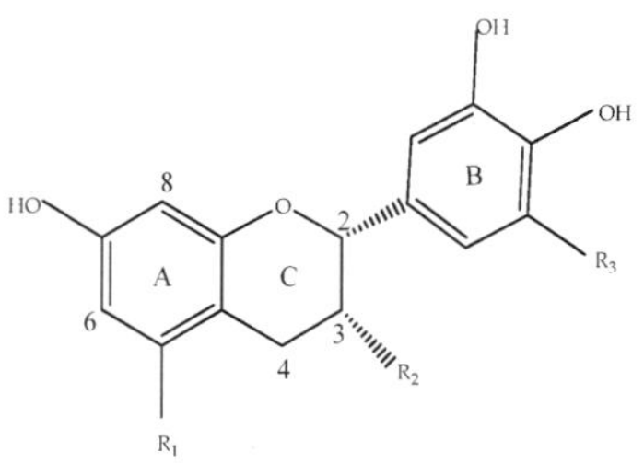

R1	R3	Class
OH	H	Proanthocyanidin
OH	OH	Prodelphynidin
H	H	Profisetinidin
H	OH	Prorobinetinidin

Fig. 6. Structure of some condensed tannins

Hydroxyl groups allow the formation of complexes with proteins, metal ions and other molecules such as polysaccharides. In general, polyphenols identified and grouped according to their basic result is a chain of six carbons (Table 1).

Atoms number	Basic carbon skeleton	Compounds
6	C_6	Simple Phenols and Benzoquinones
7	C_6 - C_1	Phenolic acids
8	C_6 - C_2	Acetofenons and fenilacetonics acids
9	C_6 - C_3	hidroxicinamics Acids, fenilpropanoids
10	C_6 - C_4	Naftoquinones
13	C_6 - C_1- C_6	Xantones
14	C_6 - C_2- C_6	Estilbens and anthraquinones
15	C_6 - C_3- C_6	Flavonoids and isoflavonoids
18	$(C_6 - C_3)_2$	Lignans and neolingnans
30		Biflavonoids
N	$(C_6 - C_3)_n$ $(C_6)_6 (C_6\text{-}C_3 - C_6)_n$	Lignins, Catecol Melanins and flavolans

Table 1. Classification of phenolics compounds in carbons atoms to base number (Garcia, 2004)

2.3 Terpenes

The *terpenoids*, constitute the largest class of secondary products, the diverse substances of this class are generally insoluble in water. The terpenes are biosynthesized from primary metabolites by at least two different routes, a route mevalonic acid, where three molecules of acetyl CoA is condensed step by step to form mevalonic acid. This six-carbon molecule is pirofosforilada and dehydrated to form isopentyl diphosphate and this is the basic unit of the terpenes active, the other route is called route metileritritol phosphate that functions in chloroplasts and other plastids. All terpenes are derived from the union of five-carbon elements that have the branched carbon skeleton of isopentane:

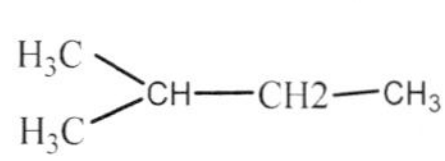

The basic structural elements of terpenes are sometimes called **isoprene units** because terpenes can decompose at high temperatures to give isoprene:

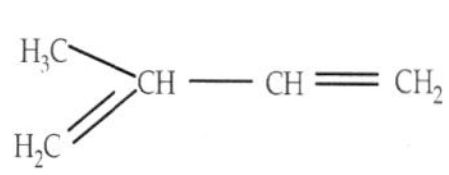

The terpenes or *isoprenoids* are classified by the number of five-carbon units they contain, example: Ten-carbon terpenes, which contain two C5 units, are called *monoterpenes;* 15-carbon terpenes (three C5 units) are *sesquiterpenes;* and 20-carbon terpenes (four C5 units) are *diterpenes.* Larger terpenes include *triterpenes* (30 carbons), *tetraterpenes* (40 carbons) and *polyterpenoids* ([C5] *n* carbons, where $n > 8$) (Taiz and Zeiger, 2002).

2.4 Nitrogenous compounds

A large variety of plant secondary metabolites have nitrogen in their structure. Included in this category are such well-known anti-defenses as alkaloids, amines, cyanogenic glycosides, non-protein amino acids, glucosinolates, alkamides and peptides (Wink & Schimmer, 2010). Most nitrogenous secondary metabolites are biosynthesized from common amino acids.

2.5 Plants with antifungal properties

It´s has studied the secondary metabolites present in various plant species, one to identify its presence, chemical structure and effect on the plant and on other organisms, so that the number of Identified Substances exceed to 100 000 at present (Wink and Schimmer, 2010) in table 2 shows a relationship of phenolic compounds in other organisms different to the plants with presence of these compounds.

Phylum	Structural patrons
Bacteria	Phenols from polyketides and quinones (occasionally present)
Fungi	Simple phenols, phenylpropanoids, quinones (usually present)
Algae	Oidados and brominated phenols, phloroglucinol derivatives from cell wall
Lichens	Anthraquinones, xanthones and depsidones
Bryophytes	Phenols in the cell wall, phenylpropanoids, stilbenes and some flavonoids
Ferns, conifers and flowering plants	Lignin in the cell wall and wide range of phenols of all kinds

Table 2. Distribution of polyphenols compounds on different phylum's in comparative to Plant phylum (Garcia, 2004, as cited in Harborne, 1990)

The number of plant species containing one or more of the major groups of compounds with anti-fungal activity is very diverse (Glasby, 1991), in Table 3 lists some of the studied plant with antifungal effect.

Specie	Compounds Identifying	Reference
Simmondsia chinensis	Glucosides	Abbassy *et al.*, 2007
Thymus zygis subsp. *sylvestris,*	Carvacrol	Gonçalves *et al.*, 2010
A. gypsicola and *A. biebersteinii*	Camphor , 1,8-cineole, piperitone , borneol and α-terpineol, *n*-eicosane , *n*-heneicosane , *n*-tricosane, linoleic acid	Kordali *et al.*, 2009
Larrea tridentata	lignans, methyl-nordihydroguaiaretic acid and nordihydroguaiaretic acid	Vargas-Arispuro *et al.*, 2005

Specie	Compounds Identifying	Reference
Chenopodium quinoa	triterpenoid saponins	Stuardo & Sn Martin, 2008
Aloe vera	Crude extracts	Jasso de Rodríguez *et al.*, 2005
Drimys winteri	essential oil	Monsálvez *et al.*, 2010,
Pimenta dioica	Essential oils	Zabka *et al.*, 2009.
Catharanthus roseus	5-hydroxy flavones	Roy & Chatterjee, 2010
Larrea tridentata, Flourensia cernua, Agave lechuguilla, Opuntia sp. and *Yucca* sp.	Condensed and hidrolizables Tannins	Castillo *et al.*, 2010,
Flourensia microphylla, Flourensia cernua and *Flourensia retinophylla*	Crude extracts	Jasso de Rodríguez *et al.*, 2007
Salvia officinalis	essential oil	Pinto *et al.*, 2007
Carya illinoensis shells and *Punica granatum*	polyphenolic extracts	Osorio *et al.*, 2010
Bulnesia sarmientoi	bulnesol, hanamyol	Rodilla *et al.*, 2011
Caesalpinia cacalaco	gallic and tannic acids	Veloz-García *et al.*, 2010
Clausena anisata	essential oils	Osei-Safo *et al.*, 2010
Ruta chalepensis	2-undecanone, 2-decanone and 2-dodecanone	Mejri *et al.*, 2010
Bucida buceras, Breonadia salicina, Harpephyllum caffrum, Olinia ventosa, Vangueria infausta and *Xylotheca kraussiana*	crude plant	Mahlo *et al.*, 2010
Agapanthus africanus	Crude extracts	Tegegne *et al.*, 2008
Reynoutria sachalinensis		Pasini *et al.*, 1997
Laurus nobilis	1.8-cineole, linalool, terpineol acetate, methyl eugenol, linalyl acetate, eugenol, sabinene, β-pinene, α-terpineol.	Corato *et al.*, 2010
Asarum heterotropoides var. *mandshuricum*	methyleugenol, eucarvone, 5-allyl-1,2,3-trimethoxybenzene and 3,7,7-trimethylbicyclo(4.1.0)hept-3-ene	Dan *et al.*, 2010
Rumex crispus	chrysophanol, parietin and nepodin	Choi *et al.*, 2004; Gyung *et al.*, 2004

Specie	Compounds Identifying	Reference
Astronium fraxinifolium, Inga marginata, Malva sylvestris, Matayba elaeagnoides, Miconia argyrophylla, Myrcia fallax, Ocimum gratissimum, Origanum vulgare, Rollinia emarginata, Siparuna arianeae, Styrax pohlii, Tabebuia serratifolia and *Trichilia pallid*	Crude extracts	Andrade *et al.*, 2010
Piper longum	Eugenol, piperine, piperlongumine and piperettine)	Lee *et al.*, 2001
Datura metel	Enzymes, peroxidase, *β*-1,3-glucanase and chitinase	Devaiah *et al.*, 2009
Calotropis procera, Nerium oleander, Eugenia jambolana, Citrullus colocynthis, Ambrosia maritima, Acacia nilotica and *Ocimum basilicum* and fruit extracts of *C. colocynthis, C. procera* and *E. jambolana*	Crude extracts	Abdel-Monaim *et al.*, 2011
Robinia pseudoacacia	Crude extracts	Zhang *et al.*, 2008
Cassia sp	cassia oil	Feng *et al.*, 2008
Reynoutria sachalinensis	Crude extracts	Konstantinidou-Doltsinis and Schmit, 1998
Hypericum perfoliatum and *Hypericum tomentosum*	α-pinene, *allo*-aromadendrene, germacrene-D, *n*-octane, α-selinene and β-selinene. Menthone, *n*-octane, β-caryophyllene, α-pinene, lauric acid and β-pinene	Hosni *et al.*, 2008
Metasequoia glyptostroboides	α-pinene, caryophyllene oxide, α-thujene, bornylene, totarol, β-caryophyllene, δ-3-carene, 2-β-pinene and α-humulene.	Bajpai *et al.*, 2007
Aegle marmelos	essential oil	Pattnaik *et al.*, 1996
Allium sativum	essential oil	Pyun and Shin 2006
Bystropogon plumosus	essential oil	Economou & Nahrstedt, 1991
Citrus aurantium	essential oil	Pattnaik *et al.*, 1996
Cryptomeria japonica	essential oil	Cheng *et al.*, 2005
Cymbopogonflexuosus	essential oil	Pattnaik *et al.*, 1996
Cymbopogon martini	essential oil	Pattnaik *et al.*, 1996

Specie	Compounds Identifying	Reference
Eucalyptus citriodora	essential oil	Pattnaik *et al.*, 1996
Melaleuca alternifolia	essential oil	Nenoff *et al.*, 1996
Mentha piperita	essential oil	Pattnaik *et al.*, 1996
Pelargonium graveolens	essential oil	Pattnaik *et al.*, (1996
Pimpinella anisum	essential oil	Kosalec *et al.*, (2005
Piper angustifolium	essential oil	Tirillini *et al.*, 1996
Salvia officinalis	essential oil	Hili *et al.*, 1997
Salvia sclarea	essential oil	Pitarokili *et al.*, 2002
Tagetes patula	essential oil	Romagnoli *et al.*, 2005
Thymbra capitata	essential oil	Salgueiro *et al.*, 2004
Thymus pulegioides	essential oil	Pinto *et al.*, 2006
Lavandula angustifolia	essential oil	D'Auria *et al.*, 2005
Dictamnus dasycarpus	Dictamnine	Zhao *et al.*, 1998
Heliotropium bursiferum	9-Angeloylretronecine, Heliotrine, Lasiocarpine, Supinine	Marquina *et al.*, 1989
Ficus septic	Antofine, Ficuseptine	Baumgartner *et al.*, 1990
Glycosmis cyanocarpa	Illukumbin B, Methylillukumbin B, Methylillukumbin A, *N*-Methylsinharine, Sinharine	Greger *et al.*, 1992, 1993
Olea europaea	Hexanal, E-2-Hexanal, E-2-Heptanal, Nonanal and E-2-Octenal	Battinelli *et al.*, 2006
Cochlospermum tinctorium	Cochloxanthin, Dihydrocochloxanthin	Diallo *et al.*, 1991
Eupatorium riparium	Methylripariochromene A	Bandara *et al.*, 1992
Apium graveolens	Angelicin, Bergapten, Columbianetin, Xanthotoxin	Afek *et al.*, 1995
Wedelia biflora	3´_-Formyl-2´_,4´_,6´_-trihydroxydihydrochalcone	Miles *et al.*, 1991
Scutellaria spp	Clerodin, Jodrellin A, Jodrellin B	Cole *et al.*, 1991
Croton sonderianus	Hardwickic acid, 3,4-Secotrachylobanoic acid	McChesney & Clark, 1991
Gomphrena martiana and *Gomphrena boliviana*	5-Hydoxy-3-methoxy-6,7-methylenedioxyflavone	Pomilio *et al.*, (1992
H. nitens	3,5,6,7,8-Pentamethoxyflavone, 3,5,6,7-Tetramethoxyflavone, 5,6,7,8-Tetramethoxyflavone, Dimethylchrysin, Trimethylgalangin	Tomas-Barberan *et al.*, 1988
H. odoratissimum	3-*O*-Methylquercetin	Van Puyvelde *et al.*, 1989
Wedelia biflora	veratrylidenehydrazide, 3,3'-di-*O*-methylquercetin, 2,7-dihydroxy-3(3t'-methoxy-4'-hydroxy)-5-methoxyisoflavone and 3',7-di-*O*-methylquercetin	Miles *et al.*, 1993
Podophyllum hexandrum	4'-*O*-demethyldehydropodophyllotoxin and picropodophyllone	Rahman *et al.*, 1995

Specie	Compounds Identifying	Reference
Piper angustifolium	Camphene	Tirillini *et al.*, 1996
Cistus incanus subsp. *creticus*	Geraniol	Chinou *et al.*, 1994
Bystropogon plumosus, B. origanifolius var. *palmensis, B. wildpretii, B. maderensis* and *B. canariensis* var. *smithianus*	Pulegone	Economou & Nahrstedt, 1991;
Zingiber officinale	Gingerenone A	Endo *et al.*, 1990
Coleonema pulchellum	Precolpuchol	Brader *et al.*, 1997
P. argentatum × *P. tomentosa*	8-oxo-Argentone, 8-oxo-15-nor-Argentone, 15-Hydroxyargentone, Argentone and 15-nor-Argentone	Maatooq *et al.*, 1996
Bidens cernua	Cernuol	Smirnov *et al.*, 1998
Garcinia mangostana	BR-xanthone A, Garcinone D, Gartanin, Mangostin, γ -Mangostin	Gopalakrishnan *et al.*, 1997
	(*E*)-3-Chloro-4-stilbenol, (*E*)-3,5-Dimethoxy-4- stilbenol, (*E*)-3,5-Dimethoxystilbene, (*E*)-3-Methoxy-4-stilbenol, (*Z*)-4-Methoxy-3-stilbenol, (*E*)-5-Methoxy-3-stilbenol, (*E*)-4-Stilbenol, (*E*)-3-Stilbenol, (*Z*)-3-Stilbenol, (*E*)-3,4-Stilbenediol, (*E*)-3,5-Stilbenediol	Schultz *et al.*, 1992
	Geraniol, Linalool, 1,8-Cineole, Citral	Pattnaik *et al.*, 1997
	Isolimonene, Isopulegol, Carvone	Naigre *et al.*, 1996
	5,7-Dihydroxy-4-hydroxyisoflavan, 6,7-Dihydroxy-4_-methoxyisoflavan, 5,7-Dihydroxy-4_-methoxyisoflavan, Biochanin A	Weidenborner *et al.*, 1990
Thymus pulegioides	Carvacrol, p-Cymene and γ -Terpinene	Pinto *et al.*, 2006
	8-Acetylheterophyllisine, Panicutin, Vilmorrianone	Rahman *et al.*, 1997
	Clausenal	Chakraborty *et al.*, 1995
	Harman, Harmine, Norharman	Quetin-Leclercq *et al.*, 1995
Calycodendron milnei	Isopsychotridine E, Hodgkinsine A, Quadrigemine C, Quadrigemine H, Psychotridine E, Vatine, Vatine A, Vatamine, Vatamidine,	Saad *et al.*, 1995

Specie	Compounds Identifying	Reference
	Dehatrine, Actinodaphnine, Anhydroushinsunine, Methoiodide, *N*-Methylactinodaphnine	Tsai *et al.*, 1989
	Anonaine	Tsai *et al.*, 1989; Simeon *et al.*, 1990
	Lanuginosine, Lysicamine	Simeon *et al.*, 1990
	Berberine	Okunade *et al.*, 1994
	Alkaloids	
	3-Methoxysampangine	Liu *et al.*, 1990
	Steroidal alkaloids	
	a-Chaconine, *a*-Solanine	Fewell & Roddick, 1993
	Polygodial	Lee *et al.*, 1999

Table 3. Chemical compounds identified with antifungal properties derived from species plants

2.6 Effect of compounds in inhibiting mycelia fungi

The most compounds have varied effects on the development of mycelia growth of fungi and the effect on sporulation rate and inhibition of germination ranging from a fungistatic effect to complete inhibition. The answer depends on the arrest of compounds derived from extracts of the species and to inhibit fungus. Table 4 shows the sensitivity of plant pathogen fungi to bioactive coumponds from plants.

Plant Specie	Plant pathogen	Fungicidal activity concentrations	References
Achillea gypsicola and *A. biebersteinii*	*Fusarium equiseti* and *F. graminearum*		Kordali *et al.*, 2009
Agapanthus africanus	*Pythium ultimum*, *F. oxysporum*, *Alternaria alternata*, *Mycosphaerella pinodes* and *Ascochyta*		Tegegne *et al.*, 2008
Aloe vera	*Rhizoctonia solani*, *F. oxysporum* and *Colletotrichum coccodes*	10^5 µl L^{-1}	Jasso de Rodríguez *et al.*, 2005
Asarum heterotropoides var. *mandshuricum*	*Alternaria humicola*, *Colletotrichum gloeosporioides*, *Rhizoctonia solani*, *Phytophthora cactorum* and *Fusarium solani*	<0.42 µg mL^{-1}	Dan *et al.*, 2010

Plant Specie	Plant pathogen	Fungicidal activity concentrations	References
Astronium fraxinifolium, Inga marginata, Malva sylvestris, Matayba elaeagnoides, Miconia argyrophylla, Myrcia fallax, Ocimum gratissimum, Origanum vulgare, Rollinia emarginata, Siparuna arianeae, Styrax pohlii, Tabebuia serratifolia and *Trichilia pallida*	*Colletotrichum lindemuthianum*	inhibition of conidial germination	Andrade *et al.*, 2010
Bucida buceras, Breonadia salicina, Harpephyllum caffrum, Olinia ventosa, Vangueria infausta and *Xylotheca kraussiana*	*Aspergillus niger, Aspergillus parasiticus, Colletotricum gloeosporioides, Penicillium janthinellum, Penicillium expansum, Trichoderma harzianum* and *Fusarium oxysporum*	0.02-0.08 mg mL^{-1}	Mahlo *et al.*, 2010
Carya illinoensis shells and *Punica granatum*	*Pythium* sp., *Colletotrichum truncatum, Colletotrichum coccodes, Alternaria alternata, Fusarium verticillioides, Fusarium solani, Fusarium sambucinum* and *Rhizoctonia solani*	0.2 mgL^{-1}	Osorio *et al.*, 2010
Cassia sp.	*Alternaria alternate*	500 µl L^{-1}	Feng *et al.*, 2008
Chenopodium quinoa	*Botrytis cinerea*	5 mg saponins ml^{-1}, 100% of conidial germination inhibition	Stuardo *et al.*, 2008
Drimys winteri	*Gaeumannomyces graminis* var *tritici*	932- 30.37mg L^{-1}	Monsálvez *et al.*, 2010,
Flourensia microphylla, Flourensia cernua and *Flourensia retinophylla*	*Alternaria* sp., *Rhizoctonia solani* and *Fusarium oxysporum*	10 to 1500µl L^{-1}	Jasso de Rodríguez *et al.*, 2007
Larrea tridentata, Flourensia cernua, Agave lechuguilla, Opuntia sp. and *Yucca* sp.,	*Rhizoctonia solani*	2000 ppm of totals polyphenols	Castillo *et al.*, 2010,

Plant Specie	Plant pathogen	Fungicidal activity concentrations	References
Larrea tridentata	*Aspergillus flavus* and *Aspergillus parasiticus*	300-500 µg mL^{-1} of NDGA	Vargas-Arispuro *et al.*, 2005
Laurus nobilis	*Botrytis cinerea, Monilinia laxa* and *Penicillium digitatum*	1, 2 and 3 mg mL^{-1}	Corato *et al.*,, 2010
Metasequoia glyptostroboides	*Fusarium oxysporum, Fusarium solani, Sclerotonia sclerotiorum, Rhizoctonia solani, Colletotricum capsici, Botrytis cinerea* and *Phytophthora capsici,*	Inhibition range of 49–70% and minimum inhibitory concentration ranging from 500 to 1000 µg mL^{-1}.	Bajpai *et al.*, 2007
Piper longum	*Pyricularia oryzae, Rhizoctonia solani, Botrytis cineria, Phytophthora infestans, Puccinia recondite* and *Erysiphe graminis*	1mg mL^{-1}	Lee *et al.*, 2001
Reynoutria sachalinensis	*Sphaerotheca pannosa* var. *rosae*		Pasini *et al.*, 1997; Konstantinidou-Doltsinis & Schmit, 1998
Robinia pseudoacacia	*Sphaerotheca fuliginea,*	80 mg mL^{-1}	Zhang *et al.*, 2008
Rumex crispus	*Blumeria graminis* f. sp. *hordei*	30 µg mL^{-1}	Choi *et al.*, 2004
Salvia officinalis	*Penicillium, Aspergillus, Cladosporium* and *Fusarium*	0.63 µl ml^{-1}	Pinto *et al.*, 2007
Thymus zygis subsp. *sylvestris*	*Aspergillus* strains	0.08- 0.16 µL mL^{-1}	Gonçalves *et al.*, 2010
Cryptomeria japonica	*Rhizoctonia solani, Collectotrichum gloeosporioides, Fusarium solani* and *Ganoderma australe*	MIC(50) values of 65, 80, 80 and 110 mg mL^{-1}	Cheng *et al.*, 2005
Melaleuca alternifolia	*Candida albicans* and *Candida sp.*	500–6000	Nenoff *et al.*, 1996
Pimpinella anisum	*Trichophyton rubrum, T. mentagrophytes, Microsporum canis* and *M. gypseum*	MIC to 1.5 and 9.0% (V/V).	Kosalec *et al.*, 2005
Piper angustifolium	*Candida albicans, Cryptococcus neoformans, Aspergillus flavus, Aspergillus fumigatus,*	10–100	Tirillini *et al.*, 1996

Plant Specie	Plant pathogen	Fungicidal activity concentrations	References
Salvia officinalis	*Torulopsis utilis, Schizosaccharomyces pombe, Candida albicans* and *Saccharomyces cerevisiae*		Hili *et al.*, 1997
Salvia sclarea	*Soil-borne pathogens*	EC50: 493–584 µL L^{-1}	Pitarokili *et al.*, 2002
Tagetes patula	*Penicillium digitatum* and *Botrytis cinerea*	1.25–10.0 µL mL^{-1}	Romagnoli *et al.*, 2005
Thymbra capitata	*Candida sp., Aspergillus* sp	0.08–0.32 µL mL^{-1}	Salgueiro *et al.*, 2004
Thymus pulegioides	*Candida, Aspergillus* and *dermatophyte* species	0.16–0.64 µL mL^{-1}	Pinto *et al.*, 2006
Lavandula angustifolia	*Candida albicans*	0.69%	D'Auria *et al.*, 2005
3-Methoxysampangine	*Candida albicans, Aspergillus fumigatus* and *Cryptococcus neoformans*	0.2–3.1	Liu *et al.*, 1990
Steroidal alkaloids			
a-Chaconine	*Ascobolus crenulatus, Alternaria brassicicola, Phoma medicaginis* and *Rhizoctonia solani*	60–100 µM	Fewell & Roddick 1993
a-Solanine	*Ascobolus crenulatus, Alternaria brassicicola, Phoma medicaginis* and *Rhizoctonia solani*	80–100 µM	Fewell & Roddick 1993
Dictamnus dasycarpus	*Cladosporium cucumerinum*	25	Zhao *et al.*, 1998
Olea europaea	*Tricophyton mentagrophytes, Microsporum canis* and *Candida* spp	1.9 -250	Battinelli *et al.*, 2006
Eupatorium riparium	*Colletotrichum gloeosporioides*		Bandara *et al.*, 1992
Wedelia biflora	*Rhizoctonia solani; Pythium ultimum;*		Miles *et al.*, 1991
Scutellaria spp	*Fusarium oxysporum* f. sp. *lycopersici* and *Verticillium tricorpus*		Cole *et al.*, 1991
Wedelia biflora	*Rhizoctonia solani; Pythium ultimum;*		Miles *et al.*, 1993
Podophyllum hexandrum	*Epidermophyton floccosum, Curvularia lunata, Nigrospora oryzae,*		Rahman *et al.*, 1995

Plant Specie	Plant pathogen	Fungicidal activity concentrations	References
	Microsporum canis, Allescheria boydii and *Pleurotus ostreatus, Drechslera rostrata*		
Piper angustifolium	*Candida albicans, Aspergillus flavus, Aspergillus fumigatus*	1.0–5.0 mM; 0.016–0.13% of oil;	Tirillini *et al.*, 1996
Cistus incanus subsp. creticus	*Candida albicans*	125–375	Chinou *et al.*, 1996
Thymus pulegioides	*Candida, Aspergillus*	1.25–20.0 µL mL^{-1}	Pinto *et al.*, 2006
Zingiber officinale	*Pyricularia oryzae*	10.0 ppm	Endo *et al.*, 1990
Coleonema pulchellum	*Cladosporium herbarum*		Brader *et al.*, 1997
Parthenium argentatum × *P. tomentosa*	*Aspergillus fumigatus* and *A. niger*	0.25 mg mL^{-1} 1.0 mg mL^{-1}	Maatooq *et al.*, 1996
Garcinia mangostana	*Fusarium oxysporum vasinfectum, Alternaria tenuis* and *Drechslera oryzae*		Gopalakrishnan *et al.*, 1997
	Aspergillus repens; A. amstelodami; A. chevalieri; A. flavus; A. petrakii;		Weidenb¨orner *et al.*, 1990a, b
	Coriolus versicolor, Gloeophyllum trabeum and *Poria placenta*	8-140	Schultz *et al.*, 1992
	Aspergillus niger	0.78–100 µL mL^{-1}	Naigre *et al.*, 1996
	Candida albicans, Trichophyton mentagrophytes, T. ruburum, Penicillium marneffei, Aspergillus fumigatus, A. flavus, P. chrysogenum, C. lipolytica and *C. tropicalis*	0.78–100.0	Lee *et al.*, 1999
Cymbopogonflexuosus		0.16–11.6	Pattnaik *et al.*, 1996
Cymbopogon martini		0.5–8.3	Pattnaik *et al.*, 1996
Eucalyptus citriodora		0.16–10.0	Pattnaik *et al.*, 1996
Bidens cernua		5.0–200	Smirnov *et al.*, 1998
Gomphrena martiana and *G. boliviana*		75	Pomilio *et al.*, 1992

Plant Specie	Plant pathogen	Fungicidal activity concentrations	References
Helichrysum nitens		1- 20 µg	Tomas-Barberan *et al.*, 1988
Allium sativum		64	Pyun and Shin 2006
Psidium acutangulum			Miles *et al.*, 1993
Croton sonderianus			McChesney & Clark, 1991
Bystropogon plumosus, B. origanifolius var. *palmensis, B. wildpretii, B. maderensis* and *B. canariensis* var. *Smithianus*		0.4–85.0% of oil	Economou & Nahrstedt, 1991; Kalodera *et al.*, 1994
Mentha piperita		0.27–10.0	Pattnaik *et al.*, 1996
Pelargonium graveolens			Pattnaik *et al.*, 1996

Table 4. Bioactive compounds from plants on fungal species.

2.7 Commercial use of natural fungicides

Currently, the commercial use of natural fungicides on the market is low, the 5th Annual Meeting of the biological control industry (Loison, 2010) reports a total of 55 biological fungicides registered in the U.S. market and in the EU the registered biopesticides are much fewer: 21 fungicides for be used in Pome fruit, vines and tomato (Table 5).

Commercial name	Active Ingradient	Company	Plant pathogen
BC 1000 ™	Bioflavonoid of Seed extracts and orange pulp	Chemie S.A.	*Botritys cinérea*
Bio save ™	Seed extracts and orange pulp	Bioland SA	*Ascochyta, Pullullaria, Fusarium, Cercospora, Botrytis, Septoria, Alternaria,* Stemphylium, *Rhizoctonia, Peronospora, Pythium, Penicilium, Sigatoka, Aspergillus.*
Agrispon ™	Plant and mineral extacts.	Agric. Sci Dallas	*Cercospora beticola*
Sincocin ™	Plant extracts	Agric. Sci Dallas	*Cercospora beticola*

Commercial name	Active Ingradient	Company	Plant pathogen
Timorex Gold	Plant extracts of *Melalueca alternifolia*	Stockton Group	*Mycosphaerella fijiensis*
Evergreen TM	Plant extracts	Aashab bio industries	
Gloves Off TM	Thymol, Carvacrol	Organozoid and Such	*Trichophyton mentagrophytes*
Garden Fungicide TM	Rosemary, thyme and clove oil	EcoSmart	
Eco Safe TM	Pongamia and Tulsi oil, Recines communis	S. K. Bio Extracts & Applications	*Root rot, Dammping off, Steam rot, leaf spot*
Gloss TM	Natural Alkaloids	S. K. Bio Extracts & Applications	fungal diseases in all field crops, vegetables and horticultural crops

Table 5. Some commercial product in the market with active ingredients from plants

3. Conclusions

The plant extracts applied in as crude state or as a fraction affect the development of fungal colonies to inhibit partially and totally in laboratory tests at low concentrations of bioactive compounds, besides affecting the incedencia and severity when applied as a treatment to increase the shelf life of products with excellent results. However, more research is needed to determine its effect on molecular changes, morphological and biochemical these compounds cause the pathogen and host.

4. References

Abbassy, M.A., Abdelgaleil, S.A.M., Belal, A.S.H. & Abdel, R.M.A.A. (2007). Insecticidal, antifeedant and antifungal activities of two glucosides isolated from the seeds of *Simmondsia chinensis*, Industrial Crops and Products, Vol. 26, No. 3, (Octuber, 2007), pp. (345-350), ISSN: 0926-6690

Abdel-Monaim, M.F., Abo-Elyousr, K.A.M. & Morsy, K.M.. (2011). Effectiveness of plant extracts on suppression of damping-off and wilt diseases of lupine (*Lupinus termis* Forsik). *Crop Protection*, Vol. *30,* No. 2, (Febrary 2011), pp. (185-191), ISSN: 0261-2194

Abou-Jawdah, Y., Sobh, H. & Salameh A. (2002) Antymicotic activities of selected plant flora, growing wild in Lebanon, against phytopathogenic fungi, Journal of Agricultural and Food Chemistry, Vol. 50, No. 11, (May 2002), pp. (3208-3213), ISSN: 0021-8561

Afek, U., Carmeli, S. & Aharoni, N. (1995) Columbianetin, a phytoalexin associated with celery resistance to pathogens during storage, *Phytochemistry,* Vol. 39, No. 6, (Augost, 1995), pp. (1347–1350), ISSN: 0031-9422

Andrade, P.J.M., Aparecida, S.E. & Ferreira. O. D. (2010). Use of plant extracts in the control of common bean anthracnose. *Crop Protection*, Vol. 29, No. 8, (Augost, 2010), pp. (838-842), ISSN: 0261-2194

Bajpai, V.K., Rahman, A. & Kang, S.Ch. 2007. Chemical composition and anti-fungal properties of the essential oil and crude extracts of Metasequoia glyptostroboides Miki ex Hu, *Industrial Crops and Products*, Vol. 26, No. 1, (January, 2007), pp. (28-35), ISSN: 0926-6690

Bandara, B.M., Hewage, C.M., Karunaratne, V., Wannigama, C.P. & Adikaram, N.K. (1992). An antifungal chromene from *Eupatorium riparium*, *Phytochemistry*, Vol. 31, No. 6, (June, 1992), pp. (1983–1985), ISSN: 0031-9422

Battinelli, L., Daniele, C., Cristiani, M., Bisignano, G., Saija, A. & Mazzanti, G. (2006). In vitro antifungal and anti-elastase activity of some aliphatic aldehydes from *Oleaeuropaea* L. fruit. *Phytomedicine*, Vol. 13, No. 8, (September, 2006), pp. (558–563), ISSN: 0944-7113

Baumgartner, B., Erdelmeier, C.A., Wright, A.D., Ralli, T. & Sticher, O. (1990) An antimicrobial alkaloid from *Ficus septica*. *Phytochemistry*, Vol. 29, No. 10, (October, 1990), pp. (3327–3330), ISSN: 0031-9422

Bautista, S.L., Barrera, N.L., Bravo, L.K. & Bermúdez, T. (2002). Antifungal activity of leaf and stem extracts from various plant species on the incidente of *Colletotrichum gloeosporioides* of papaya and mango fruits after storage, *Rev. Mex. Fitopatol*. Vol. 20, No. 1, (June, 2002), pp. (8-12), ISSN: 0185- 3309

Brader, G., Bacher, M., Hofer, O. & Greger, H. (1997) Prenylated phenylpropenes from *Coleonema pulchellum* with antimicrobial activity. *Phytochemistry*, Vol. 45, No. 6, (Jul, 1997), pp. (1207–1212), ISSN: 0031-9422

Castillo, F., Hernández, D., Gallegos, G., Méndez, M., Rodríguez, R., Reyes A. &. Aguilar, C.N. (2010). In vitro antifungal activity of plant extracts obtained with alternative organic solvents against *Rhizoctonia solani* Kühn, *Industrial Crops and Products*, Vol. 32, No. 3, (June 2010), pp. (324–328), ISSN 0926-6690

Chakraborty, A., Saba, C., Podder, G., Chowdhury, B.K. and Bhattacharyya, P. (1995) Carbazole alkaloid with antimicrobial activity from *Clausena heptaphylla*. *Phytochemistry*, Vol. 38, No. 3, (Febrary, 1995), pp. (787–789), ISSN: 0031-9422

Cheng, S.S., Lin, H.Y. & Chang, S.T. (2005). Chemical composition and antifungal activity of essential oils from different tissue of Japanese Cedar (*Cryptomeria japonica*), *Journal of Agricultural and Food Chemistry*, , Vol. 53, No. 3, (February, 2005), pp. (614–9), ISSN: 0021-8561

Chinou, I., Demetzos, C., Harvala, C., Roussakis, C. & Verbist, J.F. (1994). Cytotoxic and antibacterial labdane-type diterpenes from the aerial parts of *Cistus incanus* subsp. *creticus*, *Planta Medica*, Vol. 60, No. 1, (Febrary, 1994), pp. (34–6), ISSN: 0032-0943

Choi, G.J., Lee, S.W., Jang, K.S., Kim, J.S., Cho, K.Y. & Kim, J.Ch. (2004). Effects of chrysophanol, parietin, and nepodin of Rumex crispus on barley and cucumber powdery mildews, *Crop Protection*, Vol. 23, No. 12, (December. 2004), pp. (1215-1221), ISSN: 0261-2194

Cole, M.D., Bridge, P.D., Dellar, J.E., Fellows, L.E., Cornish, M.C. & Anderson, J.C. (1991) Antifungal activity of neo-clerodane diterpenoids from *Scutellaria*, *Phytochemistry*, Vol. 30, No. 4, (March, 1991), pp. (1125–1127), ISSN: 0031-9422

Corato, U., Maccioni, O., Trupo, M., & Di Sanzo G. (2010). Use of essential oil of *Laurus nobilis* obtained by means of a supercritical carbon dioxide technique against post harvest spoilage fungi, *Crop Protection*, Vol. *29*, No. 2, (Febrary, 2010), pp. (142-147), ISSN: 0261-2194

Cowan, M. M. (1999). Plant products as antimicrobial agents, *Clinical Microbiology Review*, Vol. 12, No. 4, (October 1999), pp. (564-582), ISSN: 1098-6618

D'Auria, F.D., Tecca, M., Strippoli, V., Salvatore, G., Battinelli, L. & Mazzanti, G. (2005) Antifungal activity of *Lavandula angustifolia* essential oil against *Candida albicans* yeast and mycelial form, *Med. Mycol.*, Vol. 43, No. 5, (Augost, 2005), pp. (391–396), ISSN: 1369-3786

Dan, Y., Hai-Yan, L., Wei-Wei, G. & Shi-Lin, Ch. (2010). Activities of essential oils from *Asarum heterotropoides* var. *mandshuricum* against five phytopathogens, *Crop Protection*, Vol. 29, No. 3, (March, 2010), pp. (295-299), ISSN: 0261-2194

Devaiah, S.P., Mahadevappa, G.H. & Shetty, H.S. (2009). Induction of systemic resistance in pearl millet (*Pennisetum glaucum*) against downy mildew (*Sclerospora graminicola*) by *Datura metel* extract, *Crop Protection*, Vol. 28, No. 9, (September, 2009), pp. (783-791), ISSN: 0261-2194

Diallo, B., Vanhaelen-Fastre, R. & Vanhaelen, M. (1991) Antimicrobial activity of two apocarotenoids isolated from *Cochlospermum tinctorium* rhizome, *Fitoterapia*, Vol. 62, No. 2, pp. (144–145), ISSN: 0367-326X

Economou, D. & Nahrstedt, A. (1991) Chemical, physiological and toxicological aspects of the essential oil of some species of the genus *Bystropogon*, *Planta Medica*, Vol. 57, No. 4, pp. (347–51), ISSN: 0032-0943

Endo, K., Kanno, E. & Oshima, Y. (1990) Structures of antifungal diarylheptenones, gingerenones A, B, C and isogingerenone B, isolated from the rhizome of *Zingiberofficinale*, *Phytochemistry*, Vol. 29, No. 3, pp. (797–799), ISSN: 0031-9422

Feng, W., Zheng, X., Chen, J. & Yang, Y. (2008). Combination of cassia oil with magnesium sulphate for control of postharvest storage rots of cherry tomatoes, *Crop Protection*, Vol. *27*, No. 1, (January, 2008), pp. (112-117), ISSN: 0261-2194

Fewell, A.M. & Roddick, J.G. (1993). Interactive antifungal activity of the glycoalkaloids, asolanine and _-chaconine, *Phytochemistry*, Vol. 33, No. 2, (May, 1993), pp. (323–328), ISSN: 0031-9422

García, D.E. (2004). Los metabolitos secundarios de las especies vegetales, *Pastos y Forrajes*, Vol. 27, No. 1, (), pp. (1-12), ISSN: ISSN 0864-0394

Glasby, J.S. (1991). Dictionary of Plants Secondary Metabolites Containing, Taylor & Francis (Eds) e-Library, 2005. 1644 p. ISBN 0-203-48987-X Master e-book.

Gonçalves, M.J., Cruz, M.T., Cavaleiro, C., Lopes, M.C. & Salgueiro, L. (2010). Chemical, antifungal and cytotoxic evaluation of the essential oil of *Thymus zygis* subsp. *sylvestris*, Industrial Crops and Products, Vol. 32, No. 1, (July, 2010), pp. (70-75), ISSN: 0926-6690

González G E, Rodríguez H R & Aguilar G C N. (2009). Biodegradación de Taninos. Cienciacierta, 17. www.postgradoeinvestigacion.uadec.mx/CienciaCierta/CC17/cc17taninos.html

Gopalakrishnan, G., Banumathi, B. & Suresh, G. (1997) Evaluation of the antifungal activity of natural xanthones from *Garcinia mangostana* and their synthetic derivatives,

Journal of Natutal Products, Vol. 60, No. 5, (Mayo, 1997), pp. (519–524), ISSN: 0163-3864

Greger, H., Hofer, O., K¨ahlig, H. & Wurz, G. (1992). Sulfur-containing cinnamides with antifungal activity from *Glycosmis cyanocarpa*, *Tetrahedron*, Vol. 48, No. 7, pp. (1209–1218), ISSN: 0040-4020

Greger, H., Zechner, G., Hadacek, F. & Wurz, G. (1993). Sulphur-containing amides from *Glycosmis* species with different antifungal activity, *Phytochemistry*, Vol. 34, No. , (), pp. (175–9), ISSN: 0031-9422

Gyung, J. Ch., Seon-Woo L., Kyoung S. J., Jin-Seog K., Kwang Y. Ch. & Jin-Cheol K. (2004). Effects of chrysophanol, parietin, and nepodin of *Rumex crispus* on barley and cucumber powdery mildews, *Crop Protection*, Vol. 23, No. 12, (December, 2004), pp. (1215-1221), ISSN: 0261-2194

Haslam, E. (1996). Natural polyphenols (vegetable tannins) as drugs: possible modes of action, *Journal of Natural Products*, Vol. 59, No. 2, (Febrero, 1996), pp. (205-215), ISSN: 0163-3864

Hernández, L. A. N., Bautista, B. S. & Velázquez del valle, M. G. (2007). Prospectiva de extractos vegetales para controlar enfermedades postcosecha hortofrutícolas, *Rev. Fitotecnia Mex.* Vol. 30, No. 2, (Abril-Junio, 2007), pp. (119-123), ISSN: 0185-3309

Hili, P., Evans, C.S. & Veness, R.G. (1997). Antimicrobial action of essential oils: the effect of dimethylsulphoxide on the activity of cinnamon oil, *Letters in Applied Microbiology*, Vol. 24, No. 4, (April, 1997), pp. (269–75), ISSN: 0266-8254

Hosni, K., Msaâda, K., Ben Taârit, M., Ouchikh, O., Kallel, M. & Marzouk, B. (2008). Essential oil composition of Hypericum perfoliatum L. and Hypericum tomentosum L. growing wild in Tunisia, *Industrial Crops and Products*, Vol. 27, No. 3, (May, 2008), pp. (308-314), ISSN: 0926-6690

Jasso de Rodríguez D., Hernández-Castillo D., Angulo-Sánchez J.L., Rodríguez-García R., Villarreal Quintanilla J.A. & Lira-Saldivar R.H. (2007). Antifungal acvtivity in vitro of Flourensia spp. extracts on Alternaria sp., Rhizoctonia solani, and Fusarium oxysporum, *Industrial Crops and Products*, Vol. 25, No. 2, (Febrary, 2007), pp. (111-116), ISSN 0926-6690

Jasso de Rodríguez D., Hernández-Castillo D., Rodríguez-García R. & Angulo-Sánchez J. L. (2005). Antifungal activity in vitro of Aloe vera pulp and liquid fraction against plant pathogenic fungi, *Industrial Crops and Products*, Vol. 21, No. 1, (January, 2005), pp. (81-87), ISSN: ISSN 0926-6690

Konstantinidou-Doltsinis, S. & Schmit, S. (1998). Impact of treatment with plant extracts from *Reynoutria sachalinensis* (F. Schmidt) Nakai on intensity of powdery mildew severity and yield in cucumber under high disease pressure, *Crop Protection*, Vol. 17, No. 8, (November, 1998), pp. (649-656), ISSN: 0261-2194

Kordali S., Cakir A, Aytas A. T, Mete E, Akcin A, Aydin T, & Kilic, H. (2009). Antifungal and herbicidal properties of essential oils and *n*-hexane extracts of *Achillea gypsicola* Hub-Mor. and *Achillea biebersteinii* Afan. (Asteraceae), *Industrial Crops and Products*, Vol. 29, No. 2-3, (March 2009), pp. (562-570), ISSN: 0926-6690

Kosalec, I., Pepeljnjak, S. & Kustrak, D. (2005). Antifungal activity of fluid extract and essential oil from anise fruits (*Pimpinella anisum* L., Apiaceae), *Acta Pharmaceutica*, Vol. 55, No. 4, (December, 2005), pp. (377–85), ISSN: 1330-0075

Lee, S.E., Park, B.S., Kim, M.K., Choi, W.S., Kim, H.T., Cho, K.Y., Lee, S.G. & Lee, H.S. (2001). Fungicidal activity of pipernonaline, a piperidine alkaloid derived from long pepper, *Piper longum* L., against phytopathogenic fungi, *Crop Protection*, Vol. *20*, No. 6, (July, 2001), pp. (523-528), ISSN: 0261-2194

Lee, S.H., Lee, J.R.L., Lunde, C.S. & Kubo, I. (1999). In vitro antifungal susceptibilities of *Candida albicans* and other fungal pathogens to polygodial, a sesquiterpene dialdehyde., *Planta Medica*, Vol. 65, No. 3, (April), pp. (204–208), ISSN: 0032-0943

Lira, S.R.H., Hernández, S.M., Chavéz, B.C., Hernández, C.F.D. & Cuellar, V.E. (2007). Biopesticides and biological control. CIQA, Monterrey, México. Pp 13-29. ISBN 968-844.054-X Mexico

Liu, S., Oguntimein, B.O., Hufford, C.D. and Clark, A.M. (1990). 3- methoxysampangine, a novel antifungal copyrine alkaloid from *Cleistopholis patens, Antimicrobial Agents and Chemotherapy*, Vol. 34, No. 4, (April, 1990), pp. (529–533), ISSN: 0066-4804

Loison, M. (2010). The plant protection corner, in: A growing industry, various business models, *New Ag Internacional magazine*, (Febrary 2010). *www.newaginternational.com/current/plantprotection*201011*.pdf*

Maatooq, G.T., Stumpf, D.K., Hoffmann, J.J., Hutter, L.K. and Timmermann, B.N. (1996) Antifungal eudesmanoids from *Parthenium argentatum x P. tomentosa. Phytochemistry*, Vol. 41, No. 2, (Febrary, 1996), pp. (519–524), ISSN: 0031-9422

Mahlo, S.M., McGaw, L.J. & Eloff, J.N. (2010). Antifungal activity of leaf extracts from South African trees against plant pathogens, *Crop Protection*, Vol. 29, No. 12, (December, 2010), pp. (1529-1533), ISSN: 0261-2194

Makkar, H P S, Siddhuraju, P & Becker, K. (2007). Plant secondary metabolites, Human press Inc, pp. 130, ISSN: 1-58829-993-7, USA.

Marquina, G., Laguna, A., Franco, P., Fernandez, L., Perez, R. & Valiente, O. (1989). Antimicrobial activity of pyrrolizidine alkaloids from *Heliotropium bursiferum, Pharmazie*, Vol. 44, No. 12, (December, 1989), pp. (870–871), ISSN: 0031-7144

McChesney, J.D. & Clark, A.M. (1991). Antimicrobial diterpenes of *Croton sonderianus*, hardwickic and 3,4-secotrachylobanoic acids, *Journal of Natural Products.*, Vol. 54, No. 6, (November, 1991), pp. (1625–1633), ISSN: 0163-3864

Meckes, M, Rivera, A.D., Nava, V. & Jimenez, A. (2004). Activity of some Mexican medicinal plant extracts on carrageenan-induced rat paw edema, *Phytomedicine*, Vol. 11, No. 5, (July, 2004), pp. (446-451), ISSN: 0944-7113

Mejri, J., Abderrabba, M. & Mejri, M. (2010). Chemical composition of the essential oil of *Ruta chalepensis* L: Influence of drying, hydro-distillation duration and plant parts, *Industrial Crops and Products*, Vol. 32, No. 3, (November, 2010), pp. (671-673), ISSN: 0926-6690

Miles, D.H., Chittawong, V., Hedin, P.A. & Kokpol, U. (1993). Potential agrochemicals from leaves of *Wedelia biflora, Phytochemistry*, Vol. 32, No. 6, (April, 1993), pp. (1427–1429), ISSN: 0031-9422

Miles, D.H., de Medeiros, J.M., Chittawong, V., Hedin, P.A., Swithenbank, C. & Lidert, Z. (1991). 3´_-Formyl-2´_4´_6´_-trihydroxydihydrochalcone from *Psidium acutangulum, Phytochemistry*, Vol. 30, No. 4, (Mayo, 1991), pp. (1131–1132), ISSN: 0031-9422

Monsálvez, M., Zapata, N., Vargas, M., Berti, M., Bittner, M., & Hernández, V. (2010). Antifungal effects of n-hexane extract and essential oil of *Drimys winteri* bark against Take-All disease, *Industrial Crops and Products*, Vol. *31*, No. 2, (March), pp. (239-244), ISSN: 0926-6690

Naigre, R., Kalck, P., Roques, C., Roux, I. & Michel, G. (1996). Comparison of antimicrobial properties of monoterpenes and their carbonylated products, *Planta Medica.*, Vol. 62, No. 3, (June, 1996), pp. (275–277), ISSN: 0032-0943

Nenoff, P., Haustein, U.F. & Brandt, W. (1996). Antifungal activity of the essential oil of *Melaleuca alternifolia* (tea tree oil) against pathogenic fungi *in vitro*, *Skin Pharmacology*, Vol. 9, No. 6, (June, 1996), pp. (388–394), ISSN: 1011-0283

Okunade, A.L., Hufford, C.D., Richardson, M.D., Peterson, I.R. & Clark, A.M. (1994). Antimicrobial properties of alkaloids from *Xanthorhiza simplicissima*, *Journal Pharmaceutical Scinces*, Vol. 83, No. 3, (March, 1994), pp. (404–6), ISSN: 0022-3549

Osei-Safo, D., Addae-Mensah, I., Garneau, F.X. & Kossi, K. H. (2010). A comparative study of the antimicrobiaa activity of the leaf essential oils of chemo-varieties of *Clausena anisata* (Willd.) Hook. f. ex Benth, *Industrial Crops and Products*, Vol. *32*, No. 3, (November. 2010), pp. (634-638), ISSN: 0926-6690

Osorio E, Flores M, Hernández D, Ventura J, Rodríguez R, & Cristóbal N. Aguilar. (2010). Biological efficiency of polyphenolic extracts from pecan nuts shell (*Carya illinoensis*), pomegranate husk (*Punica granatum*) and creosote bush leaves (*Larrea tridentata* Cov.) against plant pathogenic fungi, *Industrial Crops and Products*, Vol. *31*, No. 1, (January, 2010), pp. (153-157), ISSN: 0926-6690

Pasini, C., D'Aquila. F., Curir, P. & Lodovica, G. M. (1997). Effectiveness of antifungal compounds against rose powdery mildew (*Sphaerotheca pannosa* var. *rosae*) in glasshouses, *Crop Protection*, Vol. 16, No. 6, (May, 1997), pp. (251-256), ISSN: 0261-2194

Pattnaik, S., Subramanyam, V.R. & Kole, C.R. (1996). Antibacterial and antifungal activity of ten essential oils *in vitro*, *Microbios*, Vol. 86, No. 349, pp. (237–46), ISSN: 0026-2633

Pattnaik, S., Subramanyam, V.R., Bapaji, M. & Kole, C.R. (1997). Antibacterial and antifungal activity of aromatic constituents of essential oils, *Microbios*, Vol. 89, No. 358, pp. (39–46), ISSN: 0026-2633

Pinto, E., Pina-Vaz, C., Salgueiro, L., Goncalves, M.J., Costa-de-Oliveira, S., Cavaleiro, C., Palmeira, A., Rodrigues, A. & Martinez-de-Oliveira, J. (2006). Antifungal activity of the essential oil of *Thymus pulegioides* on *Candida*, *Aspargillus* and dermatophyte species, *Journal of Medical Microbiology*, Vol. 55, No. 10, (Octuber, 2006), pp. (1367–73), ISSN: 0022-2615

Pinto, E., Ribeiro, S.L., Cavaleiro, C., Palmeira, A. & Gonçalves, M.J. (2007). *In vitro* susceptibility of some species of yeasts and filamentous fungi to essential oils of *Salvia officinalis*, *Industrial Crops and Products*, Vol. *26*, No. 2, (August, 2007), pp. (135-141), ISSN: 0926-6690

Pitarokili, D., Couladis, M., Petsikos-Panayotarou, N. & Tzakou, O. (2002). Composition and antifungal activity on soil-borne pathogens of the essential oil of *Salvia sclarea* from Greece, *Journal of Agricultural and Food Chemistry*, Vol. 50, No. 23, (November, 2002), pp. (6688–6691), ISSN: 0021-8561

Pomilio, A.B., Buschi, C.A., Tomes, C.N. & Viale, A.A. (1992). Antimicrobial constituents of *Gomphrena martiana* and *Gomphrena boliviana*, *Journal of Ethnopharmacology*, Vol. 36, No. 2, (April, 1992), pp. (155–161), ISSN: 0378-8741

Pyun, M.S. & Shin, S. (2006). Antifungal effects of the volatile oils from *Allium* plants against *Trichophyton* species and synergism of the oils with ketoconazole, *Phytomedicine*, Vol. 13, No. 6, (june, 2006), pp. (394–400), ISSN: 0944-7113

Quetin-Leclercq, J., Favel, A., Balansard, G., Regli, P. & Angenot, L. (1995). Screening for *in vitro* antifungal activities of some indole alkaloids, *Planta Medica*, Vol. 61, No. 5, (October, 1995), pp. (475–7), ISSN: 0032-0943

Rahman, A.U., Ashraf, M., Choudhary, M.I., Rehman, H.U. & Kazmi, M.H. (1995). Antifungal aryltetralin lignans from leaves of *Podophyllum hexandrum, Phytochemistry*, Vol. 40, No. 2, (September, 1995), pp. (427–31), ISSN: 0031-9422

Rahman, A.U., Nasreen, A., Akhtar, F., Shekhani, S., Clardy, J., Parvez, M. & Choudhary, M.I. (1997). Antifungal diterpenoid alkaloids from *Delphinium denudatum, Journal of Natural Products*, Vol. 60, No. 5, (May, 1997), pp. (472–474), ISSN: 0163-3864

Rodilla, J.M., Silva, L.A., Martinez, N., Lorenzo, D., Davyt, D., Castillo, L., Giménez, C., Cabrera, R., González-Coloma, A., Zrostlíková, J. & Dellacassa, E. (2011). Advances in the identification and agrochemical importance of sesquiterpenoids from *Bulnesia sarmientoi* essential oil, *Industrial Crops and Products*, Vol. 33, No. 2, (March, 2011), pp. (497-503), ISSN: 0926-6690

Romagnoli, C., Bruni, R., Andreotti, E., Rai, M.K., Vicentini, C.B. & Mares, D. (2005). Chemical characterization and antifungal activity of essential oil of capitula from wild India *Tagetes patula* L, *Protoplasma*, Vol. 225, No. 1-2 , (April, 2005), pp. (57–65), ISSN: 0033-183X

Roy, S. & Chatterjee, P. (2010). A non-toxic antifungal compound from the leaves of *Catharanthus roseus* characterized as 5-hydroxy flavone by UV spectroscopic analysis and evaluation of its antifungal property by agar-cup method, *Industrial Crops and Products*, Vol. 32, No. 3, (November, 2010), pp. (375-380), ISSN: 0926-6690

Saad, H.E., El-Sharkawy, S.H. & Shier, W.T. (1995). Biological activities of pyrrolidinoindoline alkaloids from *Calycodendron milnei*, *Planta Medica*, Vol. 61, No. 4, (August, 1995), pp. (313–6), ISSN: 0032-0943

Salgueiro, L.R., Pinto, E., Goncalves, M.J., Pina-Vaz, C., Cavaleiro, C. Rodrigues, A.G., Palmeira, A., Costa-de-Oliveira, S. and Martinez-de-Oliveira, J. (2004) Chemical composition and antifungal activity of the essential oil of *Thymbra capitata, Planta Med*ica, Vol.70, No. 6, (June, 2004), pp. (572–575), ISSN: 0032-0943

Schultz, T.P., Boldin,W.D., Fisher, T.H., Nicholas, D.D., Murtrey, K.D. & Pobanz, K. (1992) Structure–fungicidal properties of some 3- and 4-hydroxylated stilbenes and bibenzyl analogues. *Phytochemestry*, Vol. 31, No. 11, (November, 1992), pp. (3801–6), ISSN: 0031-9422

Sepúlveda, J. G., Porta, D. H. & Rocha, S. M. (2003). La participación de los metabolitos secundarios en la defensa de las plantas, *Rev. Mex. Fitopatol.*, Vol. 21, No. 3, (Diciembre, 2003), pp. (355-363), ISSN: 0185 3309

Simeon, S., Rios, J.L. and Villar, A. (1990) Antimicrobial activity of *Annona cherimolia* stem bark alkaloids. *Pharmazie*, Vol. 45, No. 6, (June, 1990), pp. (442–3), ISSN: 0031-7144

Smirnov, V.V., Bondarenko, A.S. & Prikhodko, V.A. (1998). Antimicrobial activity of sesquiterpene phenol from *Bidens cernua, Fitoterapia*, Vol. 69, No. 1, pp. (84–5), ISSN: 0367-326X

Stuardo M & San Martín R. 2008. Antifungal properties of quinoa Chenopodium quinoa Willd) alkali treated saponins against Botrytis cinerea, *Industrial Crops and Products*, Vol. *27*, No. 3, (May, 2008), pp. (*296-302*), ISSN: 0926-6690

Swain T. (1979). Tannins and lignins. *In* Herbivores: their interactions with secondary plant metabolites (G.A. Rosenthal & D.H. Janzen, eds.). Academic Press, New York, pp. (657-682)

Swain, T. (1979). Tannins and Lignins. In: Rosenthal, G. A., Janzen, D. H. (ed.) Herbivores, their interaction with secondary plant metabolites, Academic Press. New York, p. 657- 682), ISSN:

Taiz, L & Zeiger, E. (2002). Plant physiology, Third edition. Sinauer Assoc. 690 p. ISSN: 87893-823-0, USA

Tegegne, G., Pretorius, J.C. & Swart, W.J. (2008). Antifungal properties of Agapanthus africanus L. extracts against plant pathogens, *Crop Protection*, Vol. *27*, No. 7, (July, 2008), pp. (*1052-1060*), ISSN: 0261-2194

Tirillini, B., Velaquez, E.R. & Pellegrino, R. (1996). Chemical composition and antimicrobial activity of essential oil of *Piper angustifolium*, *Planta Medica*, Vol. 62, No. 4, (Augost, 1996), pp. (372–3), ISSN: 0032-0943

Tomas-Barberan, F.A., Msonthi, J.D. & Hostettmann, K. (1988). Antifungal epicuticular methylated flavonoids from *Helichrysum nitens*. *Phytochemistry*, Vol. 27, No. 3, pp. (753–5), ISSN: 0031-9422

Tsai, I.L., Lion, Y.F. & Lu, S.T. (1989). Screening of isoquinoline alkaloids and their derivatives for antibacterial and antifungal activities, *Kaohsiung Journal of Medical Sciences*, Vol. 5, No. 3, (March, 1989), pp. (132–45), ISSN: 1607-551X

Van Puyvelde, L.V., De Kimpe, N., Costa, J., Munyjabo,V., Nyirankuliza S., Hakizamungu, E. & Niceas Schamp. (1989). Isolation of Flavonoids and a Chalcone from Helichrysum odoratissimum and Synthesis of Helichrysetin, *Journal of Natural Products*, Vol. *52*, No, 3, (May, 1989), pp. (629-633), ISSN: 1557-1676

Vargas-Arispuro, I., Reyes-Báez, R., Rivera-Castañeda, G., Martínez-Téllez, M.A. & Rivero-Espeje, I. (2005). Antifungal lignans from the creosotebush (*Larrea tridentata*), *Industrial Crops and Products*, Vol. *22*, No. 2, (September, 2005), pp. (101-107), ISSN: 0926-6690

Veloz-García, R., Marín-Martínez, R., Veloz-Rodríguez, R., Rodríguez-Guerra, R., Torres-Pacheco, I., González-Chavira, M.M., Anaya-López, J.L., Guevara-Olvera, L., Feregrino-Pérez, A.A., Loarca-Piña, G., & Guevara-González, R.G. (2010). Antimicrobial activities of cascalote (*Caesalpinia cacalaco*) phenolics-containing extract against fungus *Colletotrichum lindemuthianum*, *Industrial Crops and Products*, Vol. 31, No. 1, (January, 2010), pp. (134-138), ISSN: 0926-6690

Weidenborner, M., Hindorf, H., Jha, H.C. & Tsotsonos, P. (1990). Antifungal activity of flavonoids against storage fungi of the genus *Aspergillus*, *Phytochemistry*, Vol. 29, No. 4, (May, 1990), pp. (1103–1105), ISSN: 0031-9422

Wilson, C. L., Solar, J. M., Ghaouth, A. E. & Wisniewski, M. E. (1997). Rapid evaluation of plant extracts and essential oils for antifungal activity against *Botrytis cinerea*, *Plant Disease*, Vol. 81, No. 2, (Febrary, 1997), pp. (204-210), ISSN: 0191-2917

Wink, M & Schimmer, O. 2010. *Annual Plant Reviews* (2009) 39: 21–161: Biochemistry of Plant Secondary Metabolism. Second edition, A John Wiley & Sons, Ltd., Publication, 433 p. doi: 10.1002/9781444318876.ch2.

Zabka, M., Pavela, R. & Slezakova, L. (2009). Antifungal effect of *Pimenta dioica* essential oil against dangerous pathogenic and toxinogenic fungi, *Industrial Crops and Products*, Vol. 30, No. 2, (September, 2009), pp. (250-253), ISSN: 0926-6690

Zhang, Z.Y., Dai G.H., Zhuge Y.Y. & Li Y.B. (2008). Protective effect of *Robinia pseudoacacia* Linn1 extracts against cucumber powdery mildew fungus, *Sphaerotheca fuliginea*, *Crop Protection*, *27*, No. 6, (June, 2008), pp. (920-925), ISSN: 0261-2194

Zhao, W., Wolfender, J.L., Hostettmann, K., Xu, R. & Qin, G. (1998). Antifungal alkaloids and limonoid derivatives from *Dictamnus daysycarpus*, *Phytochemistry*, 47, No. 1, (January, 1998), pp. (7–11), ISSN: 0031-9422

5

Screening Methods in the Study of Fungicidal Property of Medicinal Plants

S. Sasidharan, L. Yoga Latha, Kwan Yuet Ping and S. Jothy Lachumy
Institute for Research in Molecular Medicine (INFORMM), Universiti Sains Malaysia, Pulau Pinang, Malaysia

1. Introduction

There is growing interest in the use of medicinal plants and herbal products for fungal infection related diseases control program because plant derived drugs are considered safe and free from adverse side effects. Plants generally produce many secondary metabolites which constitute an important source of antifungal drugs. Medicinal properties of plants are normally dependent on the presence of certain phytochemical principles such as alkaloids, anthraquinones, cardiac glycosides, saponins, tannins and polyphenols which are the bioactive bases responsible for the antimicrobial property (Ebana *et al.*, 1993). It is difficult to discover antifungal agents against yeasts and filamentous fungi compared to bacteria caused diseases. Fungal cell wall is predominantly composed chitin and other polysaccharides such as β-glucans which become hindrance to the antibiotic activity on the cells. Hence appropriate screening methods to study the antifungal activity of natural resources play an important roles in the development of fungicide.

Plant drugs still remain the principal source of pharmaceutical agents used in orthodox medicine (Khaing, 2011). The ability of plant to produce various types of phytochemicals such as alkaloid, flavonoid and saponin attract the attention of natural products researcher. Although many plant or herbal products use for the treatment of various ailments in rural communities but there is a need for scientific verification of their activities against fungi. Currently, there is little scientific evidence on the antimicrobial properties of these medicinal plants under investigation against majority of fungi (Mann et al., 2008). *In vitro* and *in vivo* antifungal activities of medicinal plants have been studied widely in the present's days. A wide variety of methods can be applied to study anti-fungal activity. However, the outcome of these studies relies on appropriate and reliable methods use by the researchers.

In this chapter we analyze and compare the various reliable methods available for the study of antifungal activity. In this chapter, we describe and discuss plant sample extraction technique, antifungal screening with determination of minimum inhibition concentration, minimum fungicidal concentration and IC_{50} value on hyphal growth inhibition. Beside that the *in situ* antifungal study method by using various microscopy techniques such as confocal laser scanning microscopy (CLSM), scanning electron microscopy (SEM) and transmission electron microscopy (TEM) also will be discussed. Fig. 1 shows the various steps involved in the evaluation of the medicinal plant's for fungicidal properties.

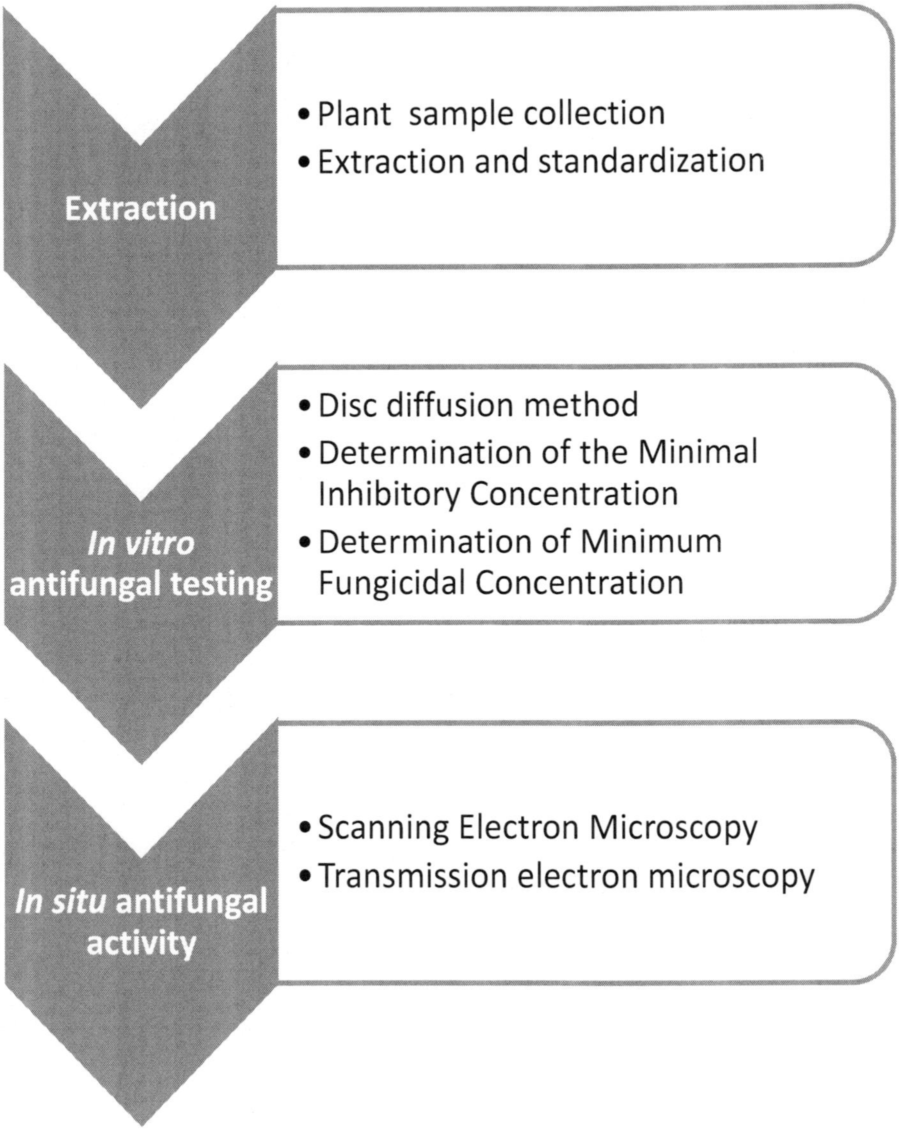

Fig. 1. Various steps involved in the development and evaluation of fungicidal property of medicinal plants

2. Extraction

First steps in the process of screening medicinal plants for antifungal activity is extraction. Extraction is the separation of medicinally active portions of plant tissues using selective solvents through standard procedures. Such extraction techniques separate the soluble plant metabolites and leave behind the insoluble cellular marc. The products so obtained from plants are relatively complex mixtures of metabolites, in liquid or semisolid state or in dry powder form, and are intended for oral or external use. These include classes of preparations known as decoctions, infusions, fluid extracts, tinctures, pilular (semisolid) extracts or powdered extracts (Sukhdev et al., 2008). The basic operations of extraction include steps, such as pre-washing, drying of plant materials or freeze drying, grinding to obtain a homogenous sample and often improving the kinetics of analytic extraction and also increasing the contact of sample surface with the solvent system. Proper actions must be taken to assure that potential active constituents are not lost, distorted or destroyed during the preparation of the extract from plant samples.

The general techniques (Fig. 2) of medicinal plant extraction include maceration (Fig. 3), infusion, percolation, digestion, decoction, hot continuous extraction (Soxhlet). Recently, modern extraction methods have been developed which includes microwave-assisted extraction (MAE), ultrasound extraction (sonication) and supercritical fluid extraction (SFE) (Table 1).

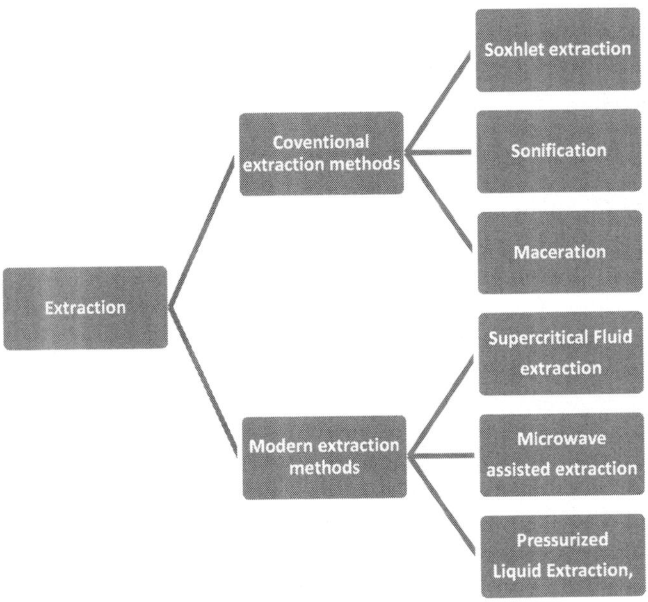

Fig. 2. Conventional and modern extraction methods

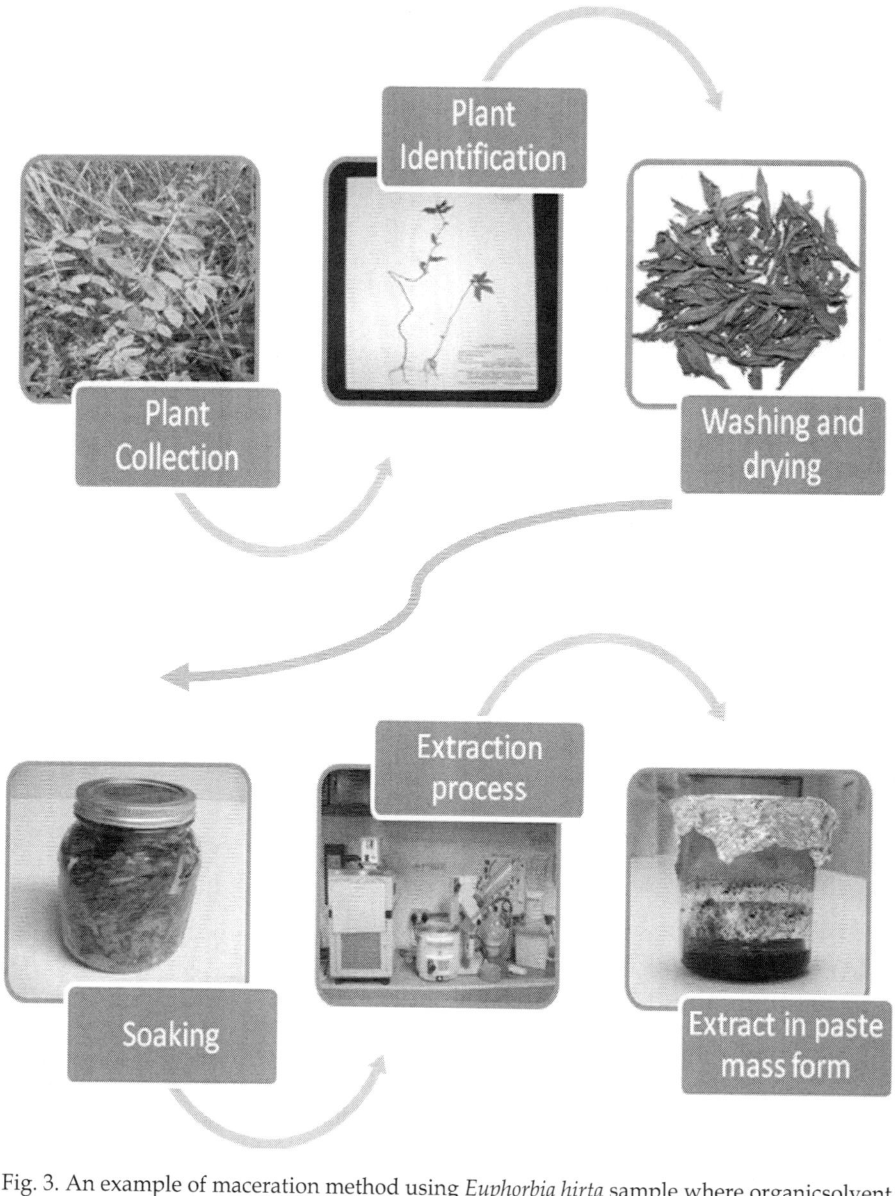

Fig. 3. An example of maceration method using *Euphorbia hirta* sample where organicsolvent extraction was performed by soaking 100g of powdered dried plant material in methanol (1.0 L) through occasional shaking and stirring for 7 days. The whole extract was then filtered and the solvent was evaporated to dryness *in vacuo* using a rotary evaporator at 40-50°C to afford a paste mass.

	Soxhlet extraction	Sonification	Maceration	Supercritical Fluid extraction (SFE)	Microwave assisted extraction (MAE)	Pressurized Liquid Extraction, (PLE)
Common solvents used	Methanol, ethanol, or mixture of alcohol and water	Methanol, ethanol, or mixture of alcohol and water	Methanol, ethanol, or mixture of alcohol and water	Carbon dioxide or carbon dioxide with modifiers, such as methanol	Methanol, ethanol, or mixture of alcohol and water	Methanol
Temperature (°C)	Depending on solvent used	Can be heated	Room temperature	40–100	80–150	80–200
Pressure applied	Not applicable	Not applicable	Not applicable	250–450 atm	Depending on if it is closed or opened vessel extraction	10–20 bar
Time required	3–18 hr	1 hr	3-4 days	30–100 min	10–40 min	20–40 min
Volume of solvent required (ml)	150–200	50–100	Depending on the sample size	Not applicable	20–50	20–30
References	Zygmunt & Namiesnik, 2003; Huie, 2002	Zygmunt & Namiesnik, 2003; Huie, 2002	Phrompittayarat et al.,2007; Sasidharan et al.,2008a; Cunha et al., 2004; Woisky et al., 1998	Zygmunt & Namiesnik, 2003; Huie, 2002; Luque de Castro et al., 2000; Liu & Wai, 2001	Zygmunt & Namiesnik, 2003; Huie, 2002; Camel, 2000; Pan et al., 2001; Pan et al., 2002 Fang et al., 2000	Ong et al., 2000; Ong & Apandi, 2001; Lee et al., 2002; Ong, 2002; Ong & Len, 2003a; 2003b; Choi et al., 2003;

Table 1. A brief summary of the experimental conditions for various methods of extraction for plants material

As the target compounds may be non-polar to polar and thermally labile, the suitability of the methods of extraction must be considered. The choice of solvent depends on several factors including the characteristics of the constituents being extracted, cost and environmental issues. SFE has been used for many years for the extraction of volatile components on an industrial scale. An important advantage of applying SFE to the extraction of active compounds from medicinal plants is that degradation as a result of lengthy exposure to elevated temperatures and atmospheric oxygen are avoided. Using MAE, the microwave energy is used for solution heating and results in significant reduction of extraction time (usually in less than 30 min) compared with conventional liquid–solid extraction methods in which a relatively long extraction time (typically 3–48 h) is required. Another advantage of MAE is that it enables a significant reduction in the consumption of organic solvents, typically less than 40 mL, compared with the 100–500 mL required in Soxhlet extraction (Huie, 2002).

3. *In vitro* antifungal testing

3.1 Microbial strain

Potato dextrose agar (PDA) medium normally used by the scientist to maintained fungal isolates and consist of extract of boiled potatoes, 200 mL; dextrose, 20 g; agar, 20 g; deionized water, 800 mL at 28°C. Spore suspensions can be prepared and diluted in sterile potato dextrose broth (PDB) to a concentration of 10^7 spores per mL. The spore population needed can be counted using a haemocytometer. Subsequent dilutions can be made from the aforementioned suspension to adjust to the required concentration, which can be used in the antifungal test.

3.2 Screening for the antifungal effect

Screening for antifungal effect can be carried out by using the disc diffusion method. The plate containing 25 mL of PDA medium will be seeded with 1 mL of fungal conidial spore suspension containing 10^5 spores per mL from a 120-h-old culture. Three Whatman filter paper No. 1 discs of 6-mm diameter can be used to screen the antifungal activity. Each sterile disk will be impregnated with 20 mL of the extract corresponding with 100 mg/mL of crude extract, myconazole 30 μg/mL, as positive control, or vehicle as negative control. The plates will be refrigerated for 2 h to allow the compounds presents in the extract diffused and then will be incubated at 28°C for 5 days. Diameter of the inhibition zone will be measured, and the mean of the three replicates are taken (Bauer et al., 1966). The disc diffusion method is a qualitative test which could provide the information whether the crude extract possessed antifungal properties

3.3 Determination of the minimal inhibitory concentration

The minimal inhibitory concentration (MIC) can be determined as the lowest concentration at which no growth occurs and is determined as follows: PDB medium will be prepared and sterilized in universal bottles, each containing 10 mL medium. Different amounts of the tested extract will be added to the broth medium to give the following concentration: 0.3125 to 100 mg/mL. To each flask 0.5 mL of Tween-80 will be added as emulsifying agent. The flasks will be inoculated with 0.5 mL fungal conidial spore suspension containing 10^5 spores per mL from a 120-h-old culture and will be incubated at 28°C for 5 days. The MIC value is determined as the lowest concentration of plant extract in the broth medium that inhibited visible growth of the test fungal strains. Each assay should be carried out in triplicate. The MIC test will be quantified the antifungal activity of plant extract.

3.4 Determination of minimum fungicidal concentration

The hyphal growth inhibition test can be used to determine the antifungal activity of the plant extract against fungal strains as previously described Picman et al. (1990). Briefly, dilutions of the test solutions dissolved in vehicle will be added to sterile melted PDA at 45°C to give final concentrations of 100, 10, 1, 0.8, 0.6, 0.4, 0.2, and 0.1 mg/mL of plants extracts. The resultant solution will be thoroughly mixed and approximately 15 mL will be poured onto the petri plate. Plugs of 1 mm of fungal mycelium cut from the edge of actively growing colonies will be inoculated in the center of the agar plate and then incubated in a humid chamber at 25°C. Control cultures will be received an equivalent amount of vehicle. Three replicates will be used for each concentration. Radial growth is measured when the control colonies almost reached 1.5 cm. Results will be expressed as the percentage of hyphal growth inhibited (Gamliel et al., 1989). Concentration response curves will be prepared in which the percentage of hyphal growth inhibition is plotted against concentration mg/mL. The concentration required to give 50% inhibition of hyphal growth IC_{50} will be calculated from the regression equation. Miconazole can be used as a positive control.

4. *In situ* antifungal activity

Electron microscopy (EM) is one of the many methods available for visual inspection of fungal strains. The effects of potential antifungal extracts from natural sources can also be evaluated by using the EM methods. Hence in this section the microscopical techniques such as Scanning (SEM) and Transmission (TEM) electron microscopy on the *in situ* antifungal activity by plant extract will be discussed.

4.1 Scanning electron microscopy

After treatment with plant extract, scanning electron microscopy SEM observation will be carried out on fungal strains. First of all, the plate containing 25 mL PDA medium will be seeded with 1 mL of the fungal conidial spore suspension containing 10^5 spores per mL from a 120-h-old culture. The extract 1mL, at the concentration of IC_{50} (obtained from the hyphal growth inhibition test), is then dropped onto the inoculated agar and will be further incubated for another 7 days at 28°C. A vehicle-treated culture can be used as a control. Five to ten mm segments will be cut from cultures growing on potato dextrose plates at various time intervals 1, 2, 3, 4, 5, 6, and 7 days for SEM examination (Sasidharan et al., 2008b). The specimen then placed on double-stick adhesive tabs on a planchette and the planchette placed in a petri plate. In a fume hood, a vial cap containing 2% osmium tetroxide in water will be placed in an unoccupied quadrant of the plate. After being covered, the plate will be sealed with parafilm, and vapor fixation of the sample proceeded for 1 h. Once the sample is vapor fixed, the planchette will be plunged into slushy nitrogen -210°C and then transferred on to the "peltier-cooled" stage of the freeze dryer, and freeze drying of the specimen will be proceeded for 10 h. Finally, the freeze dried specimen will be sputter coated with 5–10 nm gold before viewing in the SEM. The SEM is advantageous over several other microscopy methods as it is three-dimensional and almost the whole of the specimen is sharply focused. Furthermore, besides having a combination of higher magnification, larger depth of focus and greater resolution, the preparation of samples is also relatively easier, compared to the TEM method (Sasidharan et al., 2010). From the SEM micrograph (Fig. 4) we can observe the changes caused by the plant extract on fungal surface.

4.2 Transmission electron microscopy (TEM)

Further confirmation of SEM finding can be obtained from TEM study. To study the antifungal activity through TEM method the hyphal specimens (1×3 mm^2, with approximately 1 mm thickness of underlying agar blocks) of test fungal strains will be excised from the margin of actively growing SDA culture treated with plant extract using a sterilized razor blade. The specimens are then fixed with modified Karnovsky's fixative (Karnivsky, 1965) consisting of 2% (v/v) glutaraldehyde and 2% (v/v) paraformaldehyde in 0.05 M sodium cacodylate buffer solution (pH 7.2) at 4°C overnight. Subsequently, the fixed specimens are washed with the solution three times for 10 min each. The specimens were then will be post-fixed in the solution with 1% (w/v) osmium tetroxide at 4°C for 2 h and then will be washed briefly with distilled water twice each. The postfixed specimens will be *en bloc* stained with 0.5% (w/v) uranyl acetate at 4°C overnight and then will be dehydrated once in a graded ethanol series of 30, 50, 70, 80, and 95% and three times in 100% ethanol for 10 min each. The specimens will be further treated with propylene oxide twice for 30 min each as a transitional fluid and then will be embedded in Spurr's resin. Ultra-thin sections (approximately 50 nm in thickness) will be cut with a diamond/ glass knife using an ultra-microtome. The sections will be mounted on copper grids and will be stained with 2% uranyl acetate and Reynolds' lead citrate (Reynolds, 1963) for 7 min each. Finally the sections will be observed with a transmission electron microscope. From the TEM micrograph we can observe the changes caused by the plant extract on fungal cytoplasm (Fig. 5).

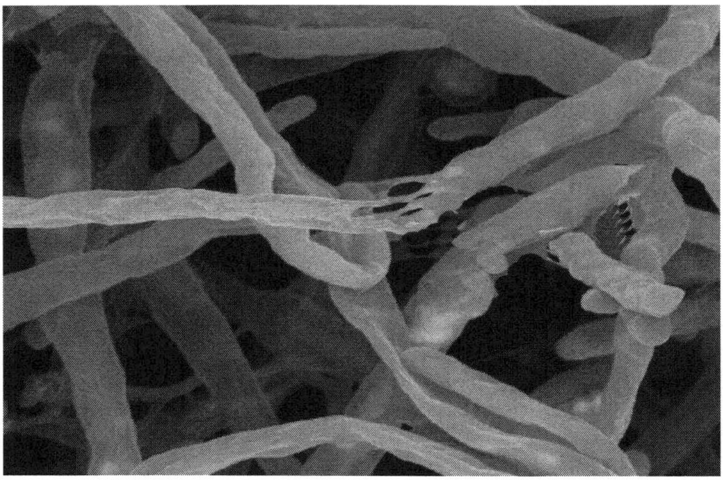

Fig. 4. SEM micrographs of *Aspergillus niger*

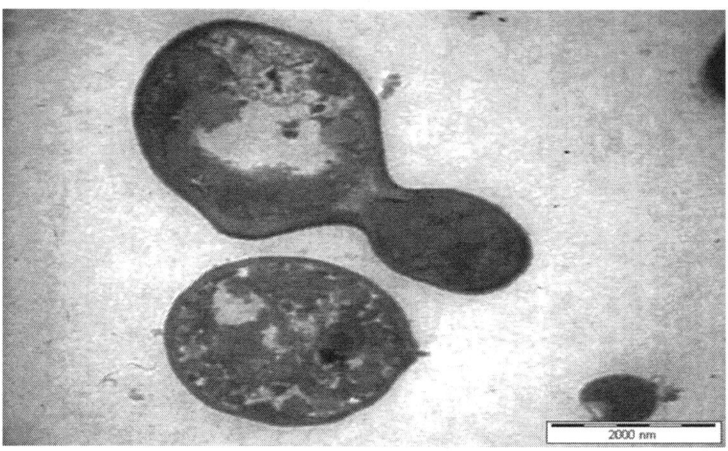

Fig. 5. TEM micrographs of *Candida albicans*

4.3 Confocal laser scanning microscopy (CLSM)

Further verification of SEM and TEM finding can be obtained from CLSM study. To study the antifungal activity through CLSM method the plant extract with MIC concentration will be prepared. 48 h fungal culture will be developed by culturing the fungal strains on SDA agar for 48 h. Controls without the plant extract or antimicrobials also will be included as control groups. The 48 h fungal culture will be gently transferred into a 12-well microtitre plate and rinsed with PBS for 15 s. The discs will be then immersed in 1 ml of the plant extract or antimicrobial agents and incubated at 37°C in an aerobic incubator for 24 h. Subsequently, the extract or antimicrobial will be removed and the viability of the fungal cells will be assessed by Molecular Probes LIVE/DEAD BacLight Bacterial viability kit which comprise SYTO-9 and propidium iodide (PI) (Molecular Probes, Eugene, OR). After incubation with the dyes, the polymethylmethacrylate discs with biofilms will be placed on glass slides and live/dead ratio of cells will be quantified using the CSLM system (Thein et al., 2007). CLSM has become a precious tool for a wide range of studies in the biological and medical sciences for imaging thin optical sections in living and fixed specimens ranging in thickness up to 100 micrometers.

5. Conclusion

The above mentions methods demonstrated the great potential in the development of antifungal testing to study the fungicidal properties of medicinal plants to develop fungicide. The main advantages of the presented methods are the following: easy; rapid; cheap and accurate. Our discussion demonstrates that the use electron microscopy is vital to reveal the cell injury caused by plants extract on fungal strains. The cell changes occurring in surface and cytoplasm of fungal cells following exposure to the plant extract could be visible using a combination of SEM and TEM studies.

6. Acknowledgment

This project was partly supported by USM Short Term Grants (304/CIPPM/639040) from Universiti Sains Malaysia. Kwan Yuet Ping was supported by MyPhD fellowship from Ministry of Higher Education, Government of Malaysia, Malaysia.

7. References

Bauer, R.W.; Kirby, M.D.K.; Sherris, J.C. & Turck, M. (1966). Antibiotic susceptibility testing by standard single disc diffusion method. *American Journal of Clinical Pathology*, 45, 493–496.

Camel, V. (2000). Microwave-assisted solvent extraction of environmental samples. *Trends in Analytical Chemistry*, 19, 229-248.

Choi, M.P.K.; Chan, K.K.C.; Leung, H.W. & Huie, C.W. (2003). Pressurized liquid extraction of active ingredients (ginsenosides) from medicinal plants using non-ionic surfactant solutions. *Journal of Chromatography A*, 983, 153-162.

Cunha, I.B.S.; Sawaya, A.C.H.F.; Caetano, F.M.; Shimizu, M.T.; Marcucci, M.C.; Drezza, F.T.; Povia, G.S. & Carvalho, P.O. (2004). Factors that influence the yield and composition of Brazilian propolis extracts. *Journal of Brazilian Chemical Society,* 15, 964–970.

Ebana, R.U.B., Madunagu, B. E. & Etok, C.A. (1993). Anti-microbial effect of *Strophantus* hipides *Secamone afzeli* on some pathogenic bacteria and their drug Research strain. *Nigeria Journal of Botany*, 6, 27-31.

Fang, Q.; Teung, H.W.; Leung, H.W. & Huie, C.W. (2000) Micelle-mediated extraction and preconcentration of ginsenosides from Chinese herbal medicine. *Journal of Chromatography A*, 904, 47-55.

Gamliel, A.; Katan, J. & Cohen, E. (1989). Toxicity of chloronitrobezenes to *Fusarium oxysporum* and *Rhizoctonia solani* as related to their structure. *Phytoparas,* 17, 101–105.

Handa, S.S.; Khanuja, S.P.S.; Longo, G. & Rakesh, D.D. (2008). Extraction Technologies for Medicinal and Aromatic Plants. International centre for science and high technology, Trieste, 21-25.

Huie, C.W. (2002). A review of modern sample-preparation techniques for the extraction and analysis of medicinal plants. *Analytical and Bioanalytical Chemistry,* 373, 23-30.

Khaing, T.A. (2011). Evaluation of the antifungal and antioxidant activities of the leaf extract of aloe Vera (Aloe barbadensis Miller). *Proceedings of World Academy of Science, Engineering and Technology,* 75, 610-612.

Karnovsky, M.J. (1965). A formaldehyde-glutaraldehyde fixative of high osmolality for use in electron microscopy. *Journal of Cell Biology,* 27, 137A-138A.

Lang, Q. & Wai, C.M. (2001). Supercritical fluid extraction in herbal and natural product studies-a practical review. *Talanta,* 53, 771-782.

Lee, H.K.; Koh, H.L.; Ong, E.S. & Woo, S.O. (2002). Determination of ginsenosides in medicinal plants and health supplements by pressurized liquid extraction (PLE) with reversed phase high performance liquid chromatography. *Journal of Separation Science,* 25,160-166.

Luque de Castro, M.D. & Jiménez-Carmona, M.M. (2000). Where Is Supercritical Fluid Extraction Going? *Trends in Analytical Chemistry,* 19, 223-228.

Ong, E. S. (2002). Chemical assay of glycyrrhizin in medicinal plants by pressurized liquid extraction (PLE) with capillary zone electrophoresis (CZE). *Journal of Separation Science*, 25, 825-831.

Ong, E.S. & Binte Apandi, S.N. (2000). Determination of Berberine and Strychnine in Medicinal Plants and Herbal Preparations by Pressurized Liquid Extraction with Capillary Zone Electrophoresis. *Electrophoresis,* 22, 2723-2729.

Ong, E.S. & Len, S.M. (2003a). Pressurized hot water extraction of berberine, baicalein, and glycyrrhizin in medicinal plants. *Analytica Chimica Acta,* 482, 81-89.

Ong, E.S. & Len, S.M. (2003b). Evaluation of surfactant assisted pressurized hot water extraction for marker compounds in Radix *Codonopsis pilosula* using liquid chromatography and liquid chromatography/electrospray ionization mass spectrometry. *Journal of Separation Science,* 26, 1533-1540.

Ong, E.S. & Len, S.M. (2004). Evaluation of pressurized liquid extraction and pressurized hot water extraction for tanshinone I and IIA in *Salvia miltiorrhiza* using LC and LC-ESI-MS. *Journal of Chromatographic Science,* 42, 211-216.

Ong, E.S. (2004). Extraction methods and chemical standardization of botanicals and herbal preparations. *Journal of Chromatography B,* 812, 23-33.

Ong, E.S.; Woo, S.O. & Yong, Y.L. (2000). Pressurized liquid extraction of berberine and aristolochic acids in medicinal plants. *Journal of Chromatography A,* 904, 57-64.

Mann, A., Banso, A. & Clifford, L.C. (2008). An antifungal property of crude plant extracts from *Anogeissus leiocarpus* and *Terminalia avicennioides.* Tanzania Journal of Health Research, 10, 34-38.

Pan, X.; Niu, G. & Liu, H. (2001). Microwave assisted extraction of tanshinones from *Salvia miltiorrhiza* bunge with analysis by high performance liquid chromatography. *Journal of Chromatography A,* 922, 371-375.

Pan, X.J.; Niu, G.G. & Liu, H.Z. (2002). Comparison of microwave-assisted extraction and conventional extraction techniques for the extraction of tanshinones from *Salvia miltiorrhiza* Bunge. *Biochemical Engineering Journal,* 12, 71-77.

Phrompittayarat, W.; Putalun, W.; Tanaka, H.; Jetiyanon, K.; Wittaya-areekul, S. & Ingkaninan, K. (2007). Comparison of various extraction methods of *Bacopa monnier. Naresuan University Journal,* 15, 29-34.

Picman, A.K.; Schneider, E.F. & Gershenzon, J. (1990). Antifungal activities of sunflower terpenoids. *Biochemical Systematics and Ecology,* 18, 325–328.

Raaman, N. (2006). *Phytochemical Techniques,* p. 306, New India Publishing, India.

Reynolds, E.S. (1963). The use of lead citrate at high pH as an electron-opaque stain in electron microscopy. *Journal of Cell Biology,* 17, 208-212.

Sasidharan, S.; Darah, I. & Jain K. (2008a). *In Vivo* and *In Vitro* toxicity study of *Gracilaria changii. Pharmaceutical Biology,* 46, 413–417.

Sasidharan, S.; Yoga Latha, L. & Angeline, T. (2010). Imaging *In vitro* Anti-biofilm Activity to Visualize the Ultrastructural Changes, In: Microscopy: Science, Technology, Applications and Education A. Méndez-Vilas & J. Díaz, (Eds.), 622-626, Formatex, Spain.

Sasidharan, S.; Zuraini, Z.; Yoga Latha, L. & Suryani, S. (2008b). Fungicidal effect and oral acute toxicity of *Psophocarpus tetragonolobus* root extract. *Pharmaceutical Biology,* 46, 261–265.

Thein, Z.M., Samaranayake, Y.H. & Samaranayake, L.P. (2007). Dietary sugars, serum and the biocide chlorhexidine digluconate modify the population and structural dynamics of mixed *Candida albicans* and *Escherichia coli biofilms. Acta Pathologica, Microbiologica et Immunologica Scandinavica,* 115: 1241–1251.

Woisky, R.G. & Salatino, A. (1998). Analysis of propolis: some parameters and procedures for chemical quality control. *Journal of Apicultural Research,* 37, 99–105.

Zygmunt, J.B. & Namiesnik, J. (2003). Preparation of samples of plant material for chromatographic analysis. *Journal of Chromatographic Science,* 41, 109–116.

6

In Vitro Multiplication of Aromatic and Medicinal Plants and Fungicide Activity

Fernanda Leal[1], Manuela Matos[1],
Ana Cláudia Coelho[2] and Olinda Pinto-Carnide[1]
[1]*IBB-Institute for Biotechnology and Bioengineering, Centre of Genomic and Biotechnology, University of Trás-os-Montes and Alto-Douro, Department of Genetics and Biotechnology*
[2]*CECAV- Center for Animal Science and Veterinary, University of Trás-os-Montes and Alto-Douro, Department of Veterinary Sciences, Portugal*

1. Introduction

Aromatic and medicinal plants, widely used as folk medicine are, beyond fruits, vegetables grains and spices, the principal source of antioxidant compounds. Several studies demonstrated that antioxidants have also antifungal activity (Jayashree & Subramanyam, 2000; Rasooli & Abyaneh, 2004). More and more, humanity try to replace synthetic metabolites by natural metabolites. Therefore, studies in aromatic and medicinal plants with the capacity to produce a different range of secondary metabolites extremely increase in late years. On the other hand, chemical products, like pesticides, fungicides or bactericides are widely used in agriculture. However, they have disadvantages to the environment, due to contamination of the soils, the final consumers or the producers. Still, the indiscriminate and recurrent use of synthetic fungicides has been found to induce resistance in several fungi, the residual toxicity of these compounds result in human health hazards and requires caution in their use for plant disease control (Singh et al., 2009). Thus, some aromatic and medicinal plants, with antifungal capacity (Soliman & Badeaa, 2002; Goun et al. 2003; Sucharita & Padma, 2010), like genus *Thymus, Mentha, Calendula* and *Catharanthus* were micropropagated for antifungal activity evaluation.

Medicinal and aromatic plants are important sources for plant secondary metabolites that are involved in many other aspects of a plant's interaction with its immediate environment. The genetic manipulation of plants together with the establishment of *in vitro* plant regeneration systems facilitates efforts to engineer secondary product metabolic pathways (Kumar & Gupta, 2007). Improvement of the yield and quality of these natural plant products through conventional breeding is still a challenge. However, recent advances in plant genomics research has generated knowledge leading to a better understanding of the complex genetics and biochemistry involved in biosynthesis of these plant secondary metabolites (Gómez-Galer et al., 2008). Advances in the cloning of genes involved in relevant pathways, the development of high throughput screening systems for chemical and biological activity, genomics tools and resources, and the recognition of a higher order of regulation of secondary plant metabolism operating at the whole plant level facilitate strategies for the effective manipulation of secondary products in plants (Kumar & Gupta, 2007).

To overcome the problem of antifungal resistance in human pathogens, plants with antimicrobial properties have been extensively studied for a possible application in food microbiology and as alternative treatments for diseases or to prevent bacterial and fungal growth. Many studies have proven very good fungicide effect of plants (Zabka et al., 2011).

The details of plants screened, their families, vernacular names and their therapeutic uses are given in Table 1.

Plant species	Family	Common name	Therapeutic use
Thymus mastichina and *Thymus zygis*	Lamiaceae	Thyme	Antiseptic, antispasmodic, antitussive (Pina-Vaz et al., 2004)
Mentha rotundifolia	Lamiaceae	Applemint	Antiseptic, antispasmodic, expectorant, vasoconstrictor (Edris et al., 2003)
Calendula sp.	Asteraceae	Marigold	Skin problems, fevers, anti-inflammatory, anti-viral, anti-bacterial and fungicide (Hänsel et al., 1992)
Catharanthus roseus	Apocynaceae	Madagascar periwinkle	Anticancer, antidiabetic, laryngitis, rheumatism, dysmenorrhea (Jaleel et al., 2009)

Table 1. Ethnomedical information of the studied species.

2. *In vitro* multiplication of plants from genus *Thymus*, *Mentha*, *Calendula* and *Catharanthus*

The multiplication by *in vitro* culture, means micropropagation, is a very important methodology to obtain a great number of homogeneous plants in a short period of time. *In vitro* culture is an important system in order to optimize and increase the secondary metabolites production. With this technique, *explants* from different species could be micropropagated under optimized condition of culture media, temperature and photoperiod. In this study, different species of plants with antifungal activity were micropropagated, and the fungicide activity of *in vitro* plants compared with the field plants.

In all the studies presented, the culture media were solidified with 0.7% of agar, and pH was adjusted to 5.6-5.8. Culture media were autoclaved at 121°C for 15 min. The cultures were maintained in a growth chamber at 24 ± 1 °C on a 16/8-h photoperiod (73 μmol $m^{-2} s^{-1}$).

Data were subjected to analysis of variance (ANOVA) using STATVIEW 5.0 program, treatment means separated using Fisher´s Least Significant Difference (LSD) test at P = 0.05.

2.1 Micropropagation of *Thymus*

Thyme is a perennial herb, a 20 to 50 cm shrub, of the Lamiaceae family, an aromatic plant native to the Mediterranean region (Miguel et al., 1999). The genus *Thymus* is exceptionally rich in species, and due to the diversity and plasticity of these plants, their geographical range is very wide. In Portugal are known, at least, 11 *Thymus* species (Afonso & McMurtrie,

1991). *Thymus* plants are of much interest owing to their use in different applications, in medicine because of their antiseptic properties, in the cosmetic industry or as a food additive for their organoleptic properties (Duke et al., 2002; Torras et al., 2007). *Thymus* species differ with regard to their morphological features and metabolism, which influences their chemical constitution. Within individual species there are chemical variations that are characterized by different plant oil compositions, usually without any morphological differences (Smolik et al., 2009). Increasingly, plant breeding has taken advantage of developments in molecular biology in order to genotype the species of interest in a way that considerably accelerates their selection. These types of approaches consist of choosing desired genotypes on the basis of molecular markers, or having prior knowledge of the genes that determine the formation of a particular trait in a plant (Pradeep-Reedy et al., 2002). There are many publications related to the antibacterial and antifungal activities of thyme essential oil (Urbanczyk et al., 2002; Priestley et al., 2003; Rasooli & Mirmostafa., 2003). It has been used as weed germination inhibitor (Angelini et al., 2003), and the different extracts from thyme leaves have shown the presence of a large number of flavonoids and vitamin E, compounds of great interest in the food industry due to their antioxidant activities (Sotomayor et al., 2004). There is also interest in using thyme essential oil for delaying the autoxidation of food lipids (Youdim et al., 2002).

2.1.1 Methodology

Branches of *Thymus* plants, species *Thymus mastichina* L. and *Thymus zygis* L., with 10 to 20 cm lenght were collected in University of Trás-os-Montes e Alto Douro (UTAD) Botanical Garden, in Vila Real, Portugal. The *explants* disinfected with commercial bleach 60% (v/v) and washed three times with sterile water, were fragmented into nodal segments and placed in different culture media. Basal culture media MS (Murashige & Skoog, 1962) were evaluated without growth regulators or supplemented with a cytokinin, 1 mg/L BAP (6-benzylaminopurine), alone or combined with an auxin, 0.2 mg/L NAA (α-naphthalene-acetic acid). Two carbon sources, sucrose and glucose, at three concentrations (2, 3 or 4 % w/v) were also tested. Number and length of shoots have been evaluated after four weeks in culture.

2.1.2 Results and discussion

The two *Thymus* species studied showed a good response to *in vitro* culture conditions (Fig. 1). After four weeks in culture, *T. mastichina* L. presented higher length of shoots (11.2 mm vs 6.0 mm), while *T. zygis* L. showed bigger shoot number per *explant* (1.9 vs 1.8). The culture media revealed a statistical significant effect ($P<0.05$) for the parameters evaluated, shoot number and length.

In what concern to carbon source, sucrose seems to be the best carbon source to thyme micropropagation. For both species, MS medium containing 3 % sucrose produced the highest shoots length (Fig. 2), similar results were obtained by Bandeira et al. (2007) in *T. vulgaris* L.. With 3 % sucrose concentration the highest shoots length (15.9 mm) was observed in *T. mastichina* L., in MS medium supplied with 1 mg/L BAP. One of the explanations for this is the presence of endogen auxin that releases the addition NAA. The biggest shoot number (9.7) appeared in *T. zygis* L. in MS medium with 2 % sucrose and 1 mg/L BAP. However, considering all the media, the ones with glucose revealed higher values in shoot number when compare with the sucrose ones, 3.6 *vs* 3.3 (though, the difference was not statistically significant $P>0.05$). Cunha and Fernandes-Ferreira (1999) and Harada e Murai (1996) obtain similar results with *Linum usitatissium* L. and *Prunus mume*, respectively.

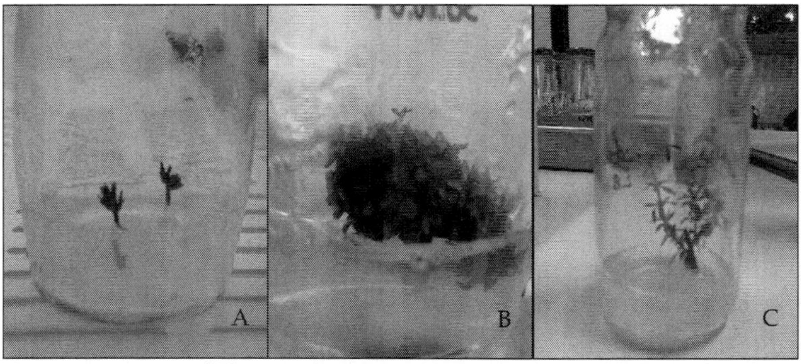

Fig. 1. Micropropagation of *Thymus* plants. A – *Explant* in the establishment medium culture; B – *In vitro* plants of *T. zygis*; C – *In vitro* plants of *T. mastichina*.

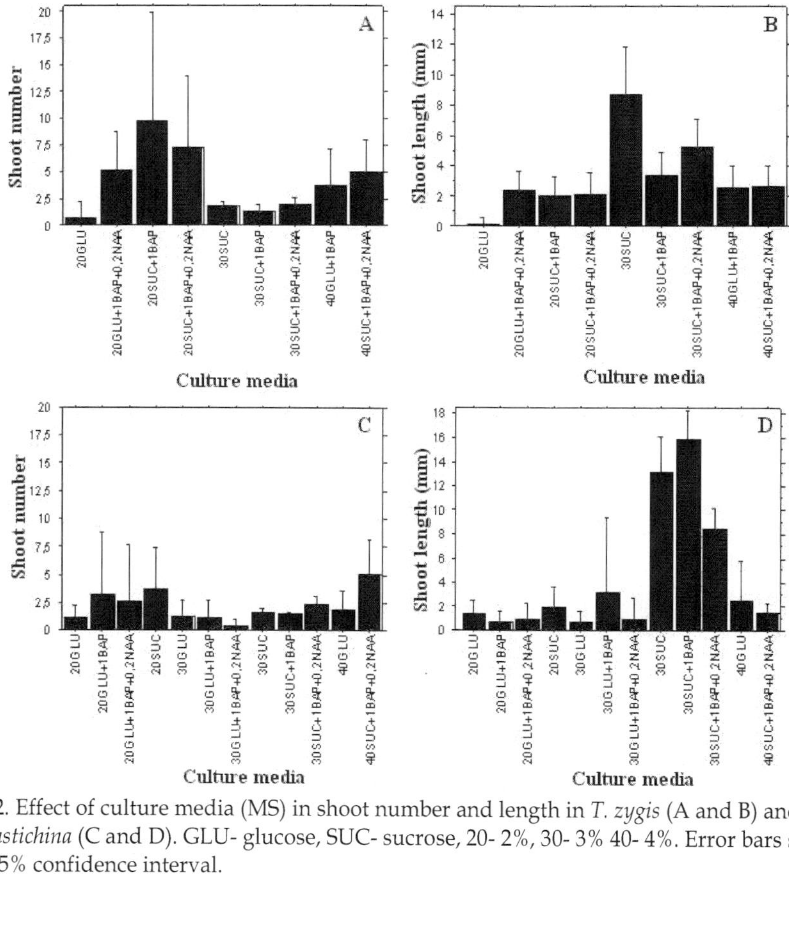

Fig. 2. Effect of culture media (MS) in shoot number and length in *T. zygis* (A and B) and *T. mastichina* (C and D). GLU- glucose, SUC- sucrose, 20- 2%, 30- 3% 40- 4%. Error bars show the 95% confidence interval.

2.2 Micropropagation of *Mentha*

The genus *Mentha* belongs to the family Lamiaceae (Labiatae) consisting of about 25 to 30 species mainly found in temperate regions of Eurasia, Australia , South Africa and North America (Shinwari et al., 2011). Mints grow 10 - 120 centimetres tall and can spread over an indeterminate sized area. Mints are used either in the herb form or as an essential oil for flavouring, perfume production and medicinal purposes (Edris et al., 2003). The *Mentha* plant produces secondary metabolites such as alkaloids, flavanoids, phenols, gummy polysaccharides. Terpens and quinines are used in food and pharmaceutical, cosmetics and pesticide industries (Khanuja et al., 2000). Some members of this genus are also used as herbal teas and condiments both in fresh and dried form due to their distinct aroma. Morphological markers (such as plant height, leaf shape, colour, etc.) are among the oldest markers used in the evaluation of genetic variability, namely in *Mentha*. However, they are not sufficiently specific and informative because different gene expression in different environments causes wide variability of phenotypic characters in individuals. In some cases congruence between morphology and molecular phylogenetics were reported (Shinwari et al., 2011 as cited in Shinwari, 1995). Genetic diversity refers to the variation at the level of individual gene and provides a mechanism for the plants to adapt in ever changing environment.

The existence of different chemotypes, based on qualitative differences within a taxon, is a common feature in most *Mentha* species and hybrids. As a result, the mint plants produce a number of commercially valuable essential oils, viz. spearmint oil, peppermint oil, pennyroyal oil, etc. (Edris et al., 2003).

2.2.1 Methodology

Nodal segments of *Mentha rotundifolia* plants were collected in the greenhouse of Botanical Garden of UTAD (Vila Real, Portugal). The *explants* were disinfected first with ethanol 70% (v/v) then with commercial bleach 40% (v/v) and washed three times with sterile water, then were fragmented into nodal segments and placed in different culture media. MS culture media with different concentrations of BAP (1.0 or 2.0 mg/L) and NAA (0.1 or 0.2 mg/L) were used.

The number and length of shoots and roots, and the presence and diameter of *calli* were evaluated during the eight weeks of *in vitro* culture. After that acclimatization was done in a mixture of peat and perlite (1:1).

2.2.2 Results and discussion

The micropropagation of Menthe had, among others, the purpose to evaluate the effect of cytokinin and auxin concentrations in the plant development. After four weeks of *in vitro* culture there was the development of plants in all culture media (Fig. 3). The results obtained in MS + 1 mg/L BAP medium with different concentrations of auxin (0.1 or 0.2 mg/L) revealed that the lowest concentration of auxin induces higher values of shoots length (27.0 mm), and the highest concentration causes higher values of the roots length (16.9 mm)(Table 2). These results are due to the fact that a low ratio of auxin / cytokinin stimulates the development of stem sprouts, and a higher ratio to promote root development (Torres et al., 1998). Pascal et al. (1991) found that the addition of a synthetic auxin (IBA) induces an increase in mulberry shoots. Analyzing the effect of BAP concentration, culture media MS + 1 mg/L BAP + 0.2 mg/L NAA and MS + 2 mg/L BAP + 0.2 mg/L NAA, it is clear that the medium with the lowest BAP concentration led to higher values for all parameters, except for *calli* induction that are similar in both medium (0.7 mm diameter). Similar results were obtained by Erig et al. (2002) with black mulberry tree by checking that an increase in BAP concentration resulted in a decrease in length of shoots.

The results showed that the composition of the culture medium influences the response of Menthe to micropropagation. Overall, MS medium without growth regulators showed better results in terms of the number of shoots (2.0 / *explant*), although shoots length was higher in MS medium with 1 mg/L BAP and 0.1 mg/L NAA (27.0 mm). The plants were successfully acclimatized with a rate of 100%.

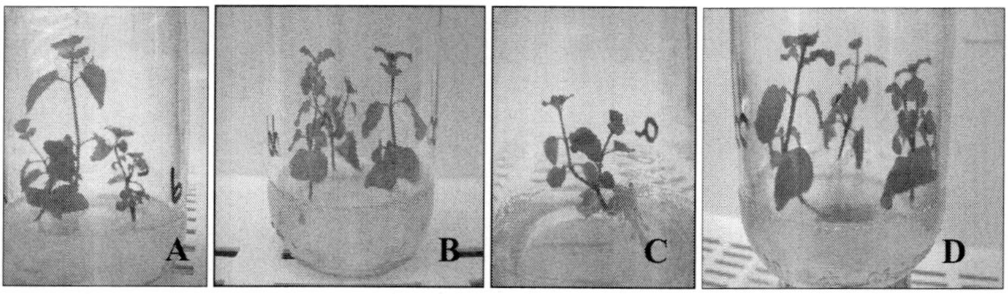

Fig. 3. Micropropagation of *M. rotundifolia*, plants after four weeks in different culture media. A - MS; B - MS + 1 mg/L BAP + 0.1 mg/L NAA; C - MS + 1 mg/L BAP + 0.2 mg/L NAA; D - MS + 2 mg/L BAP + 0.2 mg/L NAA.

Culture media	MS	MS + 1BAP+0.1NAA	MS + 1BAP+0.2NAA	MS + 2BAP+0.2NAA
Shoot number	2.0	1.9	1.8	1.5
Shoot length (mm)	18.0	27.0	22.2	11.8
Root number	3.2	2.6	2.5	0.4
Root length (mm)	18.9	14.5	16.9	4.3
Explants with *calli* (%)	0.0	15	15	6
Calli diameter (mm)	0.0	0.7	0.8	0.7

Table 2. Number and length of shoots and roots and percentage and diameter of *calli* per explant, during eight weeks of *in vitro* culture.

2.3 Micropropagation of *Calendula*

The genus *Calendula* belongs to Asteraceae family, also known as Compositae, and includes several species, namely *Calendula arvensis* and *Calendula officinalis*, which are commonly used as ornamental and medicinal plants. These plants are known to contain saponins, triterpenic alcohols and their fatty acid esters, carotenoids, flavonoids, coumarins, essential oils, hydrocarbons and fatty acids (Hänsel et al., 1992). *Calendula* is known by its medicinal properties - mainly anti-inflammatory, anti-viral, anti-bacterian and fungicide. Considering secondary metabolites, *C. arvensis* is very similar to *C. officinalis* and therefore, the majority of the traditional or folk medicinal uses are similar for both species, including for the cure of skin problems, fevers, chronic infections, wounds, bites and stings (Grieve, 1984; Chevallier, 1996). *Calendula* sp. micropropagation was achieved for the first time using several types of *explants* (Çoçu et al., 2004). Callogenesis can be useful to obtain suspension cell cultures and regeneration of plants *via* indirect organogenesis. *Calli* induction can be obtained with the addition of growth regulators, namely auxins. However, some genotypes are able to induce callus formation even in its absence. In *C. officinalis*, 2,4-D and IAA associated with cytokinins promoted an efficient callus development (Grzelak & Janiszowska, 2002). As secondary metabolites are accumulated in cell cultures from numerous species, much work is focused on their production in shoot or root cultures formed by dedifferentiation of callus (Banthorpe, 1994). Metabolites extracts can be obtained from *in vitro* material, namely *calli* (Grzelak & Janiszowska, 2002), plantlets (Schmeda-Hirschmann, 2005) and flowers (Hamburger et al., 2003).

2.3.1 Methodology

Nodal segments obtained from plants growing in greenhouse conditions were washed in running water, surface-sterilized with a 25% (v/v) sodium hypochlorite solution for 15 min. and finally rinsed 3 times with sterile distilled water (10 min./wash). The *explants* were inoculated in MS media supplemented with BAP (1 or 2 mg/L) alone or combined with 0.1 mg/L of NAA. All the media were supplemented with 30 g/L sucrose. After 4 weeks, the results were analyzed by several parameters like the number and length of shoots and induction and development of *calli*. The *calli* obtained were transferred to MS culture medium added with 3 mg/L 2,4-D, for organogenic development.

2.3.2 Results and discussion

In general *calli* induction was detected in all the media. Callogenesis occurred simultaneously with organogenesis and, after 4 weeks of *in vitro* culture, flower induction was obtained (Fig. 4). It was not observed a significant effect of the culture medium on the number and length of shoots. The highest number of shoots was obtained in the medium with 2 mg/L BAP and the larger length was obtained in the medium with 1 mg/L BAP.

Considering all the media tested, the average rate of multiplication of *C. arvensis* was 4.21 shoots/*explant*. ÇoÇu et al. (2004), using different culture media to micropropagate *C. officinalis* obtained similar values. The best results were obtained in the culture medium with 2.0 mg/L of BAP (7.8 shoots/*explant*) (Fig. 5). The presence of the auxin NAA (in combination with the cytokinin BAP), did not induce a positive effect in the number of shoots (Fig. 5). The media with 2.0 mg/L BAP and 0.1 mg/L NAA registered 4.75 shoots/*explant*. The medium with 1 mg/L BAP produced a higher number of shoots/*explant* in the presence of NAA, but not statistically different ($P>0.05$) of the medium without it.

Fig. 4. *C. arvensis* plants after four weeks of *in vitro* culture showing flower development.

The weak response of NAA was also observed by ÇoÇu et al. (2004) with hypocotyl and cotyledon *explants* of C. *officinalis* concluding that the addition of KIN and NAA to culture media reduced the frequency of shoot organogenesis. Considering the length of the shoots, they reached an average of 23.44 mm at the 4th week of *in vitro* culture.

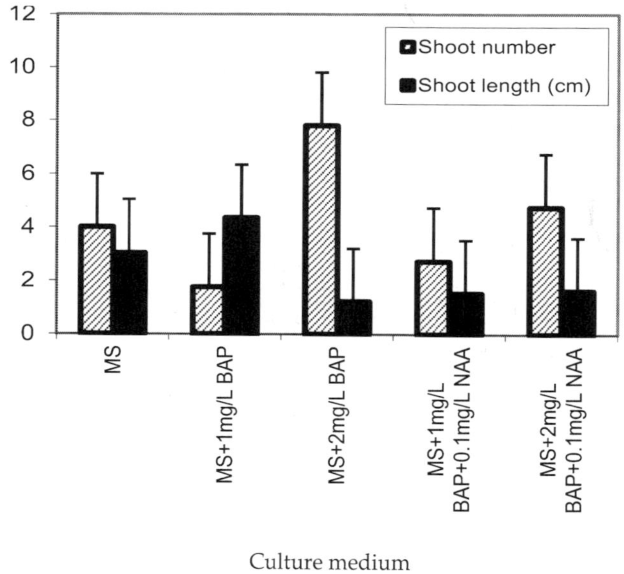

Fig. 5. Effect of different culture media on the number and length of shoots of *C. arvensis*.

2.4 Micropropagation of *Catharanthus*

Catharanthus roseus, an important medicinal plant of Apocynaceae family, is an herbaceous sub-shrub, also known as Madagascar periwinkle and *Vinca rosea* (synonym). Worldwide has been extensively studied by different groups and has been identified to be a source of numerous active principles of therapeutic importance. It is cultivated mainly for its alkaloids (Taylor et al., 1975), with anticancer activities (Jaleel et al., 2009). *Catharanthus roseus* (L.) G. Don is one of the most important medicinal plants due the accumulation in the leaves of two indolalcaloids vinblastine and vincristine, which were the first natural anticancer agents to be clinically used. The high pharmaceutical value of these secondary metabolites made of *C. roseus* one of the most studied medicinal species. A few protocols for micropropagation of *C. roseus* were reported in the last years (Junaid et al., 2007; Dhandapani et al., 2008; Ilah et al., 2009).

2.4.1 Methodology

Nodal segments of *Catharanthus roseus* were collected from plants that were potted. *Explants* were disinfected with a 40% (v/v) sodium hypochlorite solution for 15 min. and finally rinsed 3 times with sterile distilled water (10 min./wash). After that, *explants* were inoculated in MS media alone or supplemented with BAP (1 or 1.5 mg/L) alone or combined with 0.2 mg/L of IBA (indole butyric acid) or NAA. All the media were supplemented with 30 g/L sucrose. To evaluate the effect of culture medium in the development of *explants* was recorded, over eight weeks, the number and length of their shoots and roots, the number of internodes and the presence and diameter of *calli*.

2.4.2 Results and discussion

In this study a protocol was established for the micropropagation of *Catharantus roseus* (Fig. 6), MS culture medium was used because Pietrosiuk *et al* (2007) and Dhandapani *et al* (2008) described good results with this media.

Fig. 6. Different phases of the micropropagation of *C. roseus*. A – *Explants* after two weeks of in vitro culture; B - Plantlet of *C. roseus*, at six weeks of *in vitro* culture; C – Plantlet of *C. roseus* with flower, at eight weeks of *in vitro* culture.

Regarding the addition of growth regulators and their effect on the development of the *explants*, it was found that, in general, the number of shoots was favored in medium supplemented with BAP (MS medium + 1 mg/L BAP) and without auxins (2.23 shoots/*explant*) (Table 3). The length of shoots was higher in the media with the highest concentration of BAP but with the addition of the auxin NAA, MS + 1.5 mg/L BAP + 0.2 mg/L NAA (16.07 mm). It is explained, among other things, by the presence of auxin which promotes cell elongation.

Culture media	**MS**	**MS + 1BAP**	**MS + 1BAP+0.2 IBA**	**MS + 1BAP+0.2 NAA**	**MS + 1.5BAP+0.2 NAA**
Shoot number	1.33	2.23	2.08	2.16	2.15
Shoot length (mm)	9.45	13.58	15.16	11.62	16.07
Root number	3.13	0	5.94	5.79	11.26
Root length (mm)	3.50	0	8.86	12.88	35.40
Number of internodes	0.78	0.78	1.0	0.89	0.91
Explants with *calli* (%)	0	0	0.27	0.44	0.45
Calli diameter (mm)	0	0	1.23	3.35	3.41

Table 3. Number and length of shoots and roots, number of internodes and percentage and diameter of *calli per explant,* during eight weeks of *in vitro* culture.

In rooting, the results were not as expected because, as has been said, the auxin IBA, is related to the root development, so was expect that MS medium + BAP + 0.2 IBA had had a higher rate of rooting, as results obtained by Dhandapani et al. (2008). However, in our study, MS medium supplied with 1.5 BAP + 0.2 NAA revealed the largest number and length of roots, 11.3 roots/*explant* with 35.4 mm. Similar results were verified by Echeverrigaray and colleagues (2005) for thyme cultivars. With regard to the number of internodes, the medium with the addition of IBA (MS + 1 mg/L BAP + 0.2 mg/L IBA) originated the highest values.

This result was not expected since the auxin IBA is more related with rooting, as reported by Dhandapani et al. (2008) also in *Catharanthus roseus.* The culture media with the lowest values for the internodes was MS medium supplemented with NAA (MS + 1 mg/L BAP + 0.2 mg/L NAA). The culture medium MS + 1.5 mg/L BAP + 0.2 mg/L NAA proved to be the best in the number and diameter of *calli* (3.41 mm), this result was expected, since Paramageetham et al. (2007) in *Centella asiática* L. had found that the formation of *calli* occurred in response to an interaction between auxin, its concentration and type of *explant.*

3. Fungicide activity

3.1 *Aspergillus fumigatus*

Aspergillus genus is a famous taxonomic group of molds. Some of its species are very important animal and human pathogens. The disease in humans is caused mainly by *Aspergillus fumigatus, Aspergillus flavus* and *Aspergillus niger.* Other species, for example,

Aspergillus terreus or *Aspergillus nidulans* are quantitatively less prevalent (Karthaus, 2010). *A. fumigatus* is one of the most ubiquitous of the airborne saprophytic fungi that is pathogenic to plants, animals and humans. Inhalation of conidia by immunocompetent individuals rarely has any adverse effect (Latgé, 1999). However, apart from the production of mycotoxins, *A. fumigatus* is a dangerous human pathogenic, which is able to cause very serious human and animal mycoses with a high frequency of resistance to chemical antifungal drugs (Verweij et al., 2009; Lass-Florl et al., 2010; Xu et al., 2010; Zabka et al., 2011). Dramatic increases in the incidence of aspergillosis caused primarily by *A. fumigatus* have occurred in recent years. A high infection-associated death rate of up to and over 50% is attributed even today to these fungi (Karthaus, 2010). *A. fumigatus* has become the most important airborne pathogen in developed countries, causing a significant mortality in invasive aspergillosis (Latgé, 1999; Chamilos & Kontoyiannis, 2005). Patients who have been treated with steroid therapy or those with chronic obstructive pulmonary disease or severe hepatic failure are at high risk for developing pulmonary aspergillosis (Meersseman et al., 2007; Morace & Bhorgi, 2010).

The treatment of aspergillosis is most problematic and questionable owing to toxicity and side effects of the used medicines on the base of synthetic fungicides (Karthaus, 2010). Fungal infections remain a significant cause of morbidity and mortality despite advances in medicine and the emergence of new antifungal agents. Drug resistance in fungi is increasing and is becoming a serious concern and the high use and misuse of antifungal as probably the main cause of this situation. Therefore, there is an urgent need to search for effective new antifungal agents in treatment of infectious diseases at present (Xing et al., 2011).

Currently, there are four classes of antifungal agents with activity against *Aspergillus*: the polyenes, such as amphotericin B its lipid formulations, and nystatin (including liposomal nystatin); the triazoles, including itraconazole, the voriconazole and the investigational posaconazole, the echinocandins, such as caspofungin, the micafungin, and the anidulafungin; and the allylamines such as terbinafine (Groll & Kolve, 2004; Chamilos & Kontoyiannis, 2005; Meersseman et al., 2007; Shi et al., 2010). Clinical resistance of invasive aspergillosis to amphotericin B based therapy is observed frequently in clinical practice (Chamilos & Kontoyiannis, 2005). However, they are intrinsically resistant to fluconazole (Moghaddam et al., 2010).

In other way, pathogenic and toxinogenic fungi are one of the major economic problems of crop and food production (Zabka et al., 2011). In terms of food safety, species of *Fusarium*, *Aspergillus* and *Penicillium* genera are considered the most significant because they produce the great majority of known mycotoxins (Palumbo et al., 2008; Zabka et al., 2011). There has been increasing concern of the consumers about foods free or with lower level of chemical preservatives because these could be toxic for humans (Bedin et al., 1999; Souza et al., 2005). This resulted in increasing search for new technologies for use in food conservation systems which include alternatives antimicrobials (Brull & Coote, 1999; Souza et al., 2005).

3.2 Aromatic plants and antifungal activity

The spread of multidrug-resistant strains of fungus and the reduced number of drugs available led to a search for therapeutic alternatives, namely among medicinal plants and compounds isolated from them used for their antifungal properties. In these natural sources, a series of molecules with antifungal activity have been found, which are of great importance to humans and plants. Several molecules obtained from the natural environment are studied and described in bibliography with antimycotic activity. Several extracts are tested for antifungal activities like crude extracts or isolated constituents like, essential oils, terpenoids, saponins, phenolic compounds, alkaloids, peptides and proteins (Abad et al., 2007).

Aromatic plants have been widely used in folk medicine. About three quarter of the world's population relies on plants and plant extracts for healthcare (Parekh & Chanda, 2007). Several plants have been used in folklore medicine in Portugal (Pina-Vaz et al., 2004; Figueiredo et al., 2008). Spices have been used with primary purpose of enhancing the flavor of foods rather than their medicinal and antioxidant properties (Souza et al., 2005).

Plants generally produce many secondary metabolites with antifungal and microbicide activity (Bobbarala et al., 2009). According to the WHO (World Health Organization), plants would be the best source for obtaining a variety of drugs and a possible way to treat diseases caused by multidrug resistant microorganisms (Bhattacharjee et al., 2006). The medicinal value of plants lies in some chemical substances that produce a definite physiological action on the human body. The most important of these bioactive compounds of plants are alkaloids, flavanoids, tannins and phenolic compounds (Edeoga et al., 2005; Panghal et al., 2011). Some medicinal plants exert strong antifungal properties and could be conveniently used as a promising alternative source for presently problematic antifungal treatment in many areas with respect to their natural origin (Zabka et al., 2011).

Many commercially drugs used in modern medicine was initially used in crude form in traditional or folk healing practices. Benefits of using plant extracts are that they are relatively safer than synthetic alternatives, offering profound therapeutic benefits and more affordable treatment (Panghal et al., 2011).

The genus *Thymus* (Lamiaceae), widely distributed on the Iberian Peninsula, is a taxonomically complex group of aromatic plants traditionally used for medicinal purposes because of their antiseptic, antispasmodic and antitussive properties (Pina-Vaz et al., 2004). Previous studies on the antimicrobial activity of the essential oils of some *Thymus* spp., most of them possessing a large amount of phenolic monoterpenes, showed activity against fungi (Pina-Vaz et al., 2004).

Screening of medicinal plants for antimicrobial activities and phytochemicals is important for finding potential new compounds for therapeutic use (Duraipandiyan et al., 2006).

The main objective of this study was to investigate the inhibitory effects of *Thymus* and *Mentha* extracts against *Aspergillus fumigatus*.

3.3 Methodology

Fungal strain was obtained from the collection of pathogenic fungi maintained in the University of Trás-os-Montes and Alto Douro, Vila Real, Portugal. Subcultivations on Petri dishes and other manipulations with the strain were carried out in the Bio Security Level 2 (BSL 2) laboratory. The evaluation of antifungal capacity was done by the method of mycelial growth (Zhang et al., 2006). The fungus used in the assays was the fungi *Aspergillus fumigatus*. The mold was grown on potato dextrose agar (PDA). The solution was mixed with PDA culture media respectively to give a series of 5, 10, 20, and 25 mg/mL concentrations of culture media containing the compounds described above. The media were dispensed into sterilized Petri dishes (9 cm). After solidification, a mycelial disk of 4 mm diameter of the test *Aspergillus fumigatus* taken from 4 days -old fungi culture, was placed at the center of the medium.

The mycelial disks on PDA without any test constituents were performed in the same way and used as control. Radial growth of colonies was measured at two points along the diameter of the plate and the mean of these two readings was taken as the diameter of the fungal colony. After incubation at 25°C in darkness, growth zones were measured at the third, fifth and the seventh day. The growth of the colonies in control sets was compared with that of various treatments and the difference was converted into percent inhibition [$(C - T) \times 100/C$] where C and T are the radial diameters of the colony in control and treatment, respectively. The percentage of *A. fumigatus* growth inhibition is expressed as a mean of three replicate tests for each treatment. The complete antifungal analysis was carried out under strict aseptic conditions (Zhang et al., 2006).

The analyses were performed using SPSS® (Statistical Package for the Social Sciences) version 19.0. The one-way analysis of variance (ANOVA) followed by Tukey's Test with $P = 0.05$ were used to detect significant differences in inhibition fungi.

3.4 Results

Effect of four different concentrations (5 mg/mL, 10 mg/mL, 20 mg/mL and 25 mg/mL) of *Thymus* and *Mentha* extract plants was tested against *Aspergillus fumigatus*. Antifungal activity was assayed and data on effect of plant extracts on the growth of *Aspergillus fumigatus* in the third, fifth and seventh day is presented in Table 4. The data revealed that reduction in growth of *Aspergillus fumigatus* was observed with extracts of *Thymus* and *Mentha*.

Plant species	% Inhibition of *Aspergillus fumigatus*											
	Third day				Fifth day				Seventh day			
	Concentrations of aqueous plant extracts in PDA (mg/mL)											
	5	10	20	25	5	10	20	25	5	10	20	25
Thymus mastichina	—	—	—	19.1	—	—	4.6	16.7	—	0.9	7.2	18.9
Mentha rotundifolia	7.0[a]	3.9	—	—	1.2	7.4	—	—	3.9	9.9	—	—

[a]All Values are mean of three replicates.

Table 4. Inhibition effect of plant extracts on *Aspergillus fumigatus* in four different concentrations.

The results indicated that *Thymus mastichina* exhibited antifungal activity against the tested *Aspergillus fumigatus* at two different concentrations of 20 mg/mL and 25 mg/mL. The highest antifungal activity was exhibited at 25 mg/mL in *Thymus*. The percent of inhibition were statistically significant with different concentrations in *Thymus*. The lowest concentration of *Thymus mastichina* did not show any activity against *A. fumigates* in the 3 days, while the other two higher concentrations showed good antifungal activity.

Among the species tested, *Mentha* was less active. No enhancing effect was observed for *Mentha* extract against *Aspergillus fumigatus* at higher concentrations (20 mg/mL and 25 mg/mL) while the lowest concentrations i.e. 5 mg/mL, 10 mg/mL showed some inhibition activity against the mold strain. The percent of inhibition were statistically significant with different concentrations in *Mentha*.

None of the above concentrations completely inhibited the test fungus. The percent of inhibition ranged from 0.9 to 19.1%.

3.5 Discussion

Multi-drug resistance is a medical problem in world-wide and has therefore led researchers in the search for new antimicrobial drugs or resistance, particularly from natural resources (Sharma et al., 2005; Moghaddam et al., 2010). Recently, various natural products or synthetic compounds have been reported to increase the antifungal activity (Duraipandiyan et al., 2006; Bobbarala et al., 2009; Moghaddam et al., 2010; Pai et al., 2010).

Antifungal activity was exhibited by different concentrations extracts. The chronological age of the plant, percentage humidity of the harvested material, the method of extraction were possible sources of variation for the bioactivity of the extracts (Panghal et al., 2011).

The results presented indicate different spectrum of antifungal activity of the two extracts.

The antifungal activity of *Thymus mastichina* extract against the mentioned fungi was dose-dependent and increased with the increase in the plant extract concentrations. It also supports the earlier investigations of other authors (Bobbarala et al., 2009; Moghaddam et al., 2010). Previous studies have shown that *Thymus* possess antimicrobial activity (Pinto et al., 2006; Figueiredo et al., 2008).

In the other way, it was revealed in this study, that the antifungal activity of *Mentha* was enhanced in low concentrations of the extracts.

Therefore, this study suggests that plant extracts of screened plants could be helpful in treating diseases in plants caused by *Aspergillus fumigatus*.

However, there is little information about *Thymus* and *Mentha* and their derivatives in the fungal cell in order to promote fungistatic or fungicide effect (Pina-Vaz et al., 2004; Figueiredo et al., 2008). They have been empirically used as antimicrobial agents, but the mechanisms of action are still unknown (Pinto et al., 2006). Generally, inhibitory action of natural products on fungi involves cytoplasm granulation, cytoplasmic membrane lesion, and inactivation and/or inhibition of intercellular and extracellular enzymes (Cowan, 1999; Pinto et al., 2006) and might be due to various compounds, including terpenoids, phenolics and alkaloids. These compounds jointly or independently, exert different levels of antifungal effect culminating with mycelium germination inhibition (Cowan, 1999). Also, it is reported that plant lytic enzymes act in the fungal cell wall causing breakage of β-1,3 glycan, β-1,6 glycan and chitin polymers (Brull & Coote, 1999). The antimicrobial action of the aqueous extracts could be attributed to the anionic components such as thiocyanate, nitrate, chlorides and sulphates besides other water soluble components which are naturally occurring in the plant material (Darout et al., 2000).

Use of aromatic plants as microbial growth inhibitor in foods is often limited because of flavor considerations as effective antimicrobial dose may exceed the organoleptically accepted level. Nonetheless, combinations of spices and other antimicrobial barriers could enhance the food shelf stability and microbial safety even in moderated levels (Pandit & Shelef 1994; Brull & Coote, 1999; Souza et al., 2005). In the other way, the use of aromatic plants as remedies in folk medicine, provide a good reason to investigate them scientifically as potential sources of new plant drugs. It is important to prove which plant extracts have a biological activity on some specific medical conditions, *e.g.* antimicrobial and antifungal properties (Tomczykowa et al., 2008).

4. Conclusion

It was possible the establishment of a micropropagation protocol in order to multiplicate and maintain *in vitro* the aromatic and medicinal plants, to have enough material to use in future studies of antifungal activity and of genetic variability.

Considering the fact that *in vitro* cannot be directly extrapolated to *ex vitro* effects the results suggests that, the use of plant extracts such as *Thymus* and *Mentha* against *Aspergillus sp.* has potential as a topical antifungal agent as they offer a cheap and effective module for therapeutic and/or preventive purposes.

Our results showed that extracts from *Thymus and Mentha* may be particularly useful against *Aspergillus fumigatus*. These results may justify the popular use of these aromatic plants.

Compound-activity relationship for oils components against fungus organisms must be elucidated to explain its antifungal activity (Tomczykowa et al., 2008).

However, in order to evaluate possible clinical application in food microbiology and therapy of aspergillosis, further studies needed to be made.

Further phytochemical studies are required to determine the types of compounds responsible for the antifungal effects of these species.

5. Acknowledgment

Authors are grateful to Professor Mariana Sottomayor from IBMC- Institute for Molecular and Cell Biology for providing seeds for *in vitro* establishment of *Catharanthus* roseus. The authors also like to thank to Carina Alves, Luís Silva, Sandra Cabo and Tatiana Louçano, students of University of Trás-os-Montes and Alto Douro.

6. References

Abad, M.J.; Ansuategui, M. & Bermejo, P (2007). Active antifungal substances from natural sources. *ARKIVOC*, N°.7, pp.116-145, ISSN 1424-6376

Afonso, MLR. & McMurtrie, M. (1991). Plantas do Algarve Lisboa, Portugal: Serviço Nacional de Parques, Reservas e Conservação da Natureza, ISSN 0870 - 2977

Angelini, L.G.; Carpanese, G.; Cioni, P. L.; Morelli, I.; Macchia, M. & Flamini, G. (2003). Essential oil from mediterranean Lamiaceae as weed germination inhibitors. *Journal of Agricultural and Food Chemistry,*Vol. 51, N°.21, pp. 6158-6164, ISSN 0021-8561

Bandeira, J. M.; Lima, C. S.; Rubin, S.; Vaz Ribeiro, M.; Falqueto, A. R.; Peters, J. A. & Braga, E. J. (2007). Diferentes tipos de vedações dos frascos e concentrações de sacarose na micropropagação de *Thymus vulgaris* L.. *Revista Brasileira de Biociências*, Vol. 5, N° S2, pp. 472-474, ISSN 1980-4849.

Banthorpe, D.V. 1994. Secondary metabolism in plant tissue culture: scope and limitations. *Natural Product Reports*, Vol. 11, pp. 303-328, NP9941100303

Bedin, C.; Gutkoski, S. & Wiest, J. M. (1999). Atividade antimicrobiana das especiarias. *Higiene Alimentar*, Vol. 13, N°.65, pp. 26-29, ISSN 1516-8913

Bhattacharjee, I.; Chatterjee, S.; Chatterjee, S. & Chandra, G. (2006). Antibacterial potentiality of *Argemone mexicana* solvent extracts against some pathogenic bacteria. *Memórias do Instituto Oswaldo Cruz*, Vol. 101, N°.6, pp. 645-648, ISSN 0074-0276

Bobbarala, V.; Katikala, P.; Naidu, K. & Penumajji, S. (2009). Antifungal activity of selected plant extracts against phytopathogenic fungi *Aspergillus niger* F2723. *Indian Journal of Science and Technology*, Vol. 2, N°.4, pp. 87-90, ISSN 0974-5645

Brull, S. & Coote, P. (1999). Preservative agents in foods: mode of action and microbial resistance mechanisms. *International Journal of Food Microbiology*, Vol. 50, N°.1-2, pp. 1-17, ISSN 0168-1605

Chamilos, G. & Kontoyiannis, D. (2005). Update on antifungal drug resistance mechanisms of *Aspergillus fumigatus. Drug Resistance Updates*, Vol. 8, N°.6, pp. 344–358, ISSN 1368-7646

Chevallier, A. (1996). The Encyclopedia of Medicinal Plants, Dorling Kindersley. London, ISBN 9-780751-303148

ÇöÇü, S., Uranbey, S., Ìpek, A., Khawar, K.M., Sarihan, E.O., Kaya, M.D., Parmaksiz, Ì. & Özcan, S. (2004). Adventitious shoot regeneration and micropropagation in *Calendula officinalis* L. *Biologia Plantarum* Vol. 48, N°.3, pp. 449-451, ISSN 1573-8264

Cowan, M. (1999). Plant products as antimicrobial agents. *Clinical Microbiology Reviews*, Vol.12, N°.4, pp. 564-582, ISSN 0893-8512

Cunha, A. & Fernandes-Ferreira. (1999). Influence of medium parameters on somatic embryogenesis from hypocotyls *explants* and flx (*Linum usitatissium* L.). *Journal of Plant Physiology*, Vol. 155, pp. 591-597, ISSN 0176-1617.

Darout, I.; Cristy,A.; Skaug , N. & Egeberg, P. (2000). Identification and quantification of some potential antimicrobial anionic components in miswak extract. *Indian Journal of Pharmacology*, Vol. 32, N°.1, pp. 11-14, ISSN 0253-7613

Dhandapani M, Kim DH & Hong SB (2008) Efficient plant regeneration via somatic embryogenesis and organogenesis from the explants of *Catharanthus roseus. In Vitro Cell Dev Biol Plant* Vol. 44, pp. 18-25, ISSN 1054-5476

Duke, J.A.; Bogenschutz-Godwin, M.J.; De Cellier, J. & Duke, P. (2002). *Handbook of Medicinal Herbs*, Second ed. Boca Raton, CRC Press, pp. 730, ISBN 0849312841.

Duraipandiyan, V.; Ayyanar, M. & Ignacimuthu, S. (2006). Antimicrobial activity of some ethnomedicinal plants used by Paliyar tribe from Tamil Nadu, India. *BMC Complementary and Alternative Medicine*, Vol. 6, pp. 35 doi:10.1186/1472-6882-6-35, ISSN 14726882

Echeverrigaray S., Agostini G., Atti-Serfini L., Paroul N., Pauletti G.F & Santos A.C. (2001). Correlation between the Chemical and Genetic Relationships among Commercial Thyme Cultivars. *Journal of Agricultural and Food Chemistry*,Vol. 49, N° 9, pp 4220–4223, ISSN 0021-8561

Edeoga, H.O.; Okwu, D.E. & Mbaebie, B.O. (2005). Phytochemical constituents of some Nigerian medicinal plants. *African Journal of Biotechnology*, Vol. 4, N°.7, pp. 685-688, ISSN 1684–5315

Edris, A.E.; Shalaby, A.S.; Fadel, H.M. & Abdel-Wahab M.A. (2003). Evaluation of a chemotype of spearmint (*Mentha spicata* L.) grown in Siwa Oasis, Egypt. *European Food Research and Technology*, Vol. 218, N°.1, pp. 74–78, ISSN 1438-2385

Erig, A.C.; Gerson, A.R.; Fortes, R.L. (2002). 6-benzilaminopurina e ácido indolbutírico na multiplicação *in vitro* da amoreira - preta (*rubus idaeus* l.), cv. tupy. *Ciência Rural*, Santa Maria, Vol. 32, N°.5, p.765-770, ISSN 0103-8478

Figueiredo, A.C.; Barroso, J.G.; Pedro, L.G.; Salgueiro, L.; Miguel, M.G. & Faleiro, M.L. (2008). Portuguese *Thymbra* and *Thymus* species volatiles: chemical composition and biological activities. *Current Pharmaceutical Design*, Vol. 14, N°. 29, pp. 3120-40, ISSN 1381-6128

Gómez-Galera, S.; Pelacho, A.M.; Gené, A.; Capell, T. & Christou, P. (2008). The genetic manipulation of medicinal and aromatic plants. *Plant Biotechnology Reports*, Vol. 2, N°.2, pp. 93–112, ISSN 1863-5466

Goun E., Cunningham G., Chu D., Nguyen C. & Miles D. (2003). Antibacterial and antifungal activity of Indonesian ethnomedical plants. *Fitoterapia* Vol. 74, N°.6, pp. 592-596, ISSN 0367-326X

Grieve, M. 1984. A modern herbal. Penguin. London, ISBN 0-14-046-440-9

Groll, A.H. & Kolve, H. (2004). Antifungal agents: *in vivo* susceptibility testing, pharmacodynamics, and prospects for combination therapy. *European Journal of Clinical Microbiology & Infectious Diseases*, Vol. 23, N°.4, pp. 256–270, ISSN 1435-4373

Grzelak, A. & Janiszowska, W. (2002). Initiation and growth characteristics of suspension cultures of *Calendula officinalis* cells. *Plant cell, Tissue and Organ Culture*, Vol. 71, N°.1, pp. 29-40, ISSN 1573-5044.

Hamburguer, M., Adler, S., Baumann, D., Förg, A. & Weireich, B. (2003). Preparative purification of the major anti-inflammatory triterpenoid esters from Marigold (*Calendula officinalis*). *Fitoterapia*, Vol. 74, N°.4, pp. 328-338, ISSN 0367-326X

Hänsel R., Keller K., Rimpler H. & Schneider G. (1992). *Hagers handbuch der pharmazeutischen praxis*, Vol. 4. ISBN 3540626468, Berlin: Springer.

Harada, H. & Murai, Y. (1996). Micropropagation of *Prunus mume*. *Plant Cell, Tissue and Organ Culture*, Vol. 46, pp. 265-267, ISSN 0167-6857.

Ilah A, Mujib A, Junaid A, Samar F & Abdin MZ (2009) Somatic embryogenesis and two embryo specific proteins (38 and 33 kD) in *Catharanthus roseus*. *Biologia*, Vol. 64, pp. 299-304, ISSN 0006-3088.

Jaleel, C.A.; Gopi, R.; Gomathinayagam, M. & Panneerselvam, R. (2009). *Traditional and non-traditional plant growth regulators alters phytochemical constituents in Catharanthus roseus. Process Biochemistry*, Vol. 44, N°.2, pp. 205-209, ISSN 0032-9592.

Jayashree T. & Subramanyam C. (2000). Oxidative stress as a prerequisite for aflatoxin production by *Aspergillus parasiticus*. *Free Radical Biology and Medicine*, Vol. 29, N°.10, pp. 981-985, ISSN 0891-5849

Junaid A, Mujib A, Bhat MA, Sharma MP & Å amaj J (2007) Somatic embryogenesis and plant regeneration in *Catharanthus roseus*. *Biol Plant*, Vol. 51, pp. 641-646, ISSN 1054-5476.

Karthaus, M. (2010). Guideline based treatment of invasive aspergillosis. *Mycoses*, Vol. 53, Suppl.1, pp. 36-43, ISSN 0933-7407

Khanuja, S.P.S.; Shasany, A.K.A.; Srivastava, A. & Kumar, S. (2000) Assessment of genetic relationships in *Mentha* species. *Euphytica*, Vol. 111, N°.2, pp. 121-125, ISSN 0014-2336

Kumar, J. & Gupta, P.K. (2007). Molecular approaches for improvement of medicinal and aromatic plants. *Plant Cell Reports*, Vol. 26, N°.10, pp. 1689–1715, ISSN 0721-7714

Lass-Florl, C.; Perkhofer, S. & Mayr, A. (2010). *In vitro* susceptibility testing in fungi: a global perspective on a variety of methods. *Mycoses*, Vol. 53, N°.1, pp. 1–11, ISSN 0933-7407

Latgé, J.-P. (1999). *Aspergillus fumigatus* and Aspergillosis. *Clinical Microbiology Reviews*, Vol. 12, N°.2, pp. 310–350, ISSN 0893-8512

Meersseman, W.; Lagrou, K.; Maertens, J. & Wijngaerden, E. (2007). Invasive Aspergillosis in the Intensive Care Unit. *Clinical Infectious Diseases*, Vol. 45, N°.2, pp. 205–16, ISSN 1058-4838

Miguel, G., Guerrero, C., Rodrigues ,H., Brito, J., Venâncio, F., Tavares, R., Martins, A., & Duarte, F. (1999). Study of the substrate and fertilization effects on the production of essentials oils by *Thymus mastichina* (L.) L. ssp. *Mastichina* cultivated in pots. *Developments in Plant and Soil Sciences*, Vol. 86, N°.5, pp. 201-204, ISSN 0167-840X

Moghaddam, K.M.; Arfan, M.; Rafique, J.; Rezaee, S.; Fesharaki, P.J.; Gohari, A.R. & Shahverdi, A.R. (2010). The antifungal activity of *Sarcococca saligna* ethanol extract and its combination effect with fluconazole against different resistant *Aspergillus* species. *Applied Biochemistry and Biotechnology*, Vol. 162, N°.1, pp. 127–133, ISSN 1559-0291

Morace, G. & Borghi, E. (2010). Fungal infections in ICU patients: epidemiology and the role of diagnostics. *Minerva Anestesiologica*, Vol. 76, N°.11, pp. 950-956, ISSN 0026 4717

Murashige T., and F. Skoog. 1962. A revised medium for rapid growth and bioassay with tobacco tissue cultures. *Physiol. Plant.* , Vol. 15, pp. 473-497, ISSN 0032-0889.

Pai, M.B.; Prashant, G.M.; Murlikrishna, K.S.; Shivakumar, K.M. & Chandu, G.N. (2010). Antifungal efficacy of *Punica granatum, Acacia nilotica, Cuminum cyminum* and *Foeniculum vulgare* on *Candida albicans*: an *in vitro* study. *Indian Journal of Dental Research*, Vol. 21, N°.3, pp. 334-336, ISSN 09709290

Palumbo, J.D.; O'Keeffe, T.L. & Abbas, H.K. (2008). Microbial interactions with mycotoxigenic fungi and mycotoxins. *Toxin Reviews*, Vol. 27, pp. 261–285, ISSN 1556-9543

Pandit, V.A. & Shelef, L.A. (1994). Sensitivity of *Listeria monocytogenes* to rosemary (*Rosimarinus officinalis* L.). *Food Microbiology*, Vol. 11, N°.1, pp. 57-63, ISSN 0740-0020

Panghal, M.; Kaushal, V. & Yadav, J.P. (2011). *In vitro* antimicrobial activity of ten medicinal plants against clinical isolates of oral cancer cases. *Annals of Clinical Microbiology and Antimicrobials*, Vol. 10, pp. 21 doi:10.1186/1476-0711-10-21, ISSN 14760711

Paramageetham, C.; Babu, G.P. & Rao, J.V.S (2004). Somatic embryogenesis in *Centella asiática* L. Na important medicinal and neutraceutical plant of India. *Plant Cell, Tissue and Organ Culture*, Vol. 79, pp. 19-24, ISSN 0167-6857.

Parekh, J. & Chanda, S. (2007). *In vitro* antifungal activity of methanol extracts of some Indian medicinal plants against pathogenic yeast and moulds. *African Journal of Biotechnology*, Vol. 7, N°.23, pp. 4349-4353, ISSN 1684-5315

Pasqual, M.; Peixoto, P. H. P. & Santos, J. C. (1991). Propagação*in vitro* da amora-preta (*Rubus* sp) cv. Ébano: Uso de reguladores de crescimento. *Ciência e Prática*, Lavras, Vol. 3, pp.282-286, ISSN 0100-2236.

Pietrosiuk, A.; Furmanowa, M & Lata, B. (2007). *Catharantus roseus:* micropropagation and in vitro tecniques. *Phytochem Rev.* Vol. 6, N° 2, pp. 459-473, ISSN 1568-7767.

Pina-Vaz, C.; Gonçalves-Rodrigues, A.; Pinto, E.; Costa-de-Oliveira, S.; Tavares, C.; Salgueiro, L.; Cavaleiro, C.; Gonçalves, M.J. & Martinez-de-Oliveira J. (2004).

Antifungal activity of *Thymus* oils and their major compounds. *Journal of European Academy of Dermatology and Venereology,* Vol. 18, N°.1, pp. 73–78, ISSN 1468-3083

Pinto, E.; Pina-Vaz, C.; Salgueiro, L.; Gonçalves, M.J.; Costa-de-Oliveira, S.; Cavaleiro, C.; Palmeira, A.; Rodrigues, A. & Martinez-de-Oliveira, J. (2006). Antifungal activity of the essential oil of *Thymus pulegioides* on *Candida, Aspergillus* and dermatophyte species. *Journal of Medical Microbiology,* Vol. 55, pp. 1367-73, ISSN 0022-2615

Pradeep-Reddy, M., N. Sarla & E. A. Siddiq (2002). Inter simple sequence repeat (ISSR) polymorphism and its application in plant breeding. *Euphytica,* Vol. 128, N°.3, pp. 9–13, ISSN 0014-2336

Priestley, C.M.; Williamson, E.M.; Wafford, K.A. & Sattelle, D.B. (2003). Thymol, a constituent of thyme essential oil, is a positive allosteric modulator of human GABA (A) receptors and a homooligomeric GABA receptor from *Drosophila melanogaster. British Journal of Pharmacology,* Vol. 140, N°.8, pp. 1363-1372, ISSN 1476-5381

Rasooli, I. & Abyaneh, M. (2004). Inhibitory effects of Thyme oils on growth and aflatoxin production by *Aspergillus parasiticus. Food Control,* Vol. 15, N°.6, pp. 479-483, ISSN 0956-7135

Rasooli, I. & Mirmostafa, S. A. 2003. Bacterial susceptibility to and chemical composition of essential oils from *Thymus kotschyanus* and *Thymus persicus. Journal of Agricultural and Food Chemistry,* Vol. 51, N°.8, pp. 2200-2205, ISSN 0021-8561

Schmeda-Hirschmann, G., Jordan, M., Gerth, A. & Wilken, D. 2005. Secondary metabolite content in rhizomes, callus cultures and *in vitro* regenerated plantlets of *Solidago chilensis. Zeitschrift für Naturforschung C,* Vol. 60, N° 1-2, pp. 5-10, ISSN 0939-5075

Sharma, R.; Sharma, C.L. & Kapoorl, B. (2005). Antibacterial resistance: current problems and possible solutions. *Indian Journal of Medical Sciences,* Vol. 59, N°.3, pp. 120–129, ISSN 0019-5359

Shi, J.Y.; Xu, Y.C.; Shi, Y.; Lü, H.X.; Liu, Y.; Zhao, W.S.; Chen, D.M.; Xi, L.Y.; Zhou, X.; Wang, H. & Guo, L.N. (2010). *In vitro* susceptibility testing of *Aspergillus* spp. against voriconazole, itraconazole, posaconazole, amphotericin B and caspofungin. *Chinese Medical Journal,* Vol. 123, N°.19, pp. 2706-2709, ISSN 0366-6999

Shinwari, Z.K. (1995). Congruence between morphology and molecular phylogeneties in Prosartes (Liliaceae). *Pakistan Journal of Botany,* Vol. 27, N°. 2, pp. 2361-369, ISSN 2070-3368

Shinwari, Z.K.; Sultan, S. & Mahmood, T (2011). Molecular and morphological characterization of selected *Mentha* species. *Pakistan Journal of Botany,* Vol. 43, N°.3, pp. 1433-1436, ISSN 2070-3368

Singh, S., Jain, L., Pandey, M., Singh, V. & Pandey, V. (2009). Antifungal Activity of the Alkaloids from *Eschscholtzia californica. Folia Microbiologica,* Vol. 54, N°.3, pp. 204–206, ISSN 0015-5632

Smolik, M.; Jadczak, D. & Korzeniewska, S. (2009). Assessment of Morphological and Genetic Variability in some *Thymus* Accessions Using Molecular Markers. *Notulae Botanicae Horti Agrobotanici Cluj-Napoca,* Vol. 37, N°.2, pp. 234-240, ISSN 1842-4309

Soliman, K.M. & Badeaa, R.I. (2002). Effect of oil extracted from some medicinal plants on different mycotoxigenic fungi. *Food and Chemical Toxicology,* Vol. 40, N°.11, pp. 1669-1675, ISSN 0278-6915

Sotomayor, J.A., Martianez, R.M., Garcia, A.J. & Jordan, M.J. (2004). *Thymus zygis* Subsp. *Gracilis*: Watering Level Effect on Phytomass Production and Essential Oil Quality.

Journal of Agricultural and Food Chemistry, Vol. 52, N°.17, pp. 5418-5424, ISSN 0021-8561

Souza, E.; Stamford, T.; Lima, E.; Trajano, V. & Barbosa Filho, J. (2005). Antimicrobial effectiveness of spices: an approach for use in food conservation systems. *Brazilian Archives of Biology and Technology,* Vol. 48, N°.4, pp. 549-558, ISSN 516-891

Sucharita, R. & Padma, C. (2010). A non-toxic antifungal compound from the leaves of *Catharanthus* roseus characterized as 5-hydroxy flavone by UV spectroscopic analysis and evaluation of its antifungal property by agar-cup method. *Industrial Crops and Products* Vol. 32, N°.3, pp. 375-380, ISSN 0926-6690

Taylor, W.I. & Farnsworth, N.R. (1975). *The Catharanthus Alkaloids.* Marcel Dekker, New York. ISSN 0022-0957.

Tomczykowa, M.; Tomczyk, M.; Jakoniuk, P. & Tryniszewska, E. (2008). Antimicrobial and antifungal activities of the extracts and essential oils of *Bidens tripartite. Folia Histochemica et Cytobiologica,* Vol. 46, N°.3, pp. 389-93, ISSN 0239-8508

Torras, J.; Grau, M.D.; Lopez, J. & de las Heras, F.X. (2007). Analysis of essential oils from chemotypes of *Thymus vulgaris* in Catalonia , *Journal of the Science of Food and Agriculture* Vol. 87, N°.12, pp. 2327-2333, ISSN 1097-0010

Torres, A. C.; Caldas, L. S.; Buso, J. A. 1998. *Cultura de Tecidos e Transformação Genética de Plantas.* Volume I, Brasília- DF: Embrapa/CBAB. 509 p. ISBN:85-7383044-1

Urbanczyk, J.; Hanczakowska, E. & Swiatkiewicz, M. (2002). Herb mixture as an antibiotic substitute in pig feeding. *Medycyna Weterynaryjna,* Vol. 58, N°.11, pp. 887-889, ISSN 0025-8628

Verweij, P.E.; Snelders, E.; Kema, G.; Mellado, E. & Melchers, W. (2009). Azole resistance in *Aspergillus fumigatus*: a side-effect of environmental fungicide use. *The Lancet Infectious Diseases,* Vol. 9, No.12, pp.789–795, ISSN 1473-3099

Xing, Y-M.; Chen, J.; Cui, J.L.; Chen, X.-M. & Guo, S.-X. (2011). Antimicrobial activity and biodiversity of endophytic fungi in *Dendrobium devonianum* and *Dendrobium thyrsiflorum* from Vietman. , Vol. 62, N°.4, pp. 1218–1224, ISSN 1432-0991

Xu, H.; Chen, W.; Li, L.J.; Wan, Z.; Li, R.Y. & Liu, W. (2010). Clinical itraconazole resistant strains of *Aspergillus fumigatus,* isolated serially from a lung aspergilloma patient with pulmonary tuberculosis, can be detected with real-time PCR method. *Mycopathologia,* Vol. 169, N°.3, pp. 193–199, ISSN 1573-0832

Youdim, K.A.; Deans, S.G. & Finlayson, H.J. (2002). The antioxidant properties of thyme (*Thymus zygis* L.) essential oil: an inhibitor of lipid peroxidation and a free radical scavenger. *Journal of Essential Oil Research,* Vol. 14, N°.3, pp. 210-215, ISSN 1041-2905

Zabka, M.; Pavela, R. & Gabrielova-Slezakova, L. (2011). Promising antifungal effect of some Euro-Asiatic plants against dangerous pathogenic and toxinogenic fungi. *Journal of the Science and Food Agriculture* Vol. 91, N°.3, pp. 492–497, ISSN 0022-5142

Zhang, Z.Z.; Li, Y.B.; Qi, L. & Wan, X.C. (2006) Antifungal activities of major tea leaf volatile constituents toward *Colletorichum camelliae Massea. Journal of Agricultural and Food Chemistry,* Vol. 54, N°.11, pp. 3936–3940, ISSN 0021-8561.

Part 2

Biological Control

7

Biological Control Agents for Suppression of Post-Harvest Diseases of Potatoes: Strategies on Discovery and Development

Patricia J. Slininger and David A. Schisler
National Center for Agricultural Utilization Research,
United States Department of Agriculture, Agricultural Research Service, Peoria, IL
United States of America

1. Introduction

As used in plant pathology, the term "biological control" or its short form "biocontrol" commonly refers to the decrease in the inoculum or the disease-producing activity of a pathogen accomplished through one or more organisms, including the host plant but excluding man (Baker, 1987). Biological control of plant pathogens naturally occurs at some level in all agricultural ecosystems, sometimes to a degree where symptoms of disease are noticeably reduced. Thousands of potential microbial biocontrol agents have been isolated from agricultural fields and crops during research over the last 80 years, yet only a few are in commercial use. Recently, public health and safety concerns about the environmental impact of chemical pesticides have led to consideration of biological control as a natural approach to maintaining crop health. Despite environmental incentives and strong research efforts, commercialization of biocontrol agents has been slow to evolve. The momentum of the chemical industry is difficult to shift, and fermentation processes tend to be more expensive to operate than synthetic chemical processes. Yet there is a demand for biological control products, especially in the organic and agricultural niche markets, where there is no efficient chemical competitor. Indeed, the tide has been turning and a recent story in Chemical and Engineering News (Reisch, 2011) has indicated that during the last decade, the growth in sales of biological pest control agents has significantly outpaced that of chemicals. However, given this market demand, the fundamental methods of economical large-scale production and application of biological control agents are still lacking and need to be developed. Many aspects of biocontrol agent production and development represent untrodden territory in the progression of industrial fermentation technology beyond its well-established food and pharmaceuticals niche. Distinguishing them from traditional fermentation products, biocontrol agents must not only be produced in high yield but must also meet the following quality criteria: high (near 100%) retention of cell viability with maintenance of crop compatibility and consistent bioefficacy during several months of storage.

This article will focus on the control of post-harvest fungal pathogens, which present unique opportunities but also challenges. Though accurately determining the extent of losses is difficult and few reports are available, it has been estimated that post-harvest decay accounts for an approximate 25% loss of fresh commodities (USDA, 1965). Biological control using microbial antagonists can be an appropriate tool for managing post-harvest disease problems, especially in crops which are stored under controlled temperatures and high relative humidities. Such controlled storage environments represent a luxury not found when attempting to introduce microbial biological control agents into the comparatively harsh, variable environments found at the infection courts of fungal pathogens of field-grown plants. In recent years, a considerable research attention has focused on biologically controlling rots of fruits post harvest (Janisiewicz, 1988, 1991, 2002). In this chapter, research examples will be reviewed to illustrate the challenges and strategies of developing processes to manufacture and deliver biological agents for post-harvest potato disease control. Concepts to be covered will include the following: market opportunities, choosing pathosystems for biological control, enrichment techniques to enhance new strain discovery, strategies for ranking strains for commercial suitability, mode of action, production considerations, market-broadening functionality, co-cultivation of strains as the next generation biocontrol product, high-throughput screen concept for optimizing biocontrol agent performance from production to delivery, remaining knowledge gaps, and future investigations.

2. Opportunities and barriers for biopesticides on post-harvest potatoes

Market success is most likely to occur if the biological control agent is developed to combat pest problems which have no chemical pesticide solution or which exist in situations where chemical applications are prohibited. For example, in the U.S., organic farming is the fastest growing sector of agriculture. Higher commodity prices for these products and regulations restricting the use of chemical pesticides improve the chances for successful commercialization of natural biological tools in these markets (Behle et al., 1999). Once effectiveness is established in this sector of production, the transition to other sectors could follow.

Currently, a major incentive favoring the development of biopesticides is the ease of federal registration in the United States. The Environmental Protection Agency (EPA) has established a Biopesticide Pollution and Prevention Division (BPPD) to manage accelerated registration of biopesticides. In the mid 1990's, the average duration for registration of a biopesticide was 12 months compared with 36 months for all new chemical pesticide registrations (Medugno et al., 1997), and the cost of registration of a chemical was often more than eight times that of a biological (Woodhead et al., 1990). However, despite regulatory incentives, relatively few biological control agents have reached the market place, often due to one or more of the following pitfalls: (a) poor choice of pathosystem for biological control; (b) relatively few candidate microorganisms available for testing; (c) microbes are selected based on the results of an assay that does not replicate field conditions; and (d) the amenability of microbes to commercial development is excluded as a selection criterion.

3. Fusarium dry rot — An appropriate pathosystem for biological control

Characteristics of a pest problem, or "pathosystem," suitable to a biological control approach include: exploitable weakness(es), existence in an environment favorable to introduced antagonists, availability of few or no control options, and causative of significant economic loss to agriculture. Our experience on discovery and development of biological control agents first began with the need to find an alternative to thiabendazole (TBZ) for the biological control of Fusarium dry rot, an important post-harvest disease of potatoes. Dry rot is caused primarily by *Gibberella pulicaris* (Fr.:Fr.) Sacc. (anamorph: *Fusarium sambucinum* Fuckel) (Boyd, 1972). The fungus is a serious pathogen in potato tuber storages and can produce trichothecene toxins (Desjardins & Plattner, 1989) implicated in mycotoxicosis of humans and animals. Yield losses attributed to dry rot in storage range from 6 to 25% with up to 60% of tubers affected in some cases (Secor & Salas, 2001). Measures for controlling this disease in storage are limited. Resistance to TBZ, the only chemical registered for post-harvest use on tubers for human consumption, is now widespread among strains of *G. pulicaris* (Desjardins et al., 1993; Hanson et al., 1996; Kawchuk et al., 1994; Secor et al., 1994). High levels of resistance to Fusarium dry rot in potato cultivars and breeders' selections are not apparent (Pawlak et al., 1987) and all commonly grown potato cultivars are susceptible (Reiners & Petzoldt, 2004). Therefore, the potential for damage is high enough to justify the economic risk of developing a biological control agent for prevention of dry rot disease losses. A major weakness of the etiology of this pathogen is that it requires a wound in order to infect, and tubers are able to heal wounds in less than 2 weeks in storage. Additionally, the pathogen operates in an environment that is favorable to introduced antagonists in that tuber storage temperatures are uniform and relative humidities are high (>90%), a feature true for many post-harvest pathosystems.

4. Discovery of biocontrol agents amenable to commercial production

Main objectives driving the development of our techniques to discover beneficial biological control agents for dry rot suppression involved two phases: (1) rapid screening of large numbers of microbes using enrichment techniques to concentrate desirable populations and a crop-relevant bioassay to identify useful biological control agents; (2) rating potentially useful agents based on the challenges of manufacturing and delivery.

4.1 Rapid isolation from large populations via enrichment techniques

Ideally biological control agent isolation should begin in areas where biological control is naturally occurring in the field, as opposed to areas where it is not. Evaluating a maximal number of putative biocontrol agents increases the chance of discovering an effective strain. Isolating prospective biocontrol agents from appropriate tissues and under appropriate environmental conditions helps to insure that the microbial antagonists isolated will be well adapted to survival and activity on the specific tissues requiring protection. Application of these concepts resulted in our rapid isolation of 18 putative biological control agents for suppression of Fusarium dry rot. The steps of our method are illustrated in Figure 1 (Schisler & Slininger, 1994). Specifically, gamma irradiation-sterilized field soil samples were first enriched with potato tuber periderm, inoculated with a small amount of field soil obtained from potato fields with low dry rot disease incidence, and incubated for 1 week at 15°C. The microorganisms most adept at rapid growth on the nutrients found in potato periderm and at wound sites would make up the majority of microbes in each recolonized soil sample. Next, conidia of *G. pulicaris* were added to the microbially recolonized soils, and 2 days later, aqueous soil pastes of each soil were applied to wounded potato tubers to initiate a realistic disease bioassay. After incubation 4 weeks at 15°C, tubers were scored for dry rot disease development. Those wounds that developed inconsequential disease were highly likely to contain microbial communities able to survive on potato periderm, to colonize potato tissue, and to suppress disease development. Consequently, clear wounds were excavated and dilution plated on nonselective media that allowed growth of bacteria, fungi, actinomycetes, and yeasts to allow isolation of broad microbial diversity. Using this process, over 350 isolated colonies were obtained from clear wounds receiving microbial communities transferred via live soil samples from 35 locations of low disease incidence. To screen out only those strains participating in dry rot suppression, each isolate was suspended in buffer with conidia of the pathogen and inoculated to a fresh potato wound. After 3 weeks at 15°C, tubers were checked for the presence of disease and only 18 of the 350 isolates demonstrated significant dry rot suppression relative to controls inoculated only with pathogen. It is notable that all of the 18 beneficial isolates were identified as Gram-negative bacteria.

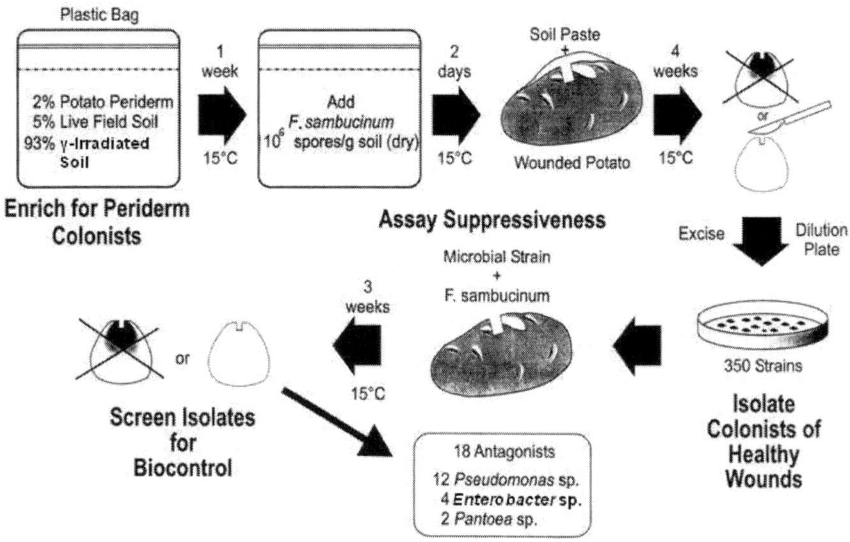

Fig. 1. Isolation of microbial antagonists effective in suppressing Fusarium dry rot of potatoes (Schisler & Slininger, 1994).

4.2 Screening potential strains for commercial suitability

Researchers involved in the discovery and development of biological control agents may speed biocontrol agent commercialization by using an end-process-oriented screening approach. This concept refers to designing the screen to select strains based on their performance under conditions simulating key challenges typically posed by mass production (Standbury & Whitaker, 1984). Since the U.S. industry standard for the manufacture of microbial products is batch liquid cultivation, it was chosen as the method of producing the 18 dry rot antagonistic bacteria for further evaluations. Compared with synthetic chemical processes, fermentation processes are relatively expensive, a circumstance which has largely limited the exploration and development of biotechnology to the food and pharmaceutical fields (Van Brunt, 1986). Primary cost factors include raw materials, utilities, labor, and capital investment. Since the culture medium is central to fermentation process design and economics, our selection of the most commercially promising strains was based on their ability to grow rapidly and to high yield on a variety of liquid culture media and then to accomplish biocontrol upon harvest and delivery to potatoes. These considerations were combined to select efficacious dry rot antagonists that could be produced with reduced fermentor volume and cultivation process costs. The steps in this screening process are illustrated in Figure 2 based on the procedure of Slininger et al. (1994).

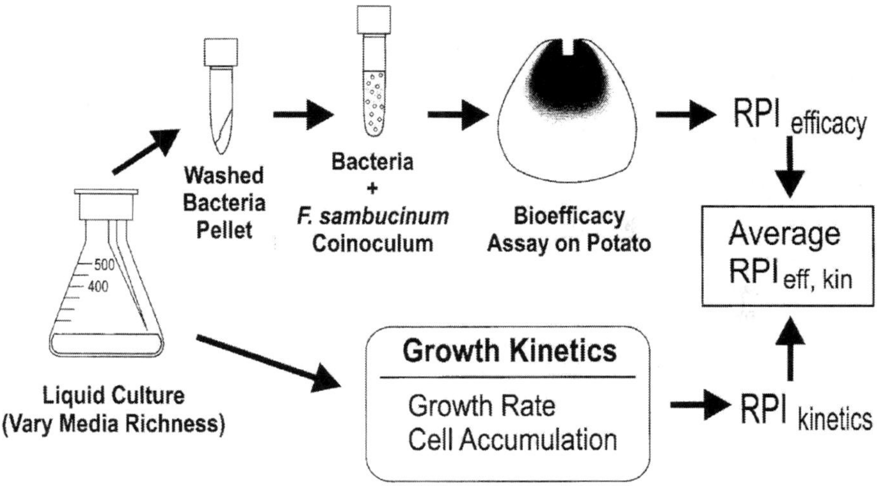

Fig. 2. Two dimensional liquid culture method of ranking commercial development potential of biological control strains using relative performance index, RPI (Slininger et al. 1994).

Strains with the nutritional flexibility to grow rapidly and achieve large bioefficacious populations were sought by challenging with glucose media ranging in richness from a minimal medium (with nitrogen supplied by urea) to a semidefined complete medium (with casamino acids and growth factors) to an undefined medium (with added yeast extract, peptone, and tryptone). Such flexibility is very desirable because it allows process optimization choices to be driven by materials cost and convenience rather than by the fastidiousness of the microorganism. Consistent with utilities considerations, shake-flask cultures were provided a low oxygen transfer coefficient (K_la~0.5 min^{-1}) and moderately warm temperature (25°C) without pH control since most soil-borne microorganisms survive and grow with temperatures ranging from 7 to 30°C and within a fairly broad pH range from 5 to 8. Harvested bacteria were then bioassayed using the wounded potato assay described above to assess efficacy.

For each bacterium, a relative performance index (RPI) was calculated based on each kinetic parameter, such as specific growth rate and cell yield. Given parameter values normally distributed across the isolate group tested, the value of $F = (X - X_{avg})/s$ ranges from –2 to +2. Here, X designates a single value observed per bacterium, and X_{avg} and s are the average and standard deviation, respectively, of all values observed for the isolate group. Using the formula $RPI = (F + 2) \times 100/4$, data corresponding to each parameter type were translated to dimensionless indices, scaled from 0 to 100, which reflected relative bacterial performance. For a given production trial, overall relative kinetic performance indices were calculated for each bacterium: $RPI_{kinetics} = (RPI_{growth\ rate} + RPI_{cell\ yield})/2$. Similarly, a relative performance index based on biocontrol efficacy was calculated for each bacterium using log (disease rating) data: $RPI_{efficacy} = (2-F) \times 100/4$. Note that the term (2-F) is used instead of (2+F) because efficacy improves as disease rating decreases. Thus, $RPI_{efficacy}$ and $RPI_{kinetics}$ are provided a common dimensionless 0-100 scale that allows both data types equal weight in the overall performance assessment. As a result of this screening method (Fig. 2), referred to as "two-dimensional liquid culture focusing" (2DLCF), the 18 bacterial dry rot antagonists were ranked with respect to potential for commercial development, 6 strains being in statistical significance group A (Table 1).

Dogma calls for screening the efficacies of prospective biocontrol agents grown under conditions as similar as possible to what is expected to be encountered in nature. Thus, "promising" isolates have been traditionally selected based on efficacy following growth on solidified media. Only after extensive laboratory, greenhouse, and field tests of these promising isolates has mass production in liquid culture become a concern. Indeed our data have shown that the traditional one-dimensional screen based on bioefficacy of agar-grown isolates selects a different set of top-performers than does the commercial process-oriented 2DLCF screen, and the traditional screen is likely to miss selection of the most commercially useful biocontrol agents because it fails to recognize that liquid culture competency varies widely among microbes. Our experiments have illustrated this by showing that the top-performing strains selected via the 2DLCF screen were often ranked the worst performing strains in the traditional one-dimensional screen of one-fifth tryptic soy broth agar-grown isolates (Table 1) (Slininger et al., 1994). If our goal is to develop bacteria with a commercial future as biocontrol agents, then early screening strategies must reflect the production requirements of the commercial setting. Since liquid culture is the industrial standard for microbial production, liquid cultivation should be the method of biocontrol agent production during early screening. In addition, a two-dimensional assay examining liquid culture growth kinetics as well as product biocontrol efficacy is needed because, our results have shown that isolate performance ranking based on kinetics is not necessarily reflective of the performance ranking based on the biocontrol efficacy, yet both of these features are critical to process economics and commercial success. The processes shown in Figures 1 and 2 have resulted in identification of strains able to suppress dry rot under commercial storage conditions (Slininger et al., 1996a; Schisler et al., 1998a; Schisler et al. 2000b), and recently have been similarly applied to find additional novel bacterial strains with commercial potential for post-harvest biocontrol of pink rot (Adiyaman et al., 2011).

Isolate number (NRRL-)	RPI[1]		Overall[2] $RPI_{Eff.Kin}$	Commercial potential group[2]	Rank[3]	
	Efficacy	Growth kinetics				
B-21050	67.3	65.0	66.2± 4.9	A	1	(18)
B-21128	66.3	64.9	65.6± 5.6	A	2	(13)
B-21133	67.2	62.3	64.8± 3.6	A	3	(11)
B-21134	66.3	60.3	63.3± 4.4	A	4	(15)
B-21132	56.9	69.4	63.1± 5.6	AB	5	(14)
B-21102	62.1	60.8	61.4± 7.4	ABC	6	(16)
B-21136	58.9	57.6	58.2± 6.9	BC	7	(8)
B-21101	56.9	58.8	57.9± 4.9	BC	8	(11)
B-21103	58.4	55.7	57.0± 4.4	C	9	(5)
B-21053	59.2	54.2	56.7± 5.7	C	10	(6)
B-21135	58.7	53.8	56.2± 6.0	C	11	(17)
B-21129	53.0	56.5	55.1± 11.6	CD	12	(8)
B-21104	63.2	35.9	49.5± 10.5	DE	13	(4)
B-21048	47.1	45.4	46.2± 11.1	DE	14	(1)
B-21137	42.8	46.7	44.7± 11.5	DE	15	(6)
B-21051	60.9	25.1	43.0± 12.2	E	16	(1)
B-21105	38.4	28.3	33.3± 7.1	F	17	(10)
B-21049	31.2	23.8	27.5± 9.6	F	18	(1)

[1]$RPI_{Efficacy}$ and $RPI_{Kinetics}$ each indicate the average of six RPI values determined from two independent productions of cells on minimal defined, semi-defined, and undefined liquid media.
[2]Commerical potential groupings were arrived at by applying a two-tailed t analysis to determine the 95% confidence intervals associated with each mean $RPI_{Eff,Kin}$ as indicated by ± values. Means that are not significantly different are designated with the same group letter.
[3]Numbers in parenthesis indicate rank based on traditional screen of efficacy of one-fifth trypticase soy agar-grown antagonists.

Table 1. Use of relative performance indices (RPI) to accomplish a two-dimensional assessment of isolate commercial potential based on growth and efficacy of cells produced in liquid culture (Slininger et al. 1994).

4.3 Multi-dimensional screens to assess commercial potential and robustness

The concept of early commercial-process-oriented screening brings us closer to rapid development of marketable biocontrol agents; however, it is likely that liquid cultivation of biocontrol agents will be followed by formulation, drying, storage, and reconstitution prior to potato application. These steps are necessary to preserve cells for convenient storage and handling in the time between production and application, and represent other features or "challenges" that could be built into an expanded multi-dimensional strategy for selecting the most commercially promising strains. Furthermore, in the natural potato storage environment, many different strains of *Fusarium sambucinum* pathogen are present to challenge biocontrol strains, and in addition, the biocontrol agent will be expected to perform well on many different potato cultivars, different crop field histories, and different wound environments. Schisler et al. (2000) examined performance variability as a function of pathogen and cultivar, and in addition to nutritional flexibility to support robustness, biocontrol strains with better overall performance against multiple strains of pathogen on multiple cultivars could be selected using the dimensionless relative performance index concept. The ability of biocontrol agents to solve multiple pest control problems is another potential screening dimension. For example, our dry rot antagonistic bacteria have also been shown to be able to suppress late blight (Slininger et al., 2007), pink rot (Schisler et al. 2009), and sprouting of stored potatoes (Slininger et al., 2000, 2003). The ability to expand the market to multiple pest control applications is expected to enhance commercial development potential of a given biocontrol product and is a recurring theme influencing the progression of our research as will be discussed at various points later in this account.

5. Bioautography as a screen for the presence of antibiotic production

A variety of potential mechanisms have generally been proposed to be involved in the biological control of plant diseases, including antibiosis, induced disease resistance, competition, parasitism, and predation. Works by Fravel (1988), Huang (1991), Loper & Buyer (1991), Schisler (1997), and Wilson et al. (1994) are useful starting points for information on mechanisms of biological control and microbial interactions potentially of relevance to dry rot disease development. Antibiosis, induced disease resistance, and competition are all possible mechanisms of control for any of our most effective strains of bacteria. However, with regard to mode of action, our studies have focused on the influence of microbial metabolites on *G. pulicaris.* In petri plate assays against the dry rot pathogen, *G. pulicaris,* our 18 bacterial cultures showed varying degrees of inhibition of fungal growth. When extracts from liquid cultures were tested by thin-layer chromatography-bioautography (TLC-BA), a useful technique which correlates antimicrobial activity with the presence of antibiotics (Lazarovits et al., 1982; Homma & Suzui, 1989), all of the cultures tested were shown to produce at least one compound which inhibited the growth of *G. pulicaris* (Burkhead et al., 1995).

Antifungal metabolites from *Enterobacter cloacae* strain S11:T:07 NRRL B-21050, which was highly ranked by the 2DLCF procedure (Table 1, Figure 2), have been isolated from Sabouraud maltose broth culture and identified as phenylacetic acid (PAA), indoleacetic acid (IAA), tyrosol (TSL), and tyrosol acetate, which are recognized to be derived from aromatic amino acids (Burkhead et al., 1998; Slininger et al., 2004). Consequently in later experiments when these compounds were assayed in cultures of strain S11:T:07 (B-21050) grown in three different growth media, it was not a surprise to learn that relative composition of the antifungal compounds produced varied as the culture nutrition, especially amino acid composition, was varied. Antifungal and sprout regulatory bioactivities of these compounds (alone and in combination) were further investigated using our wounded potato assay of dry rot suppressiveness and a cored potato eye assay of sprout inhibition. Assay results showed the antifungal activity of IAA, PAA, and TSL to suppress dry rot infection of wounded potatoes and indicated optimal efficacy when all three metabolites were applied in combination. Furthermore, dosages of IAA resulting in disease suppression, also resulted in sprout inhibition. These results suggest the potential for designing culture production and formulation conditions to achieve a dual purpose biological control agent able to suppress both dry rot and sprouting (Slininger et al., 2004).

6. Expanding the available market with broad spectrum biological control

The observation of many antifungal compounds per each dry rot suppressive isolate and the finding of diverse functional activities ranging from antifungal antibiotics to plant regulatory hormones of isolated antifungal compounds suggested the fruitfulness of exploring the spectrum of use as a means of improving the market draw for our biological control agents.

6.1 Sprout inhibition

Current practices for reducing sprouting in storage could also benefit from microbial alternatives. Because of processing demands, over 54% of the annual potato harvest must be stored at 7° to 13°C, a temperature range above that needed for ideal sprout control (ASAE, 1990). Chemical sprout inhibitors are applied to over 50% of the potato harvest to extend storage time. The potato industry has become very dependent on CIPC (1-methylethyl-3-chlorophenylcarbamate) as the most efficient sprout inhibitor with fewest detrimental side-effects on process potato quality (Lewis et al., 1996). However, recently, the tolerance for residues of CIPC has been reduced to 30 mg/kg (EPA 738-R-96-023, 1996) because of CIPC's persistence in the environment and potato tissue and concerns about its toxicity (Mondy et al., 1992). In the U.S.A., CIPC is the only synthetic chemical registered for post-harvest sprout control of stored potatoes, and it is the most widely used sprout inhibitor world-wide. Due to environmental and health safety concerns, the use of CIPC has become more restricted--opening a potential market for alternative sprout control methods. Consequently, six of our bacteria strains, exhibiting superior dry rot suppressiveness in previous research, were grown in two different liquid culture media and sprayed on Russet Burbank potatoes to assay sprout suppresiveness (Slininger et al., 2000, 2003). In growth chamber and pilot experiments repeated at two storage sites in two successive years, all six isolates demonstrated significant sprout control capabilities when applied after growth on at least one of the culture media supplied. Of the six strains tested, *Pseudomonas fluorescens* S11:P:12 (NRRL B-21133) and two strains of *Enterobacter* sp., S11:T:07 (NRRL B-21050) and S11:P:08 (NRRL B-21132), exhibited highest relative performance levels with sprout control being statistically similar to that of 16.6 ppm CIPC thermal fog after 4-5 months storage.

6.2 Late blight

Several of our top six dry rot suppressive strains have now also been found to significantly reduce late blight infection of stored potatoes (Slininger et al., 2007). Consistent with our observations of indoleacetic acid (IAA) as a major antifungal product produced by one of our dry rot suppressive strains (Slininger et al., 2004), Martiniez Noel et al. (2001) also previously showed that IAA attenuates disease severity in potato-*Phytopthora infestans* interactions and inhibits pathogen growth *in vitro*. *Phytopthora infestans*, the causative agent of the potato late blight disease, infects tubers through eyes or wounds, primarily via zoospores washed into soil from sporangia on infected leaves. Harvested tubers can become infected during washing (Fairclough et al., 1997) and during storage and handling (Lambert et al., 1998). *Phytopthora infestans* is considered to be the most significant pathogen of the crop worldwide (Fry et al., 2001) and historically was the cause of the Great Potato Famine of the late 1840's. The introduction of US-8 genotypes of *P. infestans* has coincided with an increase in severity of potato late blight in North America. As alternatives to chemical fungicides, our 18 bacterial strains patented as biological control agents of both sprouting and Fusarium dry rot were cultivated in 3 liquid media and screened in wounded potato bioassays for their ability to suppress late blight incited by *P. infestans* (US-8, mating type A2) (Slininger et al., 2007). Washed or unwashed stationary-phase bacteria were mixed with fungal zoospores to inoculate potato wounds. One-fifth of the 108 BCA treatments screened, reduced late blight by 25-60%, including among other strains *Pseudomonas fluorescens* S22:T:04 (showing most consistency), P22:Y:05 (NRRL B-21053), S11:P:12 and *Enterobacter cloacae* S11:T:07, the later known to produce IAA. Small-scale pilot testing of these four strains, alone and in combination, was conducted under conditions simulating a commercial application. All four treatments significantly reduced disease; and unwashed bacteria outperformed those washed free of culture broth, indicating a role of metabolites such as IAA. Disease suppression ranged from 35% up to 86% the first test year and from 35 to 91% the second year. Highest overall performance rankings significantly above the control were achieved by the following strains in culture broth: four-strain mix > *P. fluorescens* S22:T:04 > *P. fluorescens* S11:P:12. Combined with previous demonstrations of dry rot and sprout suppression, the consistent late blight control by these strains and strain mixtures suggests the commercial utility of a single treatment for broad spectrum suppression of post-harvest potato diseases and sprouting.

6.3 Pink rot

Pink rot disease occurs in potato growing regions around the world and is caused primarily by the oomycete *Phytophthora erythroseptica* Pethybr. Losses of over 50% of the total harvest can result from tuber contamination by either pink rot or late blight (Secor & Gudmestad, 1999). All underground portions of potato plants can be infected. Root and stem infections can result in plant wilting and death. Though some evidence indicates that there is limited genetic diversity in North American isolates of *P. erythroseptica* (Peters et al., 2005), infections initiated after tuber harvest are difficult to control. Most commercially grown potato cultivars in Canada and the United States are susceptible to pink rot and breeding efforts against this disease have been minimal (Peters et al., 2004). Mefenoxam, a phenylamide fungicide that formerly was effective in reducing the disease in storage, has lost much of its effectiveness (Taylor et al., 2006) due to widespread genetic resistance (Taylor et al., 2002) and the stability of the resistance (Abu-El Samen et al., 2005). The use of various salts (Mills et al., 2005), foliar applications of phosphorous acid (Johnson et al., 2004) and the oomycete fungicides "zoxamide" and phosphite (Miller et al., 2006) have reduced symptoms of *P. erythroseptica* on tubers. Additional disease reduction technologies are still needed for organic markets and to deter the development of resistance to chemical fungicides. Tubers generally become infected in the field via stolons previously infected by germinating oospores (a thick-walled spore resulting from sexual recombination) but zoospores (motile, asexually produced spores) or encysted zoospores of the pathogen also can infect tuber eyes, lenticels and cracks and cuts that result from tuber harvesting operations--infection courts theoretically protectable using microbial antagonists. Therefore, 10 of our bacterial antagonists that reduce Fusarium dry rot, late blight, and/or sprouting in storage were assayed for efficacy against pink rot on tubers of cultivars Russet Burbank and Russet Norkotah (Schisler et al., 2009). Antagonist strains were grown in a semidefined liquid medium, diluted to ~3 x 10^8 cfu/ml, individually combined with zoospores of *P. erythroseptica,* and used to inoculate shallow puncture wounds on tubers. Data from full factorial experimental designs with 10 levels of antagonist, 2 levels of cultivar, and 2 levels of inoculum age after inducing zoospore liberation from sporangia indicated that all factors influenced the size of pink rot lesions that developed internally around wound sites ($P \leq 0.05$). In two different sets of experiments, *Enterobacter cloacae* strain S11:T:07 reduced lesion size more than the other antagonists (19% and 32% reduction versus the control) though

Pseudomonas fluorescens S11:P:14, *Pseudomonas* sp. S22:T:04, and *Enterobacter* sp S11:P:08 also significantly reduced disease. Lesion size was greater on Russet Norkotah than Russet Burbank tubers (42.3 and 26.5 mm, respectively), but cultivar did not influence antagonist performance.

7. Co-cultivation of strains — The next generation

As reviewed above, *Pseudomonas fluorescens* strains S11:P:12, P22:Y:05, and S22:T:04 and *Enterobacter cloacae* strain S11:T:07 have been documented as top strains to suppress four important storage potato maladies—dry rot, late blight, pink rot, and sprouting. These strains are known to differ from one another in their range of antibiotic production, substrate utilization, oxygen requirement, and growth temperature optima. They are also known to differ from one another in ability to inhibit sprouting or suppress disease on various potato cultivars and when incited by various pathogens. The variety of characteristics possessed by the individuals suggests that the successful strain mixtures are likely to be more resilient and more apt to provide individuals amenable to colonize potato wounds despite the variety of environments and pathogen strains encountered. Indeed, our previous experimental results have shown that certain strain pairs applied in combination allow greater dry rot suppression than do individual strains (Schisler et al., 1997). In subsequent laboratory and field trials, we observed that formulations containing multiple strains of our dry rot antagonists performed more consistently than individual strains did when subjected to 32 storage environments varying in potato cultivar, harvest year, potato washing procedure (microflora exposure), temperature, and storage time (Slininger et al., 2001). Successful biocontrol strain mixtures often contained both *Enterobacter cloacae* and *Pseudomonas fluorescens* strains. Several other research groups have reported that mixtures of strains can enhance and/or improve the consistency of biological control (among these, Pierson & Weller, 1994; Duffy & Weller, 1995; Duffy et al., 1996; Janisiewicz, 1996; Leeman et al., 1996; Guetsky et al., 2001; Krauss & Soberanis, 2001; Hwang & Benson, 2002; Schisler et al., 2005; Cruz et al., 2006). Thus, the formulation of strain mixtures has the potential to provide better, more consistent disease control than single strain formulations. Achieving consistent efficacy at each application represents a key advancement toward commercialization of any biocontrol product. However, despite the apparent advantages of applying strain mixes, the disadvantages for the manufacturer are capital costs, operation, maintenance, registration, and management of a different fermentation for each strain used in a mix. A potential way around this obstacle is to co-culture the strains together in one fermentor. To pursue the co-culture concept, we explored the level and consistency of pest control achievable on post-harvest potatoes with the four top multi-functional biological control agents *Pseudomonas fluorescens* strains S11:P:12, P22:Y:05, and S22:T:04 and *Enterobacter cloacae* S11:T:07 (Slininger et al., 2010b). The four bacteria were applied to potatoes in the following formats: a) as co-cultures of strains, i.e. multiple strains grown together in a single culture, b) as individual strains grown separately in pure cultures, and c) as blends of individual strains grown separately in pure cultures. Consistence of biocontrol efficacy and broad pest coverage, both major factors influencing the economics of a successful product, were addressed in this research. Treatments applied in both laboratory wounded potato bioassays and small pilot trials simulating commercial storage conditions were tested, as well as treatments challenged with dry rot, late blight, pink rot, and sprouting. Experiments were designed to analyze dry rot suppression versus all strain combinations and the combination method (co-culture or blend). Results of a two-way analysis of variance of disease with strain composition and combination method showed that significantly better dry rot suppression was obtained by co-cultures (30.3 $\pm$ 2.4% relative disease) than by similar strain blends of pure cultures (41.3 $\pm$ 2.4%) ($P < 0.001$). During a 3-year study, both biocontrol efficacy and consistency were assessed in 16 laboratory and small pilot trials simulating commercial storages. The 3-strain co-culture of *Pseudomonas fluorescens* strains S11:P:12, P22:Y:05, and S22:T:04 had a lower mean disease rating than the blend in 9 of 16 experiments examining control of the 3 diseases and sprouting. The co-culture led other treatments in incidences of significant malady reduction relative to the control: 14 of 16 attempts for co-culture, 11 of 16 attempts for blend, 10 of 13 attempts for pure S11:P:12, 8 of 13 attempts for S22:T:04, and 9 of 13 attempts for P22:Y:05. Using relative performance indices to rank treatment performance across all experiments, the co-culture treatment ranked significantly higher (69th percentile) than the blend (57th percentile). A synergy analysis suggested that co-culturing strains stimulated inter-strain activities to boost biocontrol efficacy and consistency, a feature not developed in strains grown separately and mixed just prior to addition to potatoes. Although the *E. cloacae* most often dominated co-cultures which included it, the other co-inoculated *P. fluorescens* populations persisted at significantly lower levels and apparently synergized the performance of the final population in suppressing dry rot disease.

There are a number of avenues by which the unique environment fostered by co-cultivation may improve biocontrol performance, and it is possible that inter-strain communication mechanisms are involved. Gram negative bacteria partners have been reported to regulate anti-microbial metabolite production via a signaling system referred to as "quorum sensing." Quorum sensing (QS) is mediated by population size and the accumulation of acylated homoserine lactones (AHL) which stimulate the bacterial populations to express genes responsible for metabolite production (Wood & Pierson, 1996). The local fermentation environment of the co-culture may synergize the impact of such signaling on subsequent biological control performance. Arrays of AHLs are known to be produced by many common rhizosphere bacteria, and they allow not only signaling between cells within a strain population, but also between cells of different strain populations (Pierson et al., 1998). An AHL-mediated QS system was noted to regulate cell surface properties, which was different from that noted for anti-microbial phenazine production (Zhang & Pierson, 2001). Soil bacteria have also been shown to degrade AHLs, such as via lactonase activity (Molina et al., 2003), a feature suggesting the potential for curative biocontrol of bacterial diseases. In addition to metabolite regulation and disease suppression, QS has been implicated in many other aspects of biocontrol activity, including: regulation of biofilm formation and rhizosphere colonization (Wei & Zhang, 2006); pathogen virulence, motility, and fitness (Licciardello et al., 2007); indoleacetic acid (IAA) plant growth hormone synthesis (Müller et al., 2009), perhaps pertinent to IAA accumulation by our *E. cloacae* strain S11:T:07; and induced systemic resistance (Pang et al., 2009).

In addition to producing antifungal and sprout regulatory metabolites, we have recently identified the extracellular polysaccharide marginalan production by *P. fluorescens* S11:P:12 that not only improves its own survival during desiccation, but also that of co-inhabitants P22:Y:05 and S22:T:04 (Slininger et al., 2010a). This feature suggests the community benefit of one strain for others grown in association with it. In previous research by others, exopolysaccharides (EPS) have been associated with improved desiccation tolerance in

Pseudomonas sp. (Roberson & Firestone, 1992) and have been implicated as triggers to induced systemic resistance in host plants. It is believed that the EPS matrix slows the rate of water loss within the microenvironment, which enables the microbe additional time to make the necessary metabolic adjustments needed for survival. Relating to the current research, the co-culture of exopolysaccharide-producing *Paenibacillus* sp. with a *Pseudomonas* species has been reported to extend the shelf-life of the *Pseudomonas* for potential biopesticide or biofertilizer use (Kozyrovska et al., 2005). Thus, in addition to enhancing its biocontrol capacity, one of the benefits of our co-culture is that it includes an EPS-productive partner in *P. fluorescens* strain S11:P:12 that may protect bacteria against desiccation stress as they dry after application to tuber surfaces. The discovery of other mechanisms benefiting the function of the co-culture for consistent and efficacious biological control will likely be among the objectives of future research. Meanwhile, the advantages of co-culturing are compelling and spur on development efforts: economical production of multiple strains in one culture, broad disease spectrum, beneficial interactions of strains, desiccation sheltering, enhanced efficacy, and enhanced consistence.

8. Production considerations

For each strain of interest, the liquid culture production and biocontrol agent formulation processes must be designed to minimize cost and maximize production rate, yield, and quality, i.e. bioefficacy, storage stability, and host compatibility. The impact of liquid culture conditions (carbon and nitrogen sources, carbon-to-nitrogen ratio, nutrients, temperature, pH, dissolved oxygen), microbial physiology (growth state) and metabolites on the qualities of the biocontrol product will all need to be considered when designing the production processes for successful biocontrol products. To illustrate this, key findings of our research on this subject will be reviewed for a variety of our biological control agents under development, including but not exclusively our agents for post-harvest potatoes.

8.1 Manipulation of growth, metabolism, and efficacy with culture conditions

Prior research has shown that culture environment impacts metabolite accumulations and biocontrol agent quality. When strain *Pseudomonas fluorescens* 2-79 (NRRL B-15132) is efficiently delivered to the field in seed coatings, it colonizes the emerging root and produces the antibiotic, phenazine-1-carboxylic acid (PCA), as its primary means of suppressing take-all disease [incited by *Gaemannomyces graminis* var. *tritici* (Ggt)]. Our research has shown that metabolites (primarily PCA) present in liquid cultures of strain 2-79 cause significant germination losses (up to 64%) when included in seed coatings. In mass production of seed inocula, complete separation of cells from metabolites adds considerable expense and may not be feasible if metabolites are insoluble. For *Pseudomonas fluorescens* 2-79, the phytotoxicity of the cell harvest can jeopardize the most economical method of application of the biocontrol agent, which is as a wheat seed coating. Our research showed that controlling fermentor environment allowed dramatic reduction of phytotoxic metabolite production. Fermentation conditions, such as dissolved oxygen, carbon source, pH, or temperature, were controlled to allow production of cells in a phytotoxin-free culture broth which could be used directly to treat seeds without sacrifice to either seed germination or to take-all disease control via PCA production in the rhizosphere (Slininger & Shea-Wilbur, 1995; Slininger et al., 1997a, 1998).

Culture environment also impacts the metabolism and efficacy of our potato protective rhizobacteria which have been reported to produce at least one antifungal component per strain (Burkhead et al., 1995). Our process for ranking dry rot antagonists relative to commercial potential involved growing the strains on three different liquid media of varying nutritional richness and then applying them to potato wounds challenged by the pathogen. The rank of candidate strains based on the liquid culture growth and also disease suppressiveness of the product was found to vary widely with the nutritional environment provided during production of the biological control agent. The metabolite profile of *E. cloacae* strain S11:T:07 has been studied in detail, and it is known to produce indoleacetic acid, phenyl acetic acid, and tyrosol. Concentrations of these metabolites influence both disease suppression and sprouting and vary in cultures with the nutritional environment (Burkhead et al., 1998; Slininger et al., 2004). As discussed in section 7 above, the co-cultivation of multiple potato malady-suppressive strains gives rise to a unique fermentation environment that can yield a biocontrol product with improved efficacy and consistency (Slininger et al., 2010b). However, additional process optimization challenges may arise in accommodating population yields, storage stability, and efficacy. Although the performance benefits of co-cultivations to biocontrol performance had not been documented prior to our 2010 report, mixed Pseudomonad cultivations (with other *Pseudomonas* sp. or other genera) have been documented for many other applications (for example, Rodriguez & Gallardo, 1993; Kimura & Ito, 2001; Ashby et al., 2005; Kumar et al., 2006). More recently, Wu et al. (2009) examined the synergistic growth of a salt tolerant *Pseudomonas fluorescens* Rs-198 with another bacterium Rs-5 in co-culture that may have potential for application in fertilizer preparation. Co-cultures of potato-protective *P. fluorescens* strains could easily be designed to develop similar population densities even by simple adjustments in initial population densities to compensate for growth differences in strains as evaluated by a non-antibiotic selective plating technique described in Slininger et al. (2010b) (Figure 3).

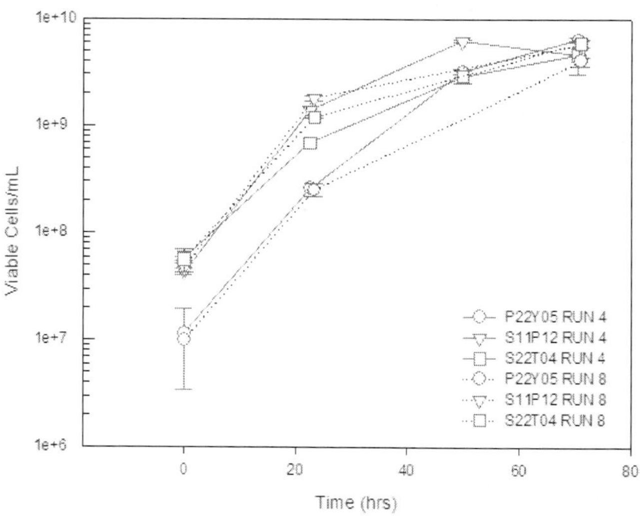

Fig. 3. Growth of P. fluorescens strains S11:P:12, P22:Y:05, and S22:T:04 in duplicate co-culture fermentor runs (Slininger et al., 2010b).

Though still synergistic to efficacy, *P. fluorescens* populations persisted at significantly lower levels than co-cultured *E. cloacae* S11:T:07 a facultative anaerobe with significant competitive advantage as dissolved oxygen is depleted. Growth suppression of multi-species bacterial populations in batch cultures by a single dominant strain has been referred to in the literature as the Jameson Effect and often involves production of specific inhibitors by one species against another. Future work may reveal inoculum population management and nutritional or physical environment management techniques that would allow testing the biocontrol efficacy of controllable population distributions of multi-genera co-cultures like *P. fluorescens* and *E. cloacae* strains, allowing investigation into the optimization of relative strain densities.

8.2 Influence of culture and physiological conditions on dry storage survival

The physiological state of cells upon harvest from the fermentation process was another factor that has been observed to influence the drying survival and shelf-life of strain 2-79 cells coated in 0.5% methylcellulose onto wheat seeds. Cells harvested from growth phase cultures (24-48 h) survived the drying process better than cells from stationary phase cultures (72-96 h), but stationary phase cells had a longer shelf-life than did growth phase cells. Our data suggest that the protective effect of residual sugar still present in growing cultures may explain why growth phase cells exhibited better survival of the drying process than did stationary cells that were deplete of sugar. The longer shelf-life of the stationary phase cells may come about via one or more of a variety of mechanisms that occur as a result of cell starvation and aging as reviewed by Slininger et al. (1996b, 1998). This feature has also been observed in our potato protective rhizobacteria and that influenced our choice of a 72-h harvest for our co-cultures in current research (Slininger et al., 2010a). The impact of cultivation conditions on biocontrol agent storage stability have been shown in virtually every other biocontrol system studied in our laboratory and not just those involving the control of plant diseases with rhizobacteria. Varying the carbon to nitrogen ratio and total carbon loading of a liquid medium for producing the bioherbicide *Colletotrichum truncatum* altered its dry storage stability as well as quantity, propagule type (conidia versus microsclerotia), and efficacy (Schisler et al., 1991; Jackson & Schisler, 1995). Yield and desiccation-tolerance of blastospores of the mycoinsecticide *Paecilomyces fumosoroseus* required appropriate concentrations of amino acids (Jackson, 1997; Jackson et al., 1997). Carbon-to-nitrogen ratio and carbon loading were found to influence the freeze-drying survival and Fusarium head blight suppressiveness of *Cryptococcus nodaensis* (now *C. flavescens*) OH 182.9 (Zhang et al., 2005); and cold shock during liquid cultivation increased the storage shelf-life of this biocontrol agent after air drying (Zhang et al., 2006).

8.3 Formulating for storage stability and delivery

Once the microbial biological control agent is harvested from the production culture, it will be necessary to formulate it. Formulations can be designed to meet a variety of objectives: to prevent biocontrol agent activity losses during storage; to facilitate convenient and efficacious delivery of the agent to the area where biocontrol is needed in the field; to promote insect biocontrol agent or plant-biocontrol agent compatibility; and to enhance the effectiveness of the biocontrol agent delivered. Examples are reviewed to illustrate formulation techniques devised by our research group to accomplish each of these goals for specific biocontrol problems.

The loss of microbial viability during storage is one of the most challenging barriers to overcome on the road to commercial success, especially for strains that do not form a resilient spore. Formulation matrices and storage temperature can play an important role in improving storage survival. For example, in the case of *Paecilomyces fumosoroseus* blastospores, calcined kaolin clay allowed significantly better drying survival and storage stability than other matrices tested (diatomaceous earth, talcs, corn starch, rice flour, and Mexican lime). Greater than 70% survival was retained after air drying and storage 42 days at 4°C, and near 20% after 21 days at 28°C was retained using 5% clay (Sandoval-Coronado et al., 2001). The trend of longer term storage survival with decreasing storage temperature has also held true for liquid formulations of *Pseudomonas fluorescens* and *Enterobacter cloacae*, our gram-negative bacterial biocontrol agents of potato dry rot and sprouting; cells frozen at -20°C in neutral buffer exhibited half-lives of 72-161 days, while those refrigerated at 4°C had half-lives of only 12-33 days (Slininger et al., 1997b). Another general finding of our work has been that the inclusion of culture broth with cells in stored formulations is often, but not always, detrimental to long term cell survival. For example, our gram-negative biocontrol agents for potatoes exhibited half-lives of 26-97 days when frozen in their culture broth and half-lives of 12-42 days when refrigerated in their culture broth. Thus, when frozen, the cells formulated in culture broth had poorer survival than cells in buffer, but when refrigerated, cell survival was similar in culture broth and buffer. When *Pseudomonas fluorescens* 2-79 was stored refrigerated in dried methylcellulose coatings of wheat seeds, the presence of the culture broth was again observed to be detrimental to drying and long-term cell survival, but the data showed that the presence of glucose in methylcellulose coatings with culture broth reduced cell losses upon drying (Slininger et al., 1996b). These examples illustrate the impact of formulation and storage conditions on biocontrol agent preservation and suggest both the aptitude and need for technology advancement in this area.

8.4 Exopolysaccharide production and use as an *in situ* formulant

Suppressive to potato diseases and sprouting, *Pseudomonas fluorescens* S11:P:12 (NRRL B-21133) produces a polysaccharide during liquid cultivation which was isolated, purified, and identified as marginalan (Slininger et al., 2010a). Dry storage results indicated that the presence of marginalan significantly reduced cell death after drying, such that the final stable viable cell density was 2.5 to 5 orders of magnitude greater, respectively, than if no marginalan were included with cells. Marginalan had no significant impact on disease or sprout suppression by strain S11:P:12, and its main benefit to biocontrol was viable cell preservation during drying and storage. When marginalan was formulated with other selected *P. fluorescens* strains P22:Y:05 and S22:T:04, as may occur in co-cultures, its benefits to drying and storage survival were again evident, though more subtle than observed for strain S11:P:12—perhaps because it was the most sensitive of the three to drying. Due to marginalan production, higher viscosity and higher fermentation power consumption for aeration and mixing will be needed to maximize viable cell yield in cultures containing S11:P:12. On the other hand, the polysaccharide offers to return value in terms of enhanced biological control as a cell desiccation protectant and should be considered in culture optimization schemes and for use in downstream formulation methodologies.

Storage Relative Humidity (%)	Time to A620=0.05 (h)[a]	Surviving Viable Cells (cfu/well)[a]	Relative Dry Rot Disease Rating (%)[a]
51	5.6 A	4.4×10^{8} A	58.6 A
86	8.7 B	4.3×10^{8} A	49.7 A
Formulation Sugar			
Lactose + BSA	5.8 A	3.9×10^{6} A	56.4 A
Lactose	7.0 AB	5.8×10^{6} A	66.9 A
Sucrose	7.2 AB	3.4×10^{6} A	57.4 A
Fructose	8.6 B	4.4×10^{6} A	35.9
Storage Time (d)			
7	3.8 A	1.1×10^{7} A	52.0 A
48	4.4 AB	3.5×10^{6} B	47.2 A
83	7.1 B	2.4×10^{6} B	55.5 A
145	13.4 C	1.3×10^{6} B	62.3 A

[a]Within each column and storage factor, means having letters in common are not significantly different ($P<0.05$).

Table 2. Impact of storage conditions on drying survival of *Enterobacter cloacae* S11:T:07 in the droplet drying screen[a] (Slininger and Schisler, 2003)

9. Filling the gaps to commercialization — A high-throughput screening concept for optimizing biocontrol performance

In conclusion, much progress has been made with methods of screening and selecting commercially useful biological control agents and in identifying key aspects of cultivation and formulation that impact biocontrol agent yield and quality. However, the development and optimization of integrated technologies to produce and deliver effective biological control agents remains a barrier to commercialization for many biocontrol agents. We have noted a myriad of variables associated with biocontrol agent cultivation, formulation, drying, storage, and reconstitution processes complicate agent quantity and quality maximization. To approach this problem, an efficient assay was conceived and applied using a 96-well microplate format to allow an integrated approach to optimizing these process variables. The basic high-throughput screening assay is depicted in Figure 4 and involves growing the BCA of interest (in flasks, fermentors, or microplates), formulating cells harvested from growth cultures, delivering microliter droplets of formulated cells to microplate wells (for rapid or slow drying), air- or freeze-drying droplets in the wells, storing plates, reconstituting dried cells, and then monitoring cell activity in terms of the rate of cell growth to a specified yield using a plate-reading spectrophotometer (Slininger & Schisler, 2003). Relevant variables (ingredients, temperature, etc.) are treated at each step of the assay process to view their individual and combined impact on resultant microbial activity, such as the speed of reaching logarithmic growth and a certain cell yield, such as in this research an optical density of 0.05 at a defined wavelength (620 nm). A kinetic activity assessment such as this is convenient to accomplish for the numerous samples that can potentially arise in multivariate experiments and is a good initial screen of biocontrol agent activity as it relates to viable cell concentration in combination with cell activity level (Figure 4).

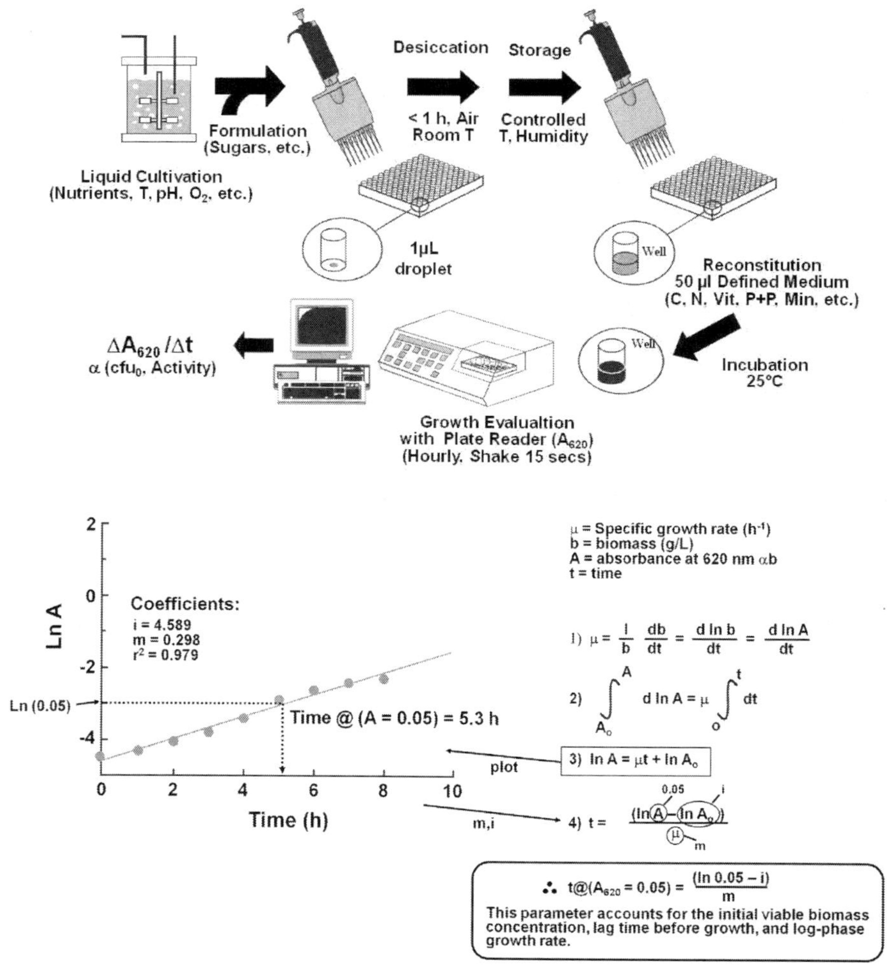

Fig. 4. Droplet drying method to screen and optimize integrated biocontrol agent production process variables (Slininger and Schisler, 2003)

Numerous variables (culture and formulation ingredients, temperature, humidity, etc.) can be tested at each step of the assay process to view their individual and combined impact on the resultant microbial activity. The results of such an assay applied to our potato dry rot protective strain S11:T:07 are shown in Table 2 along with the results of other quality assays that can be carried out on microsamples, including viable cell counts and a wounded potato disease suppressiveness assay. We are currently applying such a flexible approach to allow further optimization of an integrated production process that would support biocontrol agent commercialization in the near future.

10. Disclaimer

The mention of trade names or commercial products in this article is solely for the purpose of providing specific information and does not imply recommendation or endorsement by the U.S. Department of Agriculture. USDA is an equal opportunity provider and employer.

11. References

Abu-El Samen, F.M., Oberoi, K., Taylor, R.J., Secor, G.A. & Gudmestad, N.C. (2005). Inheritance of mefenoxam resistance in selfed populations of the homothallic oomycete *Phytophthora erythroseptica* (Pethybr.), cause of pink rot of potato. *American Journal of Potato Research* 82: 105-115

Adiyaman, T., Schisler, D.A., Slininger, P.J., Sloan, J.M., Jackson, M.A. & Rooney, A.P. (2011). Selection of biocontrol agents of pink rot based on efficacy and growth kinetics index rankings. *Plant Disease* 95(1): 24-30

ASAE (1990). Design and management of storages for bulk, fall-crop Irish potatoes. St. Joseph, MI: ASAE

Ashby, R.D., Solaiman, D.K.Y. & Foglia, T.A. (2005). Synthesis of short/medium-chain-length poly(hydroxyalkanoate) blends by mixed culture fermentation of glycerol. *Biomacromolecules* 6: 2106-2112

Baker, K.F. (1987). Evolving concepts of biological control of plant pathogens. *Annual Reviews of Phytopathology* 25: 67-85

Behle, R.W., Tamez-Guerra, P., Shasha, B.S. & McGuire, M.R. (1999). Formulating bioinsecticides to improve residual activity. *Proceedings of the Formulations Forum, March 3-5,* Orlando, FL

Boyd, A.E.W. (1972). Potato storage diseases. *Reviews of Plant Pathology* 51: 297-321

Burkhead, K.D., Schisler, D.A. & Slininger, P.J. (1995). Bioautography shows antibiotic production by soil bacterial isolates antagonistic to fungal dry rot of potatoes. *Soil Biology and Biochemistry* 27(12): 1611-1616

Burkhead, K.D., Slininger, P.J. & Schisler, D.A. (1998). Biological control bacterium *Enterobacter cloacae* S11:T:07 (NRRL B-21050) produces the antifungal compound phenylacetic acid in Sabouraud maltose broth culture. *Soil Biology and Biochemistry* 30(5): 665-667

Cruz, L.P., Gaitan, A.L. & Gongora, C.E. (2006). Exploiting the genetic diversity of *Beauveria bassiana* for improving the biological control of the coffee berry borer through the use of strain mixtures. *Applied Microbiology and Biotechnology* 71: 916-926

Desjardins, A.E., Christ-Harned, E.A., McCormick, S.P. & Secor, G.A. (1993). Population structure and genetic analysis of field resistance to thiabendazole in *Gibberella pulicaris* from potato tubers. *Phytopathology* 83: 164-170

Desjardins, A.E. & Plattner, R.D. (1989). Trichothecene toxin production by strains of *Gibberella pulicaris* (*Fusarium sambucinum*) in liquid culture and in potato tubers. *Journal of Agricultural and Food Chemistry* 37: 388-392

Duffy, B.K., Simon, A. & Weller, D.M. (1996). Combination of *Trichoderm koningii* with fluorescent Pseudomonads for control of take-all on wheat. *Phytopathology* 86: 188-194

Duffy, B.K. & Weller, D.M. (1995). Use of *Gaeumannomyces graminis* var. *graminis* alone and in combination with fluorescent *Pseudomonas* spp. to suppress take-all of wheat. *Plant Disease* 79: 907-911

EPA (1996). EPA 738-R-96-023 Registration Eligibility Decision. Environmental Protection Agency

Fairclough, R.W., Bain, R.A. & Holmes, S.J. (1997). The infection of potatoes by *Phytophthora infestans* during simulated washing and its control using a hot water treatment. *Potato Research.* 40: 91-99

Fravel, D.R. (1988). Role of antibiosis in the biocontrol of plant diseases. *Ann. Rev. Phytopathology* 26: 75-91

Fry, W.E., Thurston, H.D., & Stevenson, W.R. (2001). Late blight, In: *Compendium of Potato Diseases*, W.R. Stevenson, R. Loria, G.D. Franc & D.P. Weingartner (Eds.), 28-30, APS Press, St. Paul, MN

Geutsky, R., Shtienberg, D., Elad, Y. & Dinoor, A. (2001). Combining biocontrol agents to reduce the variability of biological control. *Phytopathology* 91: 621-627

Hanson, L.E., Schwager, S.J. & Loria, R. (1996). Sensitivity to thiabendazole in Fusarium species associated with dry rot of potato. *Phytopathology* 86: 378-384

Homma, Y. & Suzui, T. (1989). Role of antibiotic production in suppression of radish damping off by seed bacterization with *Pseudomonas cepacia. Annals of the Phytopathological Society of Japan* 55: 643-652

Huang, H.C. (1991). Control of soilborne pathogens by mycoparasites: prospects and constraints. In *The Biological Control of Plant Diseases*, J. Bay-Petersen (Ed.), 130-141, Kuo Thai Color Printing Co., Ltd., Japan

Hwang, J. & Benson, D.M. (2002). Biocontrol of rhizoctonia stem and root rot of poinsettia with *Burkholderia cepacia* and binucleate *Rhizoctonia. Plant Disease* 86: 47-53

Jackson, M.A. (1997). Optimizing nutritional conditions for the liquid culture prduction of effective fungal biological control agents. *Journal of Industrial Microbiology and Biotechnology* 19: 180-187

Jackson, M.A., McGuire, M.R., Lacey, L.A. & Wraight, S.P. (1997). Liquid culture production of desiccation tolerant blastospores of the bioinsecticidal fungus *Paecilomyces fumosoroseus. Mycological Research* 101: 35-41

Jackson, M.A. & Schisler, D.A. (1995). Liquid culture production of microsclerotia of *Colletotrichum truncatum* for use as bioherbicidal propagules. *Mycological Research* 99: 879-884

Janisiewicz, W.J. (1988). Biocontrol of postharvest diseases of apples with antagonist mixtures. *Phytopathology* 78: 194-198

Janisiewicz, W.J. (1991). Biological control of post harvest fruit diseases. In: *Handbook of Applied Mycology, Volume 1: Soil and Plants*, D.K. Arora, B. Rai, K.G. Mukerji & G.R. Knudsen, (Eds.), 301-326, Marcel Dekker, Inc., New York

Janisiewicz, W.J. (1996). Ecological diversity, niche overlap, and coexistence of antagonists used in developing mixtures for biocontrol of post-harvest diseases of apples. *Phytopathology* 86: 473-479

Janisiewicz, W.J. & Korsten, L. (2002). Biological control of postharvest diseases of fruits. *Annual Review of Phytopathology* 40: 411-441

Johnson, D.A., Inglis, D.A. & Miller, J.S. (2004). Control of potato tuber rots caused by oomycetes with foliar applications of phosphorous acid. *Plant Disease* 88: 1153-1159

Kawchuk, L.M., Holley, J.D., Lynch, D.R. & Clear, R.M. (1994). Resistance to thiabendazole and thiophanate-methyl in Canadian isolates of *Fusarium sambucinum* and *Helminthosporium solani*. *American Potato Journal* 71: 185-192

Kimura, T. & Ito, Y. (2001). Two bacterial mixed culture systems suitable for degrading terephthalate in wastewater. *Journal of Bioscience and Bioengineering* 91: 416-418

Kozyrovska, N.O., Negrutska, V.V., Kovalchuk, M.V. & Voznyuk, T.N. (2005). *Paenibacillus* sp., as a promising candidate for development of a novel technology of inoculants production. *Biopolimeri Klitina* 21:312-318

Krauss, U. & Soberanis, W. (2001). Biocontrol of cocoa pod diseases with mycoparasite mixtures. *Biological Control* 22: 149-158

Kumar, M., Leon, V., DeSisto Materano, A. & Ilzins, O.A. (2006). Enhancement of oil degradation by co-culture of hydrocarbon degrading and biosurfactant producing bacteria. *Polish Journal of Microbiology* 55: 139-146

Lambert, D.H., Currier, A.I. & Olanya, M.O. (1998). Transmission of *Phytophthora infestans* in cut potato seed. *American Journal of Potato Research.* 75: 257-263

Lazarovits, G., Brammall, R.A. & Ward, E.W.B. (1982). Bioassay of fungitoxic compounds on thin-layer chromatograms with *Pythium* and *Phytophthora* species. *Phytopathology* 72: 61-63

Leeman, M., Den Ouden, F.M., Van Pelt, J.A., Cornelissen, C., Matamala-Garros, A., Bakker, P.A.H.M. & Schippers, B. (1996). Suppression of Fusarium wilt of radish by co-inoculation of fluorescent *Pseudomonas* spp. and root-colonizing fungi. *Journal of Plant Pathology* 102: 21-31

Lewis, M.D., Thornton, M.K. & Kleinkopf, G.E. (1996). Understand why CIPC is a necessity. *Potato Grower* 25(9) : 8-9

Licciardello, G., Bertani, I., Steindler, L., Bella, P., Venturi, V. & Catara, V. (2007). Homoserine lactone quorum sensing system: its role in tomoto pathogenicity and tobacco hypersensitivity response. *FEMS Microbiology Ecology* 61: 222-234

Loper, J.E. & Buyer, J.S. (1991). Siderophores in microbial interactions on plant surfaces. *Molecular Plant-Microbe Interactions* 4: 5-13

Martiniez Noel, G.M.A., Madrid, E.A., Bottini, R. & Lamattina, L. (2001). Indole acetic acid attenuates disease severity in potato-*Phytopthora infestans* interaction and inhibits the pathogen growth *in vitro*. *Plant Physiology and Biochemistry.* 39: 815-823

Medugno, C.C., Ferraz, J.M.G., Maia, A.H.N. & Freitas, D. L. (1997). Evaluation of a wettable powder formulation for the nuclear polyhedrosis virus of *Anticarsia gemmatalis* (Lep.:Noctuidae). *Pesticide Science* 51: 153-156

Miller, J.S., Olsen, N., Woodell, L., Porter, L.D. & Clayson, S. (2006). Post-harvest applications of zoxamide and phophite for control of potato tuber rots caused by oomycetes at harvest. *American Journal of Potato Research.* 83: 269-278

Mills, A.A.S., Platt, H.W.B. & Hurta, R.A.R. (2005). Salt compounds as control agents of late blight and pink rot of potatoes in storage. *Canadian Journal of Plant Pathology.* 27: 204-209

Molina, L., Constantinescu, F., Michel, L., Reimmann, C., Duffy, B. & Defago, G. (2003). Degradation of pathogen quorum-sensing molecules by soil bacteria: a preventative and curative biological control mechanism. *FEMS Microbiology Ecology* 45: 71-81

Mondy, N.I., Sharada, D., Munshi, C.B. & Wurm, C.M. (1992). Effect of storage time, temperature, and cooking on isopropyl N-(3-chlorophenyl) carbamate levels in potatoes. *Journal of Agricultural and Food Chemistry* 40: 197-199

Müller, H., Westendorf, C., Leitner, E., Chernin, L., Riedel, K., Schmidt, S., Eberl, L. & Berg, G. (2009). Quorum-sensing effects in the antogonistic rhizosphere bacterium *Serratia plymuthica* HRO-C48. *FEMS Microbiology Ecology* 67: 468-478

Pang, Y., Liu, X., Ma, Y., Chernin, L., Berg, G. & Gao, K. (2009). Induction of systemic resistance, root colonization and biocontrol activities of the rhizospheric strain of *Serratia plymuthica* are dependent on N-acyl homoserine lactones. *Plant Pathology* 124: 261-268

Pawlak, A., Pavek, J.J., & Corsini, D.L. (1987). Resistance to storage diseases in breeding stocks. In: *The Production of New Potato Varieties: Technological Advances,* G.J. Jellis & D.E. Richardson, (Eds.), 96-98, Cambridge University Press, New York

Peters, R.D., Clark, R.J., Coffin, A.D., Sturz, A.V., Lambert, D.H. & Miller, J.S. (2005). Limited genetic diversity in North American isolates of *Phytophthora erythroseptica* pathogenic to potato based on RAPD analysis. *Plant Disease* 89: 380-384

Peters, R.D., Sturz, A.V. & Arsenault, W.J. (2004). Tuber response of six potato cultivars to inoculation with *Phytophthora erythroseptica,* the causal agent of pink rot. *Canadian Journal of Plant Pathology* 26: 63-69

Pierson, E.A. & Weller, D.M. (1994). Use of mixtures of fluorescent Pseudomonads to suppress take-all and improve the growth of wheat. *Phytopathology* 84: 940-947

Pierson, E.A., Wood, D.W., Cannon, J.A., Blachere, F.M. & Pierson III, L.S. (1998). Interpopulation signaling via N-acyl-homoserine lactones among bacteria in the wheat rhizosphere. *Molecular Plant-Microbe Interactions* 11: 1078-1084

Reiners, S. & Petzoldt, C.H. (2004). Chapter 24 Potatoes. In: *Cornell Integrated Crop and Pest Management Guidelines for Vegetable Crops,* S. Reiners & C.H. Petzoldt, (Eds.). http://www.nysaes.cornell.edu/recommends

Reisch, M.S. (2011). Going mainstream biopesticide producers look beyond organic growers. *Chemical and Engineering News* 89(16): 16

Roberson, E.B. & Firestone, M.K. (1992). Relationship between desiccation and exopolysaccharide production in a soil *Pseudomonas* sp. *Applied and Environmental Microbiology* 58: 1284-1291

Rodriguez, H. & Gallardo, R. (1993). Single-cell protein production from bagasse pith by a mixed bacterial culture. *Acta Biotechnologica* 13: 141-149

Sandoval-Coronado, C.F., Luna-Olvera, H.A., Arevalo-Nino, K., Jackson, M.A., Poprawski, T.J. & Galan-Wong, L.J. (2001). Drying and formulation of blastospores of *Paecilomyces fumosoroseus* (Hyphomycetes) produced in two different liquid media. *World Journal of Microbiology and Biotechnology* 17: 423-428

Schisler, D.A. (1997). The impact of phylloplane microorganisms on mycoherbicide efficacy and development. In: *The Mycota: Vol 5, Environmental and Microbial Relationships*, 219-235, B. Soderstrom & D.T. Wicklow, (Eds.), Springer-Verlag, New York

Schisler, D.A., Boehm, M.J. & Slininger, P.J. (2005). Dosage requirements and combinations of choline metabolizing strains and antagonist *Cryptococcus nodaensis* OH 182.9 for reducing FHB of wheat. *Phytopathology* 95: S93

Schisler, D.A., Bothast, R.J. & Slininger, P.J. (1998a).Bacterial Control of Fusarium Dry Rot of Potatoes. United States Patent No. 5783411

Schisler, D.A., Jackson, M.A. & Bothast, R.J. (1991). Influence of nutrition during conidiation of *Colletotrichum truncatum* on conidial germination and efficacy in inciting disease in *Sesbania exaltata*. *Phytopathology* 81: 587-590

Schisler, D.A. & Slininger, P.J. (1994). Selection and performance of bacterial strains for biologically controlling Fusarium dry rot of potatoes incited by *Gibberella pulicaris*. *Plant Disease* 78: 251-255

Schisler, D.A., Slininger, P.J. & Bothast, R.J. (1997). Effects of antagonist cell concentration and two-strain mixtures on biological control of Fusarium dry rot of potatoes. *Phytopathology* 87(2): 177-183

Schisler, D.A., Slininger, P.J., Hanson, L.E. & Loria, R. (2000a). Potato cultivar, pathogen isolate and antagonist cultivation medium influence the efficacy and ranking of bacterial antagonists of Fusarium dry rot. *Biocontrol Science and Technology* 10: 267-279

Schisler, D.A., Slininger, P.J., Kleinkopf, G., Bothast, R.J. & Ostrowski, R.C. (2000b). Biological control of Fusarium dry rot of potato tubers under commercial storage conditions. *American Journal of Potato Research* 77: 29-40

Schisler, D.A., Slininger, P.J., Miller, J.S., Woodell, L.K., Clayson, S. & Olsen, N. (2009). Bacterial antagonists, zoospore inoculum retention time and potato cultivar influence pink rot disease development. *American Journal of Potato Research* 86: 102-111

Secor, G.A. & Gudmestad, N.C. (1999). Managing fungal diseases of potato. *Canadian Journal of Plant Pathology* 21: 213-223

Secor, G.A., Rodriguez, D., Rodriguez, J. & Gudmestad, N.C. (1994). Distribution and incidence of benzimidazole-resistant *Fusarium sambucinum* and *Helminthosporium solani* isolated from potato in North America. *Fungicide Resistance* BCPC Monogr. No. 60: 271-274

Secor, G.A. & Salas, B. (2001). Fusarium dry rot and Fusarium wilt, In: *Compendium of Potato Diseases*, W.R. Stevenson, R. Loria, G.D. France & D.P. Weingartner, (Eds.), 23-25, APS Press, St. Paul, MN

Slininger, P.J. & Schisler, D.A. (2003). High throughput assay for optimizing microbial biological control agent production and delivery. *Phytopathology* 93: S79

Slininger, P.J., Burkhead, K.D. & Schisler, D.A. (2004). Antifungal and sprout regulatory bioactivities of phenylacetic acid, indole-3-acetic acid, and tyrosol isolated from the potato dry rot suppressive bacterium *Enterobacter cloacae* S11:T:07. *Journal of Industrial Microbiology and Biotechnology* 31: 517-524

Slininger, P.J., Burkhead, K.D., Schisler, D.A. & Bothast, R.J. (2000). Biological control of sprouting of potatoes. United States Patent No. 6107247

Slininger, P.J., Dunlap, C.A. & Schisler, D.A. (2010a). Polysaccharide production benefits dry storage survival of the biocontrol agent *Pseudomonas fluorescens* S11:P:12 effective against several maladies of stored potatoes. *Biocontrol Science and Technology* 20(3): 227-244

Slininger, P.J., Schisler, D.A. & Bothast, R.J. (1994).Two-dimensional liquid culture focusing: A method of selecting commercially promising microbial isolates with demonstrated biological control capability, In: *Improving Plant Productivity with Rhizosphere Bacteria, Proceedings of the Third International Workshop on Plant Growth-Promoting Rhizobacteria,* M.H. Ryder, P.M. Stephens & G.D. Bowen (Eds.), 29-32, Adelaide, South Australia: CSIRO

Slininger, P.J., Schisler, D.A. & Bothast, R.J. (1996a). Bacteria for the control of Fusarium dry rot to potatoes. United States Patent No. 5552315.

Slininger, P.J., Schisler, D.A., Burkhead, K.D. & Bothast, R.J. (2003). Postharvest biological control of potato sprouting by Fusarium dry rot suppressive bacteria. *Biocontrol Science and Technology* 13(5): 477-494

Slininger, P.J., Schisler, D.A., Burkhead, K.D., Ostrowski, R.C., & Bothast, R.J. (1997b). Development of biological agents for the control of Fusarium dry rot disease of stored potatoes. *American Potato Journal* 74: 467-468

Slininger, P.J., Schisler, D.A., Ericsson, L.D., Brandt, T.L., Frazier, M.J., Woodell, L.K., Olsen, N.L. & Kleinkopf, G.E. (2007). Biological control of post-harvest late blight of potatoes. *Biocontrol Science and Technology* 17(5/6): 647-663

Slininger, P.J., Schisler, D.A. & Kleinkopf, G. 2001. Combinations of dry rot antagonistic bacteria enhance biological control consistency in stored potatoes. *Phytopathology* 91: S83

Slininger, P.J., Schisler, D.A., Shea-Andersh, M.A., Sloan, J.M., Woodell, L.K., Frazier, M.J. & Olsen, N. (2010b). Multi-strain co-cultures surpass blends for broad spectrum biological control of maladies of potatoes in storage. *Biocontrol Science and Technology* 20(8): 763-786

Slininger, P.J. & Shea-Wilbur, M.A. (1995). Liquid-culture pH, temperature, and carbon (not nitrogen) source regulate phenazine productivity of the take-all biocontrol agent *Pseudomonas fluorescens* 2-79. *Applied Microbiology and Biotechnology* 43: 794-800

Slininger, P.J., VanCauwenberge, J.E., Bothast, R.J., Weller, D.M., Thomashow, L.S. & Cook, R.J. (1996b). Effect of growth culture physiological state, metabolites, and formulation on the viability, phytotoxicity, and efficacy of the take-all biocontrol agent *Pseudomonas fluorescens* 2-79 stored encapsulated on wheat seeds. *Applied Microbiology and Biotechnology* 45: 391-398

Slininger, P.J., VanCauwenberge, J.E., Shea-Wilbur, M.A. & Bothast, R.J. (1998). Impact of liquid culture physiology, environment, and metabolites on biocontrol agent qualities: *Pseudomonas fluorescens* 2-79 versus wheat take-all, In: *Plant-Microbe Interactions and Biological Control*, G.G. Boland & L.D. Kuykendall (Eds.), 329-353, Marcel Dekker, Inc., New York, NY

Slininger, P.J., VanCauwenberge, M.A., Shea-Wilbur, M.A., Burkhead, K.D., Schisler, D.A. & Bothast, R.J. (1997a). Reduction of phenazine-1-carboxylic acid accumulation in growth cultures of the biocontrol agent *Pseudomonas fluorescens* 2-79 eliminates phytotoxic effects of wheat seed inocula without sacrifice to take-all suppressiveness, In: *Plant Growth-promoting Rhizobacteria: Present Status and Future Prospects, Proceedings of the Fourth International Workshop on Plant Growth-promoting Rhizobacteria Japan-OECD Joint Workshop*, A. Ogoshi, K. Kobayashi, Y. Homma, F. Kodama, N. Kondo & S. Akino, (Eds.), 464-467, The 4th PGPR International Workshop Organizing Committee, Faculty of Agriculture, Hokkaido University, Sapporo, Japan

Standbury, P.F. & Whitaker, A. (1984). *Principles of Fermentation Technology.* Oxford: Pergamon Press.

Taylor, R.J., Pasche, J.S. & Gudmestad, N.C. (2006). Biological significance of mefenoxam resistance in *Phytophthora erythroseptica* and its implications for the management of pink rot of potato. *Plant Disease* 90: 927-934

Taylor, R.J., Salas, B., Secor, G.A., Rivera, V. & Gudmestad, N.C. (2002). Sensitivity of North American isolates of *Phytophthora erythroseptica* and *Pythium ultimum* to mefenoxam (metalaxyl). *Plant Disease* 86: 797-802

USDA (1965). Losses in Agriculture, U.S. Department of Agriculture Handbook 291. U.S. Department of Agriculture, Agricultural Research Service.

Van Brunt, J. (1986). Fermentation economics. *Biotechnology* 4: 395-401

Wei, H.L. & Zhang, L.Q. (2006). Quorum-sensing system influences root colonization and biological control ability in *Pseudomonas fluorescens* 2P24. *Antonie van Leeuwenhoek, International Journal of General and Molecular Microbiology* 89: 267-280

Wilson, C.L., Ghaouth, A.E., Chalutz, E., Droby, S., Stevens, C., Lu, J.Y., Khan, V. & Arul, J. (1994). Potential of induced resistance to control postharvest diseases of fruits and vegetables. *Plant Disease* 78: 837-843

Wood, D.W. & Pierson III, L.S. (1996). The phzI gene of *Pseudomonas aereofaciens* 30-84 is responsible for the production of a diffusible signal required for phenazine antibiotic production. *Gene* 168: 49-53

Woodhead, S.H., O'Leary, D.J. & Rabatin, S.C. (1990). Discovery, development, and registration of a biocontrol agent from an industrial perspective. *Canadian Journal of Plant Pathology* 12: 328-331

Wu, Z., Ma, J., Wang, Y., Yang, L. & Lou, K. (2009). Identification of salt tolerant promoting growth bacteria Rs-198 and study on co-culture with Rs-5'. *Shengwu Jishu* 19: 63-66

Zhang, S., Schisler, D.A., Boehm, M.J. & Slininger, P.J. (2005). Carbon-to-nitrogen ratio and carbon loading of production media influence freeze-drying survival and biocontrol efficacy of *Cryptococcus nodaenisis* OH 182.9. *Phytopathology* 95: 626-632

Zhang, S., Schisler, D.A., Jackson, M.A., Boehm, M.J., Slininger, P.J. & Liu, Z. (2006). Cold shock during liquid production increases storage shelf-life of *Cryptococcus nodaensis* OH 192.9 after air drying. *Biocontrol Science and Technology* 16: 281-293

Zhang, Z. & Pierson III, L.S. (2001). A second quorum-sensing system regulates cell surface properties but not phenazine antibiotic production in *Pseudomonas aureofaciens*. *Applied and Environmental Microbiology* 67: 4305-4315

8

Control of Major Diseases in Horticulture

Emma W. Gachomo[1,2], Jose C. Jimenez-Lopez[3], Adéchola P. P. Kayodé[4], Lamine Baba-Moussa[5] and Simeon O. Kotchoni[1,2]

[1]Department of Biology, Rutgers University, Camden,
[2]Center for Computational and Integrative Biology, Rutgers University, Camden,
[3]Department of Biochemistry, Cell and Molecular Biology of Plant, Estacion Experimental del Zaidin, Consejo Superior de Investigaciones Cientificas, Granada,
[4]Département de Nutrition et Sciences Alimentaires, Faculté des Sciences Agronomiques, Université d'Abomey-Calavi, Cotonou,
[5]Laboratoire de Biologie et de Typage Moléculaire en Microbiologie, Faculté des Sciences et Techniques, Université d'Abomey-Calavi, Cotonou,
[1,2]USA
[3]Spain
[4,5]Benin

1. Introduction

Cultivation of fruits, vegetables, flowers, and/or ornamental plants is one of the economically important sectors of crop production. However cultivation of these plants plagued with wide range of most destructive, reoccurring and widespread stress factors both biotic (diseases) and abiotic (environmental) (Gachomo et al., 2009). Plant diseases are caused by many different types of organism that include fungi, bacteria, viruses etc. While plant disease may be caused by many factors, majority of them are caused by fungi and fungi like organisms (FLO). Over 8,000 species of fungi and FLOs have been reported to cause disease (Ellis et al., 2008). Disease caused by fungi and FLOs have been recorded to cause damage that has had far reaching effects on human population as a result of the suffering that follow crop loss (Gachomo, 2005) e.g. the great famine of 1845-1852 that occurred in Ireland as a result of potato crop failure due to potato blight infection. During this famine about a million people died of hunger and another million emigrated out of Ireland. Control of plant disease is both expensive and time consuming and represents a large segment of the pesticide market (Gachomo et al., 2009; Killian & Steiner, 2003), despite continuous efforts to control their detrimental effect on crop production. For example cost of control and damage by *Phytopthora infestans* (that causes late blight of potatoes) in Europe is estimated at more than €1,000,000,000 ($1,400,000,000) per year (Haverkort et al., 2008). The typical symptoms of these fungal induced diseases cover all color spectra (from white to black) and form different types of symptoms both internal and external, microscopic and macroscopic, local and systemic which appear (but are not limited to) as leaf spots, galls, lesions, blights and cankers incase of local symptoms or as wilting, yellowing, and dwarfing incase of systemic symptoms because they involve most or all of the plant (Gachomo et al., 2006; Gachomo & Kotchoni 2007; Gachomo et al., 2010). Common fungal diseases include leaf spots, cankers, powdery mildews, downey mildews, smuts, blights, damping off, blasts, anthracnose, wilts, fruit rots, root rots, stem rots, rusts and molds among any others (Bowen & Roark, 2001; Walker et al., 1995).

Once the fungal diseases are established, they are generally difficult to control despite a combination of practices including sanitation measures and fungicide applications (Behe et al., 1993). Management of the diseases relies often on intensive use of fungicides (Bowen & Roark, 2001; Reddy et al., 1992; Walker et al., 1995). Fungicides used in fungal disease management include dormant sprays, protectant and systemic fungicides (Bowen & Roark, 2001; Killian & Steiner, 2003). Protectants, also known as contact fungicides, remain on the outside of the plant and kill parts of fungal structures growing on the surface of the plant (Gachomo et al., 2009). However, the activity of protectant fungicides decreases with time because of exposure to environmental factors such as ultraviolet (UV) radiation and wash-off caused by the rain, and the expansion of foliage surface in growing plants leading to a breaking effect on protective layer. Consequently, protectant fungicides need frequent applications to be effective. On the other hand, systemic fungicides are absorbed into the plant tissue and can minimize disease symptoms after infection has occurred (Gachomo, 2005). One group of systemic fungicides, the demethylation inhibitors (DMI), has been widely used to control fungal diseases in horticulture (Killian & Steiner, 2003). DMI common mode of action is centered on the fungal sterol biosynthesis, i.e. the inhibition of demethylation at position 14 of lanosterol or 24-methylene dihydrolanosterol, the precursors of sterols in fungi (Kuck et al., 1996). However, strobilurins constitute the novel promising group of fungicides to control a wide range of pathogens infecting cereals, potatoes, fruit, grapevine, vegetables, turf grass and ornamentals (Ebeling et al., 2003; Stark-Urnau et al., 1997; Wojdyla & Orlikowski, 1999). The strobilurins inhibit the mitochondrial respiratory chain by blocking the electron transport at the Qo site of cytochrome b. Forthermore, the strobilurins used either alone or in combination with other fungicides control a wide array of fungal diseases, including diseases caused by water moulds, downy mildews, powdery mildews, fruit rotters, rusts, leaf spotting and blight fungi (Ammermann et al., 1992, Margot et al., 1998; Reuveni, 2001; Ypema & Gold 1999).

In this chapter, we give a detail overview of disease symptoms, their causal agents and fungicide dependent control of selected major fungal diseases in horticulture such as powdery mildew, gray mold, blackspot, late and early blight diseases, and Fusarium wilt disease. Optimum number and timing of fungicide sprays for an efficient disease control in horticulture are here discussed. In addition, we report several aspects of fungicide activity, including the effects on fungal establishment within the host.

2. Diseases of flowers

2.1 Powdery mildew

Powdery mildew as the name suggests initially appears as a white grayish coating on leaves (Figure 1 A). When the disease is severe it produces a white gray coating on stems, petals and buds. This white coating spreads over time to cover the whole leaf (Figure 1 B, C). The affected leaves or petals get distorted, shrivel and may fall to the ground. Several disease cycles occur within the plants' growth season. The disease makes the plants unattractive and repeated disease cycles cause reduced vigor of the plants (Gachomo, 2005). Among the most often affected flowers and shrubs are columbine, dahlia, delphiniums, honeysuckle, ivy, lilac, phlox, privet, snapdragons, lungwort, bee balm, garden phlox, verbena, rose, and zinnias.

2.2 Life cycle of powdery mildew causal agents

The powdery mildews are caused by a group of more than 300 related fungal species which are host specific, meaning that a fungal species that attacks one type of plant species might not infect another plant species. Examples **of fungi causing powdery mildew**include *Sphaerotheca pannosa* that infects roses; *Erysiphe cichoracearm* that infects Chrysanthemums. The primary inoculum comes from dormant infections that survive the winter. The overwintering structures called cleistothecia produce sexual spores, which infect young leaves and blossoms when the weather is favorable. Mycelia formed in the infected tissues form conidia, which infected other leaves, shoots and buds. Several generations of conidia are formed in the growing season as long as the weather remains favorable. Towards the end of the season when the conditions become unfavorable for disease development, cleistothecia are formed in the infected tissue and will serve as a source of inoculum during the next growing season.

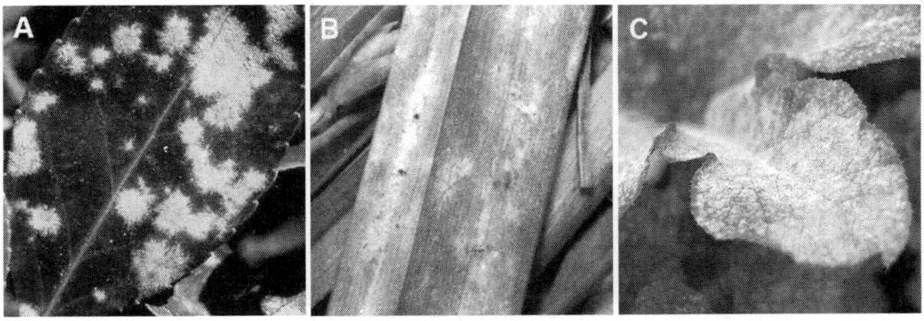

Fig. 1. Disease symptom of powdery mildew on the upper leaf surface of Euonymous plant (A), grass blades: Photo, courtesy of P. Flynn (B) and pumpkin leaves: Photo, courtesy of J. Kunbino (C).

2.3 Control and prevention of powdery mildew disease

Spores of the fungal causing disease are spread by wind and water splash; it is therefore crucial to avoid wetting the leaves during watering as one of the most fundamental strategies of powdery mildew control. In addition, avoiding the use of sprinkling irrigation but rather watering at the base of the plants or use of drip irrigation are the general rules in controlling this disease. In order for the disease to develop, the spores require plants wetness for several hours to be able to infect the plants, therefore water the plants preferentially in the morning rather than in the evening so that they have time to dry during the day, therefore reducing time of leaf wetness. Varieties that are susceptible should not be planted under the shade but in areas exposed to the sun because shading does not allow dew to dry quickly. The plants should be well spaced and kept free of weed to increase aeration around the plants and reduce moisture retention around the base of the plant. Plants should be pruned to reduce dense foliage and increase aeration. The fungus overwinters in fallen plant debris, therefore it is very important to clean up all fallen plant materials and prune infected tissues at the end of the season, which will ultimately reduce the source of inoculum for the next season. Up to date, the efficient control of powdery mildew disease relies on fungicide application. Several fungicides have been developed for the management of this disease. In Table 1, we show some of the widely used fungicides in powdery mildew disease control.

2.4 Gray mold disease

Generally, the gray mold disease appears as a gray to brown mold on dying plant tissue such as spent flowers, leaves and stems (Figure 2 A, B). The symptoms are observed as light spot surrounded by maroon halos on petals. These spots enlarge and turn brown over time. Infected petals may fall off. Cankers and dieback may be observed on stems. A wide range of plant varieties are affected by this disease. Susceptible plant species to gray mold include house plants such as African violets, amaryllis, Amazon-lily, azaleas, begonias, cacti, caladium, calla lily, camellias, castorbeans, chrysanthemums, cinerarias, coleus, corn flowers, dalias, dracaenas, dusty millers, ferns, fig, fuchsias, ardenias, gloxinias, heliotrope, orchids, passion flower, and poinsettias. In addition, members of herbaceous annual plants susceptible to gray mold are ageratum, begonia, caldium, carnation, celosia, chicory, geranium, gerbera, gladiola, impatiens, marigold, nasturtium, pansy, petunia, snapdragon, statice, stock, sunflower, sweet pea, verbena, and zinnia. Furthermore herbaceous perennials including anemone, aster, baby's-breath, bellflower, bleeding heart, bloodroot, bluebell, buttercup, calendula, candytuft, carnation, chrysanthemum, chicory, coralberry, cranesbill, dandelion, daylily, delphinium, Dutchman's- pipe, foxglove, globe-amaranth, hyacinth, iris, Jack-in-the-pulpit, liy, lily-of-the-valley, lupine, Maltese cross, narcissus, peony, phlox, pinks, plantain lily, poppy, primrose, purple coneflower, rose, snowdrop, tulip, vinca, and violet are susceptible to gray mold disease.

Fig. 2. Disease symptom of Gray mold disease on red raspberry fruit (A) and on tomato fruit (B). Photo, courtesy of P. R. Bristout.

2.5 Life cycle of gray mold causal agents

Gray mold disease is caused by many strains of the fungus *Botrytis cinerea*. This fungus is not host specific and therefore infects many different types of plant species. Symptoms vary with the plant species attacked. The fungus needs moisture to cause infection and a favorable temperature ranging from 18° C to 25° C.

2.5.1 Control of gray mold disease

To contain the disease all infected tissue should be remove as soon as possible; infected debris should be collected. The plants should be well spaced and weeded to allow good air circulation; sprinkler irrigation and wetting of the foliage should be avoided because this provides the humidity needed for the proliferation of the disease. Application of fungicides in combination with the good sanitation measures will ultimately help to contain the disease. It is imperative to apply preventive fungicides as soon as disease is detected. Fungicides with the active ingredients such as chlorothalonil, dichloran, fludioxonil, trifloxystrobin, iprodione, mancozeb, copper sulfate pentahydrate, fenhexamid, azoxystrobin, and thiophanate methyl are registered for *Botrytis* control and therefore recommended to use in case of gray mold disease. Be sure to rotate applications among chemical classes as fungicide resistant strains of *Botrytis* have been reported.

2.6 Blackspot disease

Blackspot is a disease of roses. It appears as small black spots on the upper surface of the leaves (Figure 3 A), which first appear on the lowest leaves and may first appear as purple spots on stems that eventually turn black. The area around the spots turns yellow and the spot may coalesce to form black blotches (Figure 3 A, B, C). The yellow leaves easily fall off the plants. The disease spreads from lower leaves to younger upper leaves leading to further defoliation. Severe defoliation reduces vigor of the plants and decrease flower production (Gachomo et al., 2010).

Fig. 3. Photographs of rose leaves infected with *Diplocarpon rosae* showing, (A) the symptoms on leaves followed by the yellowing of the leaves (B) a close up of a sporulating spot showing the dome shaped-unopened acervuli that have pushed the cuticle upwards, (C) a close up of a sporulating spot where a mass of white conidia oozes out of the acervuli (Adopted from Gachomo, et al., 2010)

2.6.1 Life cycle of blackspot causal agents

Blackspot disease is caused by a fungal pathogen, *Diplocarpon rosae*. The fungus overwinters on infected canes and fallen debris (Gachomo, 2005). During the favorable weather conditions the spores are splashed from infected plant parts to young leaves by rain splash and irrigation water. The fungus produces conidia within 10 to 14 days (Figure 4 A, B) which are splashed to other young leaves. Several disease cycles can occur within a growing season. Once established the disease is difficult to control.

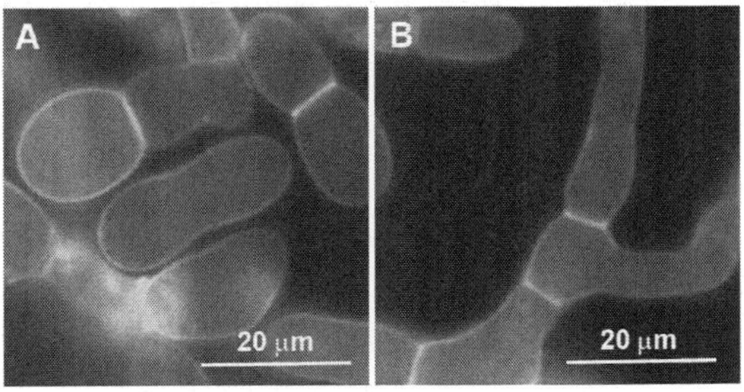

Fig. 4. Light microscope photographs of *Diplocarpon rosae* growing on artificial malt agar medium: (A) *D. rosae* two-celled conidial structures before and during germination. (B) Three day-old conidium germination. (Courtesy of E. W. Gachomo).

2.6.2 Control of blackspot disease

Control practices start with planting resistant rose varieties where available. Good sanitation is key to keeping the rose disease free. Recommended cultural practices are: All infected debris should be collected and burnt or buried; all infected canes must be pruned; overhead irrigation must be avoided because it tends to splash conidia from infected to non-infected parts of the plants. It recommended to water plants at the base; the plants should be preferably watered in the morning as opposed to the evening, because the conidia require several hours of wetness to cause infection, therefore watering in the morning reduces the hours of leaf wetness. In addition, plants should be well spaced and kept weed free to allow for aeration. Furthermore, one must avoid planting susceptible plants under the shade. When blackspot disease is establish, its control relies heavily on fungicides. In Table 1, the fungicides recommended in blacksopt disease management are listed.

Mode of Action	Target site and code	Group name	Chemical group	Common name	FRAC* code
sterol biosynthesis in membranes	C14-Demethylase in sterol biosynthesis *(erg11/cyp51)*	**DMI**-fungicides (**De**M**ethylation** **I**nhibitors) (SBI: Class I)	Triazoles	Myclobutanil (Immunox) Propiconazole (Banner Maxx) tebuconazole	3
sterol biosynthesis in membranes	C14-Demethylase in sterol biosynthesis *(erg11/cyp51)*	**DMI**-fungicides (**DeM**ethylation **I**nhibitors) (SBI: Class I)	Piperazines	Triforine	3
mitosis and cell division	ß-tubuline assembly in mitosis	**MBC**-Fungicides (**M**ethyl **B**enzimidazole **C**arbamates)	Thiophanates	Thiophanate-methyl 3336 4.5 F 3336 50W Halt	1
Multi-site contact activity	multi-site contact activity	Inorganic	Inorganic	Copper (different salts)	M1
Multi-site contact activity	multi-site contact activity	Inorganic	Inorganic	Sulphur	M2

*FRAC (Fungicides Resistance Action Committee)

Table 1. Fungicides labeled for the control of powdery mildew on roses.

3. Diseases of vegetables

3.1 Bottom rot disease of lettuce

Bottom rot disease of lettuce can be recognized by brown spots that initially appear on the midribs of the lower leaves that are in contact with the soil (Figure 5 A). The rot spreads rapidly under favorable conditions to affect larger sections of the midrib and leaf blades, and may affect the inner leaves of the head. Symptoms are more severe during heading.

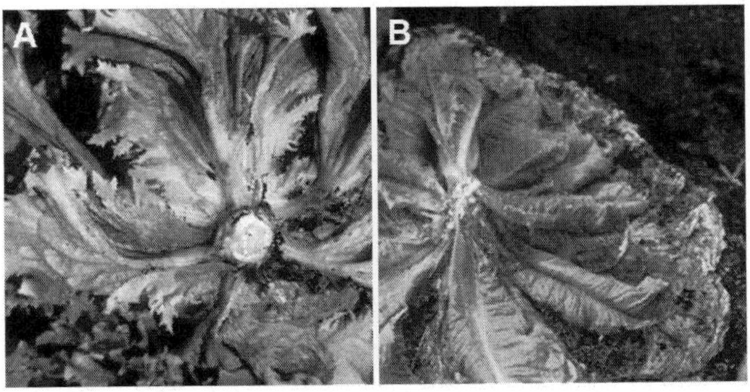

Fig. 5. Disease symptom of bottom rot (A) and Fusarium wilt of lettuce (B). Photos courtesy of A. F. Sherf (A) and T. A. Zitter (B).

3.1.1 Fungal agent of bottom rot disease

Bottom rot is caused by a soilborne fungal pathogen *Rhizoctonia solani*. The fungus overwinters in the soil or in crop debris as sclerotia or mycelia. It may survive in alternate hosts and serve as a source of inoculum, sexual spores. It is disseminated by wind or rain splash in the next growing season. *R. Solani* has a wide host range e.g. eggplant, soybean, potato, cotton, alfafa, maize, wheat and several weed species.

3.1.2 Control of bottom rot disease

Cultural measures includes three year rotations with non-host plants; collecting plant debris and burying it or plowing it deep in the soil; planting varieties that have an upright architecture to reduce contact with the soil; keeping the fields weed free and removing volunteer crops to reduce possible alternate hosts. Since *R. Solani* is able to survive on non decomposed organic matter, it is important to avoid planting lettuce in a field that has high amounts of organic matter that is not decomposed; avoid overhead irrigation during heading of the plants; plant lettuce on ridges which increases aeration and helps avoid plants contact with the soil. Fungicides (Table 2) are the most effective means to control bottom rot disease. However, fungicide control is only satisfactory when used in combination with cultural control strategies. Proper placement and timing of fungicide applications are key elements for effective disease management.

Mode of Action	Target site and code	Group name	Chemical group	Common name	FRAC code
Respiration	complex III: cytochrome bc1 (ubiquinol oxidase) at Qo site *(cyt b gene)*	**QoI**-fungicides (**Q**uinone **o**utside **I**nhibitors	Methoxy acrylates	Azoxystrobin (Amistar) (Quadris Flowable)	11
Respiration	complex II: succinatedehydrogene	Carboxamides	Pyridine-carboxamides	Boscalid (Endura)	7
Signal transduction	MAP/Histidine-Kinase in osmotic signal transduction *(os-1, Daf1)*	Dicarboximides	Dicarboximies	Iprodione (Rovral 50 W) Vinclozolin (Ronilan DF)	2

Table 2. Fungicides Recommended for control of bottom rot on lettuce.

3.2 Fusarium wilt of lettuce

Lettuce seedlings affected by this disease wilt and ultimately die, while in mature plants the symptoms include red-brown to black discoloration of internal taproot and crown tissue, yellowing of leaves, tipburn of heads (Figure 5 B) and when infection is severe plants are stunted and may fail to form heads.

3.2.1 Fungal pathogen of Fusarium wilt disease of lettuce

Fusarium wilt of lettuce is caused by a soil-borne fungus, *Fusarium oxysporum* f.sp. *lactucae* forma specialis nov. This pathogen can remain viable in the soil for many years. *Fusarium oxysporum* f.sp. *lactucae* forma specialis nov is host specific to lettuce and therefore only affect/grow on lettuce.

3.2.2 Fusarium wilt disease control

Recommended cultural practices include clean of farm equipment, avoiding to plant lettuce in infected field and planting resistant/tolerant lettuce varieties.

4. Diseases of potato and tomato

4.1 Late blight disease

Late blight is one of the most destructive diseases of potato and tomato. It is found wherever these crops are grown. On potatoes it appears as small light green water soaked spots at the edges of leaves. During favorable weather conditions, cool and moist, the lesions enlarge rapidly, and turn brown to black (Figure 6 A, B). The lesions coalesce to cover entire leaves and even affect the stem. Infected tissue dries up when the weather is dry. The disease spreads rapidly and all the leaves may be killed in a few days. On tubers, the disease appears as irregular, dry, brown depressions. Copper brown, granular lesions are found underneath the skin (Figure 6 A). Potatoes infected with the late blight pathogen are generally susceptible to secondary infection from other fungi and/or bacteria.

4.1.1 Fungal pathogen of late blight disease

Late blight disease is caused by a fungal pathogen, *Phytophthora infestans*. The primary sources of inoculum are infected seed tubers, volunteer plants and plant debris. Spores are dispersed by wind and water splash from infected to non-infected plants. The disease spreads rapidly at temperatures between 10 and 21°C in combination with high humidity. Several strains of the fungus have been reported and strains recombination increases the chance of having novel strains that are either resistant to fungicides or more tolerant to harsh environmental conditions. *P. infestans* also infects tomatoes and causes mild infections on eggplants, peppers and related weed species.

4.1.2 Late blight disease management

It is recommended to destroy all volunteer potato and other susceptible plants because *P. infestans* survives on these volunteer plants that represent the primary sources of inoculum during the next season. Potato growers should only use certified seed potatoes and avoid using their own grown tubers as seed in order to contain the devastating effect of late blight disease. It is advisable to make sure that other crops that can also be infected by *P. infestans* are disease free. Cull piles of infected potatoes should be destroyed because they serve as a source of inoculum. The fields should be scouted for late blight on a regular basis, paying close attention to low lying areas, areas under shade, or near water sources. It is important to avoid overhead irrigation in the evening because this provides long periods of leaf wetness that favors disease development. Potato tubers should be harvested after the vines die, which also kills the spores on them and avoids transmission of spores to the tubers. Infected tubers should be removed before storage in order to avoid spreading the disease to the healthy tubers. Planting resistant or moderately resistant potato varieties where available is advisable.

4.1.3 Chemical control of late blight disease

The fungicides recommended for use against late blight disease vary from region to region because strains of *P. infestans* found in one region might not be present in another, and fungicide sensitivity might be different among fungal isolates. Genotypes of *P. infestans* have been reported to recombine to produce new genotypes that are resistant to the recommended systemic fungicides, but resistance to protectant fungicides has not been reported.

In fields that have already been reported to have late blight, the first application of a protectant fungicide is recommended before row closure and a second application should follow within 7-10 days. Further applications of protectants should be done when the weather conditions are conducive for late blight development. A late blight epidemic is difficult to control, therefore regular applications of protectants during the growing season is important to keep new foliage covered. Applications should be made even late in the season as long as parts of the vines are still green to avoid tuber infections. For a complete list of fungicides recommend in a region, it is advisable to consult the area extension office. However, we highlight in Table 3 some of, the recommended fungicides used to control late blight disease on potatoes.

4.2 Early blight disease of potato and tomato

On potato and tomato foliage early blight appears as brown to black spots, which coalesce to form lesions that are restricted by large veins and therefore having an angular shape (Figure 6 C, D). Occasionally, a chlorotic border may be formed around the lesions. When stems are infected the disease appears as small dark spots. On tubers there are dark sunken lesions that are surrounded by raised margins. The tissue underneath the lesions is dry, reddish brown in color, and leathery in texture.

FRAC code	Mode of action	Group name	Chemical group	Common name
M3	Multi-site inhibitor	Dithiocarbamates and relatives	Dithiocarbamates and relatives	**Manebs:** (Maneb 75 DF;Maneb 80; Maneb + Zinc; Manex)
M3	Multi-site inhibitor	Dithiocarbamates and relatives	Dithiocarbamates and relatives	**Mancozebs:** (Dithane M-45; Dithane F-45; Dithane DF; Penncozeb 80 WP; Penncozeb 75 DF)
M5	Multi-site contact activity	Chloronitriles	Chloronitriles	**Chlorothalonil:**(Bravo 500; Terranil Excell; Bravo Ultrex; Terranil 6L; Bravo Weatherstik; Bravo Zn)
11	Respiration	**QoI** - fungicides (**Q**uinone **o**utside **I**nhibitors)	Methoxy acrylates	**Azoxystrobin:**(Quadris)
40	Lipids and membrane synthesis	**CAA**-fungicides (**C**arboxylic **A**cid **A**mides)	Cinnamic acid amide	**Dimethomorph:**(Acrobat MZ)
22	mitosis ß-tubulin assembly		Benzamides	**Gavel 75 DF**
27	Unknown mode of action	Cyanoacetamide-oxime	Cyanoacetamide-oxime	**Cymoxanil:** (Curzate 60 DF)

Table 3. Fungicides listed for control of late blight on potatoes.

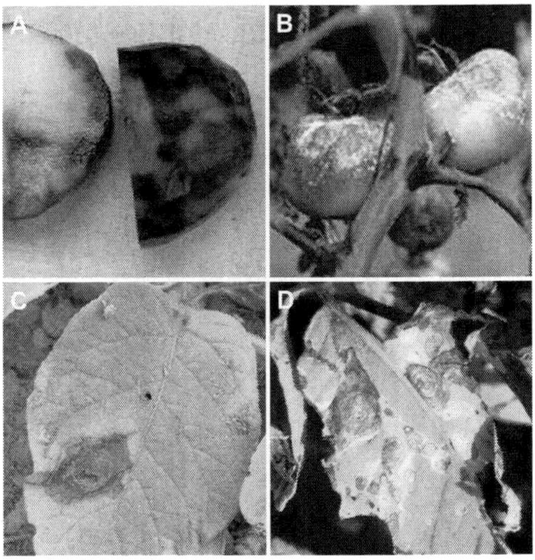

Fig. 6. Late blight (A-B) and early blight (C-D) disease symptom on potato (A, D) and tomato plants (B, D) respectively. Late blight disease is depicted on potato (A) and tomato fruit (B), while early blight disease is depicted on potato leaf (C) and tomato leaf (D). Photos: courtesy of B. Millett (A); W. R. Stevenson (B); S. R. Rideout (C); and R. Mulrooney (D).

4.2.1 Fungal pathogen of early blight disease

Early blight disease is caused by a fungus, *Alternaria solani*. The fungus overwinters in plant debris, infected tubers, soil and on other host species. Disease development is favored by temperatures between 20°C and 30°C; long periods of leaf wetness, high relative humidity under alternating wet and dry conditions. Spores are dispersed by wind, water splash, insects, machinery and animals. The disease occurs late in the season and increases rapidly during flowering and senescence. Both biotic and abiotic stresses favor disease development. Bruising or wounding of tubers during harvest leads to infection with early blight.

On tomato the disease symptom is characterized by lesions with dark concentric rings. Diseased leaves wither, dry and fall off. Severe defoliation reduces plant vigor and exposes tomato fruits to sunscald. Disease is first observed on the lower leaves and spreads to the upper leaves. Other symptoms include damping-off, collar rot, stem cankers, leaf blight, and fruit rot.

4.2.2 Early blight disease management

The following cultural practices that promote a healthy crop and therefore hinder early blight disease establishment include: three year crop rotations with non-susceptible crops; removing volunteer crops and keeping the field weed free; planting resistant/tolerant varieties; removing plant debris or burying it in the soil; irrigating in the morning so that the plant have enough time to dry; keeping the plants healthy so that they are less susceptible to disease; having proper spacing between the plants and rows to provide for good air circulation; using certified disease-free tomato seed and transplants; planting potatoes away from previous season potato fields; avoiding bruising and wounding of tubers during harvesting.

4.2.3 Fungicides use in management of early blight

On potatoes it is recommended to apply protectant fungicides at beginning of flowering or at the earliest symptoms of early blight. On tomatoes fungicide application is recommended soon after transplanting or two to three weeks after emergence. In Table 4, the recommended fungicides used in early blight disease control are summarized.

4.3 Black scurf disease of potato

On underground stems and stolons the disease appears as brown to black sunken lesions that cause the plants to look weak. These lesions may girdle the stolons and cut them off from the rest of the plant. Lesions that girdle the main stem cause the leaves to turn purplish or yellowish and curl upwards. Other symptoms include formation of aerial tubers and formation of whitish mold on the stems at the soil line. On tubers the disease causes tubers to crack or get deformed. Overwintering structures formed on surface of tubers appears as dark masses or as netted residues.

4.3.1 Fungal causal agent of black scurf disease

Black scurf of potatoes is cause by a fungal pathogen, *Rhizoctonia solani* Kuhn. The fungus overwinters in the soil on plant debris or inform of sclerotia. Sclerotia may also survive on tubers. Initial infection occurs when sclerotia germinate to infect stem and sprouts. Tubers are most susceptible to infection when left in the soil after the vines die. Infection is favored by cool (12-16°C) moist soils.

Mode of Action	Target site and code	Group name	Chemical group	Common name	FRAC code
Respiration	complex III: cytochrome bc1 (ubiquinol oxidase) at Qo site *(cyt b gene)*	**QoI**-fungicides (**Q**uinone **o**utside **I**nhibitors	Methoxy-acrylates	azoxystrobin,	11
Respiration	complex III: cytochrome bc1 (ubiquinol oxidase) at Qo site *(cyt b gene)*	**QoI**-fungicides (**Q**uinone **o**utside Inhibitors	Methoxy-carbamates	pyraclostrobin	11
Multi-site contact activity	Multi-site contact activity	Chloronitriles	Chloronitriles	Chlorothalonil (Daconil, Bravo, Echo, Fungonil)	M5
Multi-site contact activity	Multi-site contact activity	Inorganic	Inorganic	Copper (Bordeaux Mixture, Kocide, Tenn-Cop,Liqui-cop, Basicop, Camelot)	M1
Multi-site contact activity	Multi-site contact activity	Dithio carbamates and relatives	Dithio carbamates and relatives	mancozeb maneb Ziram	M3
not classified	unknown	diverse	diverse	Mineral oils, organic oils,potassium bicarbonate (Armicarb 100, Firststep), hydrogen dioxide (Oxidate) material of biological origin (*Bacillus subtilis*).	NC

Table 4. Fungicides for early blight control in tomato

4.3.2 Disease management

Recommended cultural practices in management of black scurf of potatoes include planting certified disease free seed, planting in warm soils (16°C); warming the seed before planting; rotation with non-host plants such as grasses; avoiding field with a history of disease because the fungal population builds in the soil when potatoes are grown in the same field.

5. Acknowledgement

We gratefully acknowledge data availability by colleagues from Research units and Extension services from the University of Maine, Idaho, Pennsylvania State University,

University of Illinois extension, and Cornell University extension services. In addition, we wish to sincerely apologize to colleagues whose data are not here acknowledged. We wish to thank anonymous reviewers of the manuscript for their valuable and useful comments and suggestions.

6. References

Ammermann, E., Lorenz, G., Schelberger, K., Wenderoth, B., Sauter, H. & Rentzea C. (1992). BAS 490F: A broad spectrum-fungicide with a new mode of action. In Proceedings of the Brighton Crop Protection Conference-Pest and Diseases. pp. 403-410. Eds BCPC. Surrey, UK: BCPC Publications.

Behe, B.K., Williams, J.D., Cobb, P., Hagan, A.K. & Stritikus G. (1993). Growing roses. Alabama Cooperative Extension Service, Circular ANR-157.

Bowen, K.L. & Roark, R.S. (2001). Management of black spot of rose with winter fungicide treatment. *Plant Disease* 85: 393-398.

Ebeling, M., Heimann, K.-G., Schoefer, S. & Sonder K. (2003). The human and environmental safety aspects of trifloxystrobin. *Pflanzenschutz-Nachrichten Bayer* 56: 231-245.

Ellis, S.D., Boehm, M.J. & Mitchell, T.K. (2008). Fungal and fungal-like diseases of plants: Ohio State University Extension.

Gachomo, E.W. & Kotchoni, S.O. (2007). Detailed description of developmental growth stages of *Diplocarpon rosae* Wolf: a core building block for efficient disease management. *Annals of Applied Biology* 151: 233-243.

Gachomo, E.W. (2005). Study of the Life cycle of Diplocarpon rosae Wolf and the Effects of Fungicides on Pathogenesis. Goettingen, Germany: Cuvillier Verlag.

Gachomo, E.W., Dehne, H.-W. & Steiner, U. (2006). Microscopic evidence for the hemibiotrophic nature of Diplocarpon rosae, cause of black spot disease of rose. *Physiological and Molecular Plant Pathology* 69: 86-92.

Gachomo, E.W., Dehne, H-W. & Steiner, U. (2009). Efficacy of triazoles and strobilurins in controlling black spot disease of roses caused by *Diplocarpon rosae. Annals of Applied Biology* 154: 259-267.

Gachomo, E.W., Manfredo, J., Seufferheld, M.J. & Kotchoni S.O. (2010). Melanization of appressoria is critical for the pathogenicity of *Diplocarpon rosae. Molecular Bioliology Reports* 37: 3583-3591.

Haverkort, A.J., Boonekamp, P.M., Hutten, R., Jacobsen, E., Lotz, L.A.P., Kessel, G.J.T., Visser, R.G.F. & van der Vossen, E.A.G. (2008). Societal costs of late blight in potato and prospects of durable resistance through cisgenic modification. *Potato Research* 51: 47-57.

Killian, M. & Steiner, U. (2003). Bactericides and fungicides. In Encyclopaedia of Rose Science, pp. 190-198. Eds A.V. Roberts, T. Debener and S. Gudin. Amsterdam, the Netherlands: Elsevier Academic Press.

Kuck, K.H., Scheinpflug, H. & Pontzen, R. (1996). DMI fungicides. In Modern Selective Fungicides; Properties, Applications, Mechanisms of Action. 2nd revised and enlarged edn, pp. 205–258. Ed. H. Lyr. Stuttgart, Germany: Gustav Fisher Verlag.

Margot, P., Huggenberger, F., Amrein, J. & Weiss B. (1998). CGA279202: a new broad-spectrum strobilurin fungicide. Proceedings of the Brighton Crop Protection Conference-Pest and Diseases, 2, 375-382.

Reddy, S., Spencer, J.A. & Newman S.E. (1992). Leaflet surfaces of blackspot-resistant and susceptible roses and their reactions to fungal invasion. *HortScience* 27: 133-135.

Reuveni, M. (2001). Activity of trifloxystrobin against powdery and downy mildew diseases of grapevines. *Canadian Journal of Pathology* 23: 52-59.

Stark-Urnau, M., Gold, R., Guggenheim, R. & Dueggelin M. (1997). Sensitivity of different mildew and rust fungi against kresoxim-methyl. In Proceedings of the 9th European Mediterranean Cereal Rusts Powdery Mildews Conference, pp. 268-271. Eds G.H.J. Kema, R.E. Niks & R.A. Daamen. Wageningen, The Netherlands: Research Institute for Plant Protection.

Walker, S., Mandegaran, Z. & Roberts A.M. (1995). Screening roses for resistance to *Diplocarpon rosae*. *Acta Horticulturae* 424: 209-213.

Wojdyla, A.D. & Orlikowski, L.B. (1999). Strobilurin compounds in the control of rust, powdery mildew and blackspot on some ornamental plants. *International Symposium on Crop Protection (Gent)* 64: 539-545.

Ypema, H.L. & Gold R.E. (1999). Kresoxim-methyl: modification of a naturally occurring compound to produce a new fungicide. *Plant Disease* 83: 4-19.

9

Target-Site-Specific Screening System for Antifungal Compounds

Consuelo Corrales-Maldonado,
Miguel Ángel Martínez-Téllez and Irasema Vargas-Arispuro
Centro de Investigación en Alimentación y Desarrollo, A.C.
Mexico

1. Introduction

The rapid emergence of fungicide resistance has brought a strong demand for crop protection agents with a new mode of action. One of the challenges for modern plant pathology research is the discovery of new modes of action that provide improved activity of fungicides against commercially important target, combined with assures environmental and public safety, is a critical step in safeguarding food security. This chapter reviews biochemical and molecular biological approaches that have revealed new insights into fungal growth and morphogenesis and offer potential target sites for the development of new fungicides. It also discusses prospects for exploiting these modern technologies for the development of fungicides.

Agriculture will always face crop losses caused by microorganisms. Since bible the first pages talk about pest, nowadays, in the XXI century must be added to pest, the losses caused by the effect of global climate change (Gustafson, 2011). Particularly, fungal plant pathogen that comprises an important group of microorganisms that causes significant economic losses in agriculture around the world. They are able to infect any tissue at any stage of plant growth (Garrido *et al.*, 2010). Control of plant diseases typically depends upon the application of chemical fungicides, despite their potentially toxic effects on non-target organisms and the environment (Santos *et al,* 2008; Ferrer-Alcón, *et al,* 2009). Although effective, their extensive use for several decades has disrupted biological control by natural enemies and has led to new pathogen races that are resistant to fungicides (Fernandez-Acero *et al.,* 2006). In spite of the incredible amount of biological information about fungal plant pathogens, there is a scarce commercial fungicide developed from a new knowledge approach. The absence of fungicides that are capable of acting in more than one site of action is a direct consequence of resistance to fungicides, which is common among currently used agrochemicals (Brent & Hollomon, 2007). For example, chlorhexidine, quaternary ammonium compounds, organic acids, esters and alcohols that acts by interfering with the structure or permeability of the cell membrane, subsequently altering its barrier function (Russell, 2003). Pyribencarb cause an inhibition of the electron transport system in fungi, it has been suggested that pyribencarb inhibited succinate-cytocrome C reductase in *Botrytis cinerea* and *Corynespora cassicola* and decylubiquinol-cytocrome C reductase in *B. cinerea* in the same way as strobilurin fungicides . Benzimidazole fungicides, such as benomyl, act through specific binding of the β-tubulin subunit of fungal tubulin, which consequently interferes with microtubules assembly, which in turn is essential for numerous cellular processes, such as mitosis and cytoskeleton formation. Metal ions such copper and silver have been proposed to interact strongly with thiol groups in fungal enzymes and proteins. The inhibitory activity of these compounds may be caused by enzyme damage through binding to key functional groups, particularly sulphydril groups in plasma membrane and cytosol. Flucytosine (pyrimidine analog) the sites of action are nucleic acids, this agent is taken up by fungal cells via the enzyme cytosine permease. Biocides exhibit a multiplicity of antifungal mechanisms. The knowledge of their mechanism of action, combined with an understanding of quantitative structure activity relationships, provides an important platform from which novel biocides may emerge, offering enhanced activity and environmental acceptability (Fernandez- Acero *et al.,* 2011).

Nowadays, new approaches based on graph-theoretical descriptors have emerged as powerful tools for the design of bioactive agents (Marrero -Ponce et al., 2008). The purpose of these approaches is to perform a massive screening of databases of heterogeneous series of compounds and to extract as much structural information as possible at different levels of chemical diversity. So, the use of methodologies and promising approaches may enable the discovery and identification of new candidates as potential fungicides. The new agrochemicals that can be designed will have a wide range of action against different species. Also, they will be able to act by different mechanisms of action and thus avoid the problems of cross-resistance (Speck-Planche *et al,* 2011).

2. The use of genomics, proteomics and bioinformatics in fungicide design

During decades, prior to the development of functional genomics, target discovery was relied on the "observation based" approach. That is, the target strategy involved screening of large numbers of small molecules against particular and desired phenotypes. From this approach, libraries of compounds were constructed with biologically derived or chemically synthesized agents who were used in a systematic manner. However, the result of this approach produced a low number of drugs (Ferrer-Alcon *et al.,* 2009, Steffens *et al.,* 1996). It can be stated that the pace of natural product research and the level of global interest in the particular area of our environment as risen dramatically in the past few years. This period is projected to continue for the future as the interface between biology and chemistry becomes even more blurred and public demand rises for the cost effective medications and biological agents from sustainable resources. The research approach should focus on how to discover novel plant derived natural products through molecular docking as new lead compounds for potential agents and to modify these compounds to find still more potent agents with focus being on the application of homology modeling (Singh & Sharma, 2011). Significant improvements in the era of genomics and proteomics and concurrent progresses in bioinformatics techniques, have given rise to the expectation that the three dimensional structure or reliable homology modeling of target proteins can be achieve in a reasonably short time. Traditionally, the medical plants have provided lead for antifungal compound. Most of fungicides available today were discovered from the screening of synthetic or natural product libraries. Natural products, either as pure compounds or as standardized plant extracts, provide unlimited opportunities for new fungicides leads because of the unmatched availability of chemical diversity. Newer molecular structures as isolated from natural products may be suitably modified to obtain designer molecules for fungicide. Pharmacological testing, modifying, derivatising and research on these natural products represent a good strategy for discovering and developing new fungicides. The combinatorial chemistry has helped in the development of a series of similar but homologous structural compounds for testing (Singh & Sharma, 2011).

At present, structure based-drug design and ligand-based drug design are two great strategies that can be applied for the discovery and/or development of new fungicides. Structure based-drug design relies on a knowledge of the three dimensional structure of the biological receptor, obtained through experimental methods such as X-ray crystallography or NMR spectroscopy. When the experimental structure of a target is not available, it may be possible to create a homology model of the biological receptor on the basis of the experimental structure of another known material (mostly a related protein). The use of various tools like automated computational procedures has provided a means of suggesting new drug candidates and optimizing time and resources. Sometimes the information about the three dimensional structure of the receptor is not available. In this sense, ligand-based drug design is focused on the knowleage of other molecules can be used to derive the minimum necessary structural characteristics that a molecule must present in order to bind to the receptor. Ligand-based drugs design can be applied in cases where the structure of the receptors is uknown but a series of compouds have been identified that exert the fungicide activity. It is necessary to have several compounds structurally similar with high activity, with no activity and with a range of intermediate activities. These other compounds that bind to the biological receptor of interest provides us information the minimum necessary structural characteristics that a molecule must present in order to bind to the receptor (Speck-Planche *et al.,* 2011). Both strategies of drug discovery can be extended to and applied in the design of more effective agrochemicals, and specifically fungicides. Example of applying new technologies towards the rational design of fungicides to control phytopathogenic fungi of commercial crops was used by Fernández-Acero *et al,* (2006). They found substrates with antifungal properties against oomycetes, they screened compounds analogous to various phytoalexins and to flavanes derivatives which display antifungal activity against *Phytophthora* fungi.

The use of bioinformatics techniques to biological systems was demonstrated in the Structural Proteomics In Europe (SPINE) project, which was established to develop new methods and technology for high throughput structural biology. Developments covers target selection, target registration, wet and dry laboratory data management and structure annotation as they pertain to high throughput study (Albeck *et al.,* 2006). How this program, there are now many databases which is constantly being updated with the latest data of groups to seek new targets, new fungicides and relevant information like new virulence factors of some fungi, some of these pages are: (www.broadinstitute.org/science/projects/fungal-genomeinitiative, http://cogeme.ex.ac.uk. www.phi-base.org, http://www.expasy.org/ and http://bioinformatics.charite.de/supernatural2/.

In the post-genomic era, new terms related with chemical "-omics" have appeared. The term "genetic chemical" describes the use of small molecules to selectively perturb gene function. When this concept is applied on a genome-wide scale it is named "chemogenomics". The application of chemogenomics to protein targets is named "chemoproteomics"; although a more explicit definition is TRAP (targeted related affinity profiling) defined as the use of biology to inform chemistry (Xu *et al.*, 2007). The accumulation of proteomic information of fungal plant pathogens may be an incentive to the development of new and environmentally friendly fungicides. Particularly, Proteomics is another is a high-throughput technology that allows an in depth study of the sets of proteins synthesized in a specific sample at any specific moment. By protein profile comparison between samples, the proteins involved in specific biological processes may be revealed. One of the most interesting applications of the proteomics is its use in discovering new protein targets for drug design including fungicides (Fernandez-Acero *et al.*, 2010; Ferrer-Alarcon *et al.*; 2009). It involves the identification and early validation of disease-associated targets.

The accumulation of information over the last decades, relating to a) fungal molecular genetic data, b) pathogenicity/virulence factors and c) proteomic approaches, has led to the appearance of several web-accessible databases which contribute to the fungal scientific community's development in this field. More than 50 genomes of pathogenic fungi are published in the Broad Institute Database for public perusal (www.broadinstitute.org/science/projects/fungal-genomeinitiative); and further data in the Phytopathogenic Fungi and Oomycete EST Database, COGEME, http://cogeme.ex.ac.uk/). In spite of the incredible amount of biological information about fungal plant pathogens, there is no commercial fungicide developed from a molecular approach.

3. Pathogenesis

As fungal pathogens have an enormous impact on plant production worldwide, the strategies they use to infect plants and to cause disease are a topic of great interest (Van De Wouw & Howlett, 2011). Knowledge of the pathogenicity/virulence factors essential for fungal infections is very important because it represent the targets that researches must attack in the fight against these pathogens (Fernandez-Acero *et al.*, 2011).

We define pathogenicity gene as those necessary for disease development but not essential for pathogen to complete its lifecycle *in vitro*. Pathogenicity genes are of interest not only to increase our overall knowledge of disease process, but also because any such gene could became a target for disease control.

The types of genes essential for pathogenesis depend on the infection process of a particular fungus. Some fungi degrade the cuticle and cell wall to enter the plant; others form specialized structures, such as appressoria, to penetrate the epidermis, while others enter the host through wounds or natural openings (Idnurm & Howlett, 2001). Once the fungus has colonized plant, it may grow obtaining nutrients from its host without killing cells (as a biotroph); some fungi produce specialized infection structures (e.g. haustoria) during biotrophic stage. Other fungi species act killing host cells with the use of toxins (as necrotrophs) or act as hemibiotroph (biotroph and necrotroph at different stages if infection). Toxins often are major components of the arsenal or virulence determinants by necrotrophic fungi. They can be host specific or non host specific, and they kill or disable functions of host cells (Van De Wouw & Howlett, 2011).

A number of fungal mechanisms and molecules have been shown to contribute to fungal pathogenicity or virulence, understood as the capacity to cause damage in a host, in absolute or relative terms. Among them, cell wall degrading proteins, inhibitory proteins, and enzymes involved in the synthesis of toxins are included. These virulence factors are typically involved in evolutionary arms races between plants and pathogens (Gonzalez Fernandez *et al.*, 2010).

Knowledge of the pathogenic cycle and virulence factors of the fungus is crucial for designing effective crop protection strategies, including the development of resistant plant genotypes through classical plant breeding or genetic engineering, fungicides or the use of biological control strategies (Gonzalez-Fernandez *et al.*, 2010). The determination of a specific factor as virulence or pathogenicity has been achieved by constructing defective mutants in the specific genes. The infection power of the analyzed mutants should at least decrease or disappear compared to the wild type, if the deflections of these genes in mutants produce a loss of vegetative lesion, it is logical to assume that the inhibition of this enzyme or set of enzymes by targeted strategies, should produce new fungicides. In this context, the use of natural products or related compounds as specific enzymes inhibitors is an archetype, as they would be species specific and the environmental impact would be reduced to a minimum (Fernandez-Acero *et al.*, 2011).

A diversity of fungi, oomycetes secrete proteins and other molecules to different cellular compartments of their hosts to modulate plant defense circuitry and enable parasitic colonization, these molecules have been called "effectors". The usage of the term "effector" became popular in the field of plant-microbe interactions with the discovery that plant pathogenic gram-negative bacteria utilize a specialized machinery to deliver proteins inside host cells. More recently, a broader range of plant microbiologists have adopted the term effector and its associated concepts (Abramovitch et al., 2006). This term is now also routinely used in the fungal and oomycete literature and is becoming increasingly popular in nematology to describe secreted proteins that exert some effect on plant cells (Hogenhout et al., 2009).

Some effectors are avirulence proteins and have a 'gene-for gene' relationship with resistance proteins in the host. When a fungal avirulence gene is mutated, hosts with the corresponding resistance gene no longer detect the pathogen; this leads to a compatible interaction. Host-specific proteinaceous toxins that have an 'inverse' gene-for-gene relationship with the host, whereby the interaction leads to disease such genes would be classified as pathogenicity genes (Oliver & Solomon, 2010). Small proteins encoded by fungal genes involved at various stage of infection, alter host cell structure and function facilitate infection. These proteins are often cysteine rich (hogenhout *et al.*, 2009).

Fungi use signaling cascades to respond to changes in the environment by altering their gene expression. The interruption of these signaling genes results in the loss and/ or reduction in pathogenicity, as well as pleiotrophic effects on cellular processes, including mating, conidiation, growth rate and toxin production. Therefore, it is difficult to determine which aspect of fungal physiology is responsible for the loss of pathogenicity. The components of these signal transduction cascades may represent targets for the development of fungicides (Van De Wouw & Howlett, 2011). *Phytophtora infestans,* one of the most destructive pathogen of potato in the history, have a remarkable speed of adaptation to control strategies such as genetically resistant cultivars, comparison with two other *Phytophthora* genomes showed rapid turnover and extensive expansion of specific families of secreted disease effector proteins, including many genes that are induced during infection stage. These fast-evolving effector genes are localized to highly dynamic and expanded regions of the *P. infestans* genome. This probably plays a crucial part in the rapid adaptability of the pathogen to host plants and underpins its evolutionary potential (Haas *et al.*, 2009). Other example of pathogen trainer of virulence factors is *Botrytis cinerea* to date have been identified a wide range of compounds and enzymes that uses the fungus to exerts its pathogenicity, thanks to the arsenal of degrading enzymes, *B. cinerea* is able to feed on different plant tissues, this fungus shares conserved virulence factors with other phytopathogens (Choquer *et al.*, 2007). *Botrytis cinerea* is a ascomycete necrothrofic, this fungi alone is responsible for 10% of the global fungicide market (Fernandez-Acero *et al.*, 2011), is thought to enter the host mainly by producing degrading enzymes and causing an oxidative burst, specifically because secretes nonspecific phytotoxins to kill cells from a large spectrum of plants. Among the numerous metabolites isolated from fermentation broths, the most well known is the sesquiterpene botrydial (Deighton *et al.*, 2001). This fungus is notably equipped with multiple cell wall-degrading enzymes that allow plant tissue colonization and the release of carbohydrates for consumption. Pectin, the major host cell wall component, can be degraded by a set of fungal pectinases (Ten Have *et al*, 2001; Choquer *et al*, 2007). Xilanase (Xyn11A) from *Botrytis cinerea* contributes to the infection process with the necrotizing and not with the xylan hydrolyzing activity. The main contribution of the xylanase Xyn11A to the infection process of *B. cinerea* is to induce necrosis of the infected plant tissue. A conserved 30-amino acids region on the enzyme surface, away from the xylanase active site, is responsible for this effect and mediates binding to plant cells (Noda *et al*, 2010).

New technologies like proteomics are very good tools for obtain information about proteins secreted by pathogenic fungi. Nowadays, there is lot of candidate pathogenicity genes, but there is stagnation for their functional analysis, since experiments are time-consuming and difficult to realize for some fungi. However, genomic tools are already providing a much more integrated picture of pathogenicity mechanisms, compared with the previous focus on individual genes. Many fungal genes affecting disease progression are involved in growth and development, and there are few genes for which the only effect is on disease, proteomics allowed to identify between 10 and 100 different biological functions from each gene (Van De Wouw & Howlett, 2011; Fernandez-Acero *et al.*, 2011). This is an important progress in the new fungicide discovery because pathogenic genes with function in growth and development could be target site for new fungicides.

4. Morphogenesis

Since fungi are a eukaryotic organism, they have diverse metabolic profiles similar to many mammalian and plant. Hence, several antifungal agents discovered to be potentially active against plant pathogenic fungi have failed to survive during the testing process because the target site of the fungicide is found in another organism (Thines *et al.*, 2004).

The most successful fungicide in the market today acts relatively broadly by targeting fungal vegetative growth and thus the entire fungal life cycle. Examples of a successful mode of action classes interfering with biochemical progresses essential in fungi include compounds targeting respiration and sterol biosynthesis. An example of the former mode of action class is given by compounds which target mitochondrial electron transport within the respiration chain. Fungicides of the strobilurin class have this mode of action. These compounds are structurally based on natural products and interfere with the ubiquinol cytochrome C oxidoreductase (Foster & Thines, 2009).

There are numerous antifungal compounds that were discovered from diverse microbial sources using traditional activity-based screening techniques. These microbial compounds showed potent control efficacy against various plant diseases, including chronic diseases which are difficult to control with conventional synthetic fungicides. Advances in screening systems directed to specific targets of fungal metabolism have increased the opportunities to discover novel antifungal agents with selectivity over non-target organisms. Microbial metabolites have also been exploited as a source for non-fungicidal disease control agents that do not inhibit vegetative hyphal growth, but rather interfere specifically with the infection process of pathogenic fungi, such as spore germination, the formation of penetration structures and sporulation (Seok & Kook, 2007). Infection structures of phytopathogenic fungi are modified hyphae specialized for the invasion of plant tissue, initial events are adhesion to the cuticule and directed growth of the germ tube on the plant surface (Mendgen *et al.,* 1996). The specificity of microbial fungicides is a highly preferred characteristic in terms of impacting the environment, where it is closely related to the occurrence of fungicide resistance. The most recently developed fungicides from microbial metabolites, the strobilurins, provide a cue for the high risk of resistance development of site-specific fungicides.

These compounds have long been applied to control fungal diseases of rice, vegetables and fruits, and their effectiveness in controlling these diseases has been tested in the field and proven over many years. The importance of microbial fungicides, compared to synthetic compounds, may have been under evaluated in the past due to the limitation in their activity spectrum and in certain instances, the development of resistance (Knight *et al.,* 1997). Nevertheless, the excellent fungicidal activity of these microbial metabolites and their potential as lead candidates for further fungicide development continue to stimulate research and screening for antifungal microbial metabolites, some examples of this compounds are:

Kasugamycin is an amino-sugar compound discovered from the metabolites of *Streptomyces kasugaensis, in vivo* studies have shown that kasugamycin efficiently suppresses the development of *M. grisea* mycelia on rice plants in both preventive and curative treatments, to overcome potential resistance problems, mixtures of kasugamycin with different synthetic fungicides having different modes of action are currently in use.

Polyoxins are peptidylpyrimidine nucleoside antibiotics isolated from the culture broth of *Streptomyces cacaoi* var. *asoensis,* such excellent characteristics come from the fact that polyoxins selectively inhibit the synthesis of cell-wall chitin in sensitive fungi but have no adverse effects on organisms lacking chitinous cell walls.

Validamycin A produced by *Streptomyces hygroscopicus* var. *limoneus* has been effective in controlling rice sheath blight caused by *R. solani.* Validamycin A is converted within fungal cells to validoxylamine A, an extremely strong inhibitor of trehalose. The mode of action of validamycin A is favorable for biological selectivity, because vertebrates do not depend on the hydrolysis of trehalose (Doumbou *et al.,* 2001).

Promising microbial metabolites continue to be discovered using traditional activity-based screening procedures against various plant pathogenic fungi. In particular, many of the site-specific antifungal metabolites have recently been discovered from microbial metabolites. These microbial metabolites include non-fungicidal compounds that interfere with the infection process of pathogenic fungi, and specific inhibitors of the fungal biosynthetic pathways for chitin, fatty acids and nucleic acids (Seok & Kook, 2007).

Other kind of antifungal agents are proteins; antifungal proteins have been isolated from various organisms ranging from bacteria, plants, insects and amphibians to human beings. Both their fungal target site and their mode of action are extremely diverse. In order for it to be applied, an antifungal protein needs to fulfill several prerequisites such as antifungal activity *in vivo* and lack of effects on the host cells. Furthermore, resistance mechanisms need to be excluded as far as possible. Therefore, investigation of the target site and the mode of action of an antifungal protein should reveal whether the protein is suitable for an application (Theis *et al.,* 2005). The antifungal protein (AFP) are abundantly secreted by the filamentous fungus *Aspergillus giganteus,* this cysteine-rich protein have ability to disturb the integrity of fungal cell walls and plasma membranes but does not interfere with the viability of other eukaryotic systems (Barakat *et al.,* 2010; Meyer, 2008).

Severe membrane alterations in *A. niger* were observed, whereas the membrane of *P. chrysogenum* was not affected after treatment with AFP. The protein localized predominantly to a cell wall attached outer layer which is probably composed of glycoproteins, as well as to the cell wall of *A. niger.* It was found to accumulate within defined areas of the cell wall, pointing towards a specific interaction of AFP with cell wall components. In contrast, very little protein was bound to the outer layer and cell wall of *P. chrysogenum.* The protein was found to act in a dose-dependent manner: it was fungistatic when applied at concentrations below the minimal inhibitory concentration, but fungicidal at higher concentrations. Using an *in vivo* model system was demonstrated that AFP indeed prevented the infection of tomato roots (*Lycopersicon esculentum*) by the plant-pathogenic fungus *Fusarium oxysporum* f. sp. *Lycopersici* (Theis *et al.,* 2005).

The fungal profilins, small actin-binding proteins that share limited homology to human profilin, can operate as a potential drug target, since these proteins are essential for the growth of most eukaryotic cells, including *S. cerevisiae* (Witke, 2004). Addition, the existence of structural information can support the design of structure-based ligands for profilins. Peptides are generally used as lead compounds in drug development, to design a novel peptide ligand, an *in vitro* evolution approach has often been used. Although this approach can be used without three-dimensional (3D) structural information about the target protein, it requires laborious experimental procedures, including library constructions and the screening of bioactive peptide ligands. In this respect, if information on the structure and the active site of the target protein is known, an *in silico* approach based on the 3D structure of the target protein is a useful approach to designing the peptide ligand. The validity of the profilin as antifungal drug target was evaluated by Ueno *et al.* 2010, amino acid alignments showed the low homology between human and fungal profilins. This implies that the fungal profilin could be a target with high selectivity. Furthermore, a mouse infection study showed that the suppression of profilin expression attenuated the fungal burden in the kidney and indicated that the profilin was required for survival in the host's body (Ueno *et al.,* 2010).

A major challenge for drug development today is how best to employ the new genome wide technologies to identify which metabolic pathways and gene products are critical for disease establishment and progression and thus to increase the probability of finding novel fungicide targets. Novel targets can be located either in biochemical or signaling pathways essential for vegetative growth or for pathogenic development of the fungus. Targets with high specificity can thereby be expected in pathways involved in adhesion, host-recognition/pre-penetration processes, host colonization and the final reproductive differentiation processes during pathogenic development. An attractive proposition is the development of fungicides interfering with pathogenic development but not with vegetative growth. Such a strategy could prevent or cure infections by fungal pathogens without affecting neutral or benign species (Foster & Thines, 2009).

5. Conclusion

We would like to conclude by stating that antifungal targets-site are extremely diverse. However, substances that acts on these target-sites needs to fulfill several prerequisites such as antifungal activity *in vivo* and lack of effects on the host cells. Furthermore, resistance mechanisms need to be excluded as far as possible. Therefore, investigation of the target site and the mode of action of an antifungal compound can be explored by statistical learning algorithms. Performance and applicability of the statistical learning methods in studying "fungal-target likeness" may be further improved by incorporation of new information from advances in genomic, proteomics, pathogenesis and morphogenesis studies. Efficiency and accuracy of statistical learning methods in the prediction of fungal-target like proteins can also be enhanced from new progress in learning algorithms and sequence descriptors.

6. References

Abramovitch, R. B., Anderson, J. C. & Martin, G. B. (2006). Bacterial elicitation and evasion of plant innate immunity. *Nature Reviews Molecular Cell Biology*. 7:601-611.

Albeck, S., Alzari, P., Andreini, C. , Banci, L., Berry, M., Bertini, I.,Cambillau, C., Canard, B., Carter, L., Cohen, S. X., Diprose, J. M., Dym, O., Esnouf, R. M., Felder, C., Ferron, F., Guillemot, F., Hamer, R., Ben Jelloul, M., Laskowski, R. A., Laurent, T., Longhi, S. Lopez, R., Luchinat, C., Malet, H., Mochel, T., Morris, R. J., Moulinier, L., Oinn, T.,Pajon, A., Peleg, Y., Perrakis, A., Poch, O., Prilusky, J., Rachedi, A., Ripp, R., Rosato, A., Silman, I., Stuart, D. I., Sussman, J. L., Thierry, J.C., Thompson, J. D.,Thornton, J. M., Unger, T.,Vaughan, B., Vranken, W., Watson, J. D., Whamondg, G. & Henrickg K. (2006). SPINE bioinformatics and data management aspects of high throughput structural biology. *Acta Chrystallographic*. D62:1184-1195.

Barakat, H. Spielvogel, A., Hassan, M., El-Desouky A., El-Mansy H., Rath, F. Meyer, V. & Stahl, U. (2010). The antifungal protein AFP from *Aspergillus giganteus* prevents secondary growth of different *Fusarium* species on barley. *Applied Microbiology and Biotechnology*. 87:617–624.

Brent, K.J. and Hollomon, D.W. *Fungicide Resistance in Crop Pathogens: How Can it be Managed?*. Manuscript, Fungicide Resistance Action Committee, CROPLIFE INTERNATIONAL, Brussels, Belgium (2007). Available from http://www. frac.info/frac/publication/publication.htm. Reviewed on June 24, 2011.

Choquer, M., Fournier, E., Kunz, C., Levis, C., Pradier, J., Simon, A. & Viaud, M. (2007). *Botrytis cinerea* virulence factors: new insights into a necrotrophic and polyphageous pathogen. Minireview. Journal compilation. Federation of European Microbiological Societies, *Microbiological letters.* 277:1-10.

Deighton, N., Muckenschnabel, I., Colmenares, A.J., Collado, I.G. & Williamson, B. (2001). Botrydial is produced in plant tissues infected by *Botrytis cinerea. Phytochemistry* 57: 689–692.

Doumbou, C.L., Hamby, S. M., Crawford, D. L. & Beaulieu, C. (2001). Actinomycetes, promising tools to control plant diseases and to promote plant growth. *Phytoprotection.* 82: 85-102.

Fernandez-Acero, F., Carbú, M., El-Akhal, M., Garrido, C., Gonzalez-Rodriguez, V. & Cantoral, J. (2011). Development of proteomics based fungicides: New strategies for environmentally friendly control of fungal plant disease. *International Journal of Molecular Science.* 12:795-816.

Fernández-Acero, F.J., Colby, T., Harzen, A., Carbú, M., Wieneke, U., Cantoral, J.M. & Schmidt, J. (2010). 2-DE proteomic approach to the *Botrytis cinerea* secretome induced with different carbon sources and plant-based elicitors. *Proteomics.* 10:2270–2280.

Fernandez-Acero, F.J., Carbú, M., Garrido, C., Collado, I. G., Cantoral, J. & Vallejo, I. (2006). Screening study of potential lead compounds for natural products based fungicides against *Phytophthora* species. *Journal of phytopathology.* 154, 616-621.

Foster, A. & Thines, E. (2009). Identification of fungicide targets in pathogenic fungi. The Mycota. A comprehensive treatise on fungi as experimental system for basic and applied research. Physiology and Genetics XV. Springer. Germany. Pp.232.

Ferrer-Alcon, M., Arteta, D., Guerrero, M.J., Fernandez-Orth, D., Simon, L. & Martinez, A. (2009). The use of gene array technology and proteomics in the search of new targets of diseases for therapeutics. *Toxicology letters.* 186: 45-51.

Garrido, C., Cantoral, J.M., Carbu, M., Gonzalez-Rodriguez, & V. Fernandez-Acero, F. (2010). New proteomic approaches to plant pathogenic fungi. *Current Proteomics.* 7:306-315

Gonzalez-Fernandez, R., Prats, E. & Jorrın-Novo, J. V. (2010). Proteomics of Plant Pathogenic Fungi. *Journal of Biomedicine and Biotechnology.* 2010;ID 932527, 36 p.

Gustafson, D. I. (2011). Climate change: a crop protection challenge for the twenty-first century. *Pest Management Science.* 67:691-696.

Haas, B. J., Kamoun, S., Zody, M., Jiang, R., Handsaker R., Cano, L., Grabherr M., Kodira, C., Raffaele, S. & Torto-Alalibo, T. (2009). Genome sequence and analysis of the Irish potato famine pathogen Phytophthora infestans. *Nature.* 461:393-398.

Hogenhout, S., Van der Hoorn, R., Terauchi, R. & Kamoun S. (2009). Emerging concepts in effector biology of plant-associated organisms. *Molecular Plant–Microbe Interaction.* 22: 115–122.

Idnurm, A. & Howlett, B. (2001). Phatogenicity genes of phytopathogenic fungi. *Molecular Plant Pathology.* 2:241-255.

Knight, S.C., Anthony, V.M., Brady A. M., Greenland, A.J., Heaney, S.P., Murray, D.C., Powell, K. A., Schulz, C.A., Spinks, C. A., Worthington, P. A. & Youle, D. (1997). Rationale and perspectives on the development of fungicides. *Annual Review of Phytopathology.* 35:349–372.

Marrero-Ponce, Y., Meneses-Marcel, A., Rivera-Borroto, O.M., Garcia-Domenech, R., De Julian-Ortiz, J.V., Montero, A. (2008). Bond-based linear indices in QSAR: computational discovery of novel anti-trichomonal compounds. *Journal of Computer Aided Molecular Desing.* 22:523–540.

Mendgen, K., Hahn, M. & Deising, H. (1996). Morphogenesis and Mechanisms of penetration by plant pathogenic fungi. *Annual Review of Phytopathology.* 34:367-386.

Meyer, V. (2008). A small protein that fights fungi: AFP as a new promising antifungal agent of biotechnological value. *Applied Microbiology and Biotechnology.* 78:17–28.

Noda, J., Brito, N. & Gonzalez, M. (2010). The *Botrytis cinerea* xylanase Xyn11A contributes to virulence with its necrotizing activity, not with its catalytic activity. *BMC Plant Biology.* 10:38-47.

Oliver, R.P. & Solomon, P.S. (2010). New developments in pathogenicity and virulence of necrotrophs. *Current Opinion in Plant Biology.* 13, 1–5.

Russell, D. (2003). Similarities and differences in the responses of microorganisms to biocides. *Journal of Antimicrobial Chemotherapy.* 52: 750–763

Santos, M., Diánez, F., de Cara, M. & Tello, J.C.(2008). Possibilities of the use of vinasses in the control of fungi phytopathogens. *Bioresource Technology.* 99: 9040–9043.

Seok, B. & Kook, B. (2007). Microbial Fungicides in the Control of Plant Diseases. *Journal of Phytopathology.* 155, 641–653

Singh, S. & Sharma, D. (2011). In silico modeling in conjunction with natural products: Paving the way for drug design. *Biotechnology and Molecular Biology Review.* 6:88-93.

Speck-Planche, A., Guilarte-Montero.L., Yera-Bueno. R., Rojas-Vargas, J., Garcia-López, L.A., Uriarte, E. & Molina-Pérez, E. (2011). Rational design of new agrochemicals fungicides using substructural descriptors. *Pest Management Science.* 67:438-445.

Steffens, J. J., Pellt, J.E. & Tien, M. (1996). Mechanisms of fungicide resistence in phytopathogenic fungi. *Current Opinion in Biotechnology.* 7:348-355

Ten Have, A., Breuil, W.O., Wubben, J.P., Visser, J. & Van Kan, J.A. (2001). *Botrytis cinerea* endopolygalacturonase genes are differentially expressed in various plant tissues. *Fungal Genetics and Biology.* 33: 97–105.

Theis, T., Marx, F., Salvenmoser W., Stahl U., & Meyer, V. (2005). New insights into the target site and mode of action of the antifungal protein of *Aspergillus giganteus. Research in Microbiology.* 156: 47–56.

Thines, E., Anke, H. & Weber, W.S. (2004). Fungal secondary metabolites as inhibitors of infection-related morphogenesis in phytopathogenic fungi. *Mycological Research.* 108:14–25.

Ueno, K. Tamura, Y. & Chibana, H. (2010). Target validation and ligand development for a pathogenic fungal profilin, using a knock-down strain of pathogenic yeast *Candida glabrata* and structure-based ligand design. *Yeast*. 27: 369–378.

Van De Wouw, A. & Howlett, B. (2011). Fungal pathogenicity genes in the age of 'omics'. *Molecular Plant Pathology*. 12: 507-514.

Witke, W. (2004). The role of profilin complexes in cell motility and other cellular processes. *Trends Cell Biology*. 14: 461–469.

Xu, H., Xu, H., Lin, M., Wang, W., Li, Z., Huang, J., Chen, Y., Chen, X. (2007). Learning the drug target-likeness of a protein. *Proteomics. 7: 4255–4263.*

10

An Integrated Approach to Control the Black Leaf Streak Disease (BLSD) of Bananas, while Reducing Fungicide Use and Environmental Impact

Jacky Ganry[1], Eric Fouré[2], Luc de Lapeyre de Bellaire[2] and Thierry Lescot[2]
[1]*International Cooperation, Agreenium, CIRAD,*
[2]*CIRAD- Persyst, UPR 'Systèmes de culture bananiers, plantains et ananas', Montpellier, France*

1. Introduction

This chapter aims at describing an integrated approach to control the development of Black Leaf Streak Disease(BLSD) of bananas and plantains plants with a minimum of fungicide applications resulting in lower cost of control, lower risk of fungicide resistance and lower environmental negative impact.

Bananas are one of the main agricultural productions in the world, and the main fruit crop, with an annual production of about 117 Mt (Lescot, 2011). International trade of bananas represents 14Mt·year-1 with a value of more than 4 billion US$ (Loeillet, 2005). The production of bananas for this international trade is grown in a small number of tropical countries mainly located in Latin America (80%), in African and Asian countries, and in the European community (French West Indies, Canary Islands). In all these countries, this industry is economically important and a source of direct and indirect employment. But one of the main problems is that it relies on a very narrow genetic base, the Cavendish sub-group, which is exposing the industry to important pest and disease threats.

Photo credit: CARBAP (J. Ngando)

Photo 1. Banana plants with necrosis of BLSD .

The Banana leaf spot diseases are a major constraint in most large production areas, resulting in very significant losses in the absence of suitable control system (Stover, 1986; Stover & Simmonds, 1987). There are two major impacts of the disease on the production.

Photos credit: CARBAP (J. Ngando)

Photo 2. Very severe attack of BLSD on a banana plant with a bunch

Photos credit: CARBAP (J. Ngando)

Photo 3. Early ripening of the fruits due to severe attack of BLSD

The first one is a reduction of yield due to a negative effect on the photosynthetic assimilation (up to 50 to 100%) and thus on dry matter production. Bunch weight is strongly affected (Ramsey & Daniells, 1990), and significant yield losses (20% to 50%) were observed on plantains (Stover, 1983; Mobambo et al., 1996).

A second one is a reduction of the conservation potential of the fruit, expressed by its "greenlife", which is the time between harvest and ripening, arguing for a direct effect of these foliar diseases on fruit physiology by mechanisms that remain to be discovered. (Abadie et al., 2008; Chillet et al., 2009). The impact on export quality is thus very critical as there are high risks of advanced maturation (Stover, 1972). Therefore banana growers must control BLSD to a level that does not alter fruit exportation.

The economic incidence of BLSD is important since the cost of control can reach more than 25% of the production cost (Stover & Simmonds, 1987) and the BLSD incidence in the banana industry is regularly increasing.

Two species are responsible for the current epidemics, *Mycosphaerella musicola* Leach, causal agent of the Sigatoka Leaf Spot Disease (SLSD) and *Mycosphaerella fijiensis*, causal agent of the Black Leaf Streak Disease (BLSD). The latter is much more virulent than the former resulting in more intense damage and more difficult control. In addition the spectrum of susceptible varieties is much wider. The group of plantains is particularly susceptible to BLSD when it is considered resistant to SLSD, except in special geographical conditions (altitude zones) (Fouré & Lescot, 1988).

BLSD is regarded as the most economically important disease (Jones, 2000) for the banana industry and it is a major constraint for small farmers growing banana and plantains for domestic markets (Marin et al., 2003). It is considered as one of the most important crop diseases in the world (Pennisi, 2010).

Such a situation is mainly related to the fact that the banana trade relies on a unique group of banana cultivars (*Musa acuminata,* AAA, Cavendish subgroup) which are highly susceptible to this disease.

BLSD is present in Latin America since 1972, in Cameroon since 1980 and has appeared in Ivory Coast in 1985. Its expansion in the Caribbean is more recent having started from two independent introductions in the north (Cuba in 1990, and then Jamaica, the Dominican Republic, Haiti and Puerto Rico) and in the Lesser Antilles (through Trinidad in 2003) with a very recent arrival in Martinique (2010).

The life cycle of BLSD has been widely described and starts with leaf infection by either ascospores or conidia (Fig. 1) It is important to highlight the main characteristics of this life cycle that are useful for BLSD control.

After a period of epiphyllic growth of generally 2-3 days, germ tubes penetrate stomata. In good conditions, the first symptoms appear generally 10-14 days after. The symptoms then gradually evolve from stage 1 to stage 6 (Fouré, 1982)

Conidia are produced on young stages of the disease (stage2-4) and have been considered as water dispersed at short distances, even if they are also present in the "airspora". Ascospores are produced at later stage and wind-dispersed after perithecia burst. They are considered to be dispersed at longer distances than conidia.

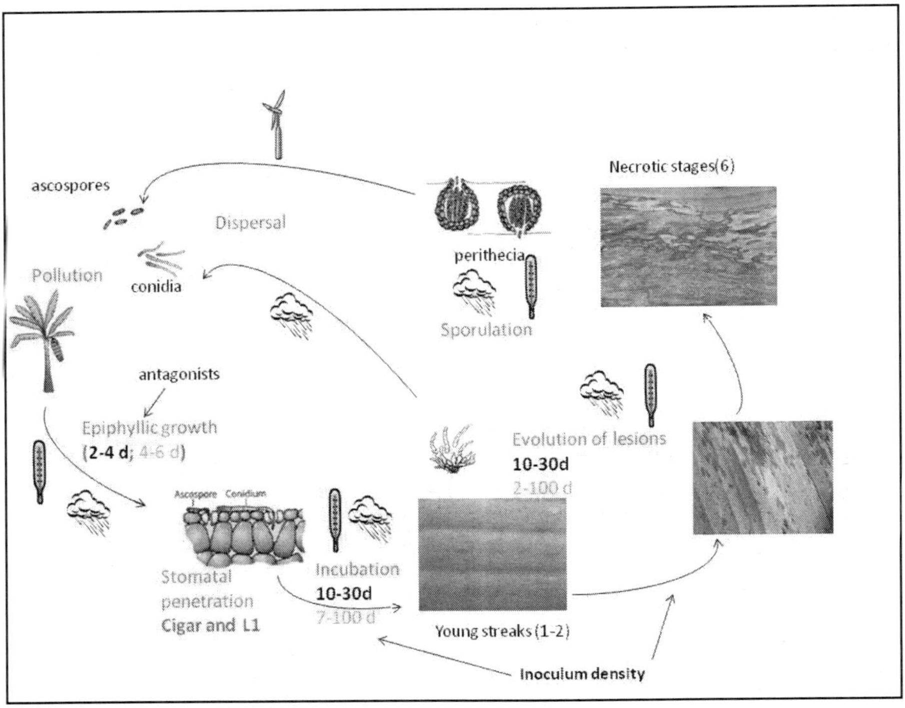

(in yellow: data for the Sigatoka Disease and in black, data for the BLSD)

Fig. 1. The biological cycle of the banana leaf spot diseases

2. Principle and various components of an integrated approach to control BLSD of bananas

The principle and various components of this integrated approach are presented. They are based on a holistic approach including technical aspects and logistical, economic and political aspects as well.

2.1 Historical features

It is useful to look at some historical features to better understand the genesis of this integrated control of BLSD.

An effective chemical control of SLSD, based on the use of mineral oils, was developed by a French team of IFAC in Guadeloupe in the fifties (Guyot &Cuillé, 1954; Guyot &Cuillé, 1955).

The work done at that time quickly showed that the foliar emission rate of the banana plant would impose a very large number of sprays to ensure that each new leaf is protected in time. This should lead to a minimum of forty sprays per year with an additional processing constraint to apply about 2000 liters of aqueous formulation per hectare, which was quite intolerable.

Photo 4. Stages 1 and 2 of the disease (adaxial surface)

Photo 5. Stages 2 and 3 of the disease (adaxial surface)

Photo 6. Stages 4 (adaxial surface)

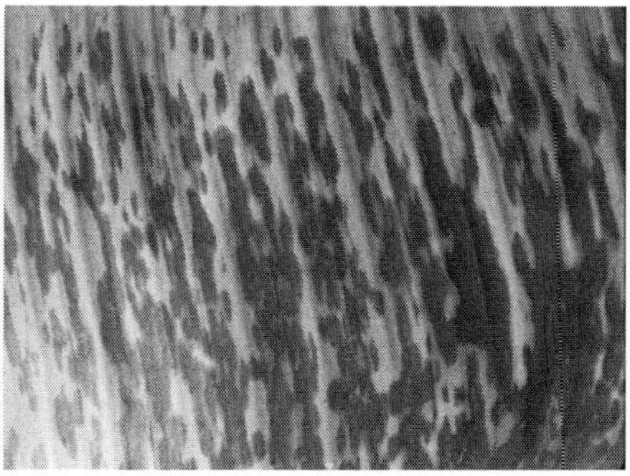

Photo 7. Stages 5 of the disease (abaxial surface)

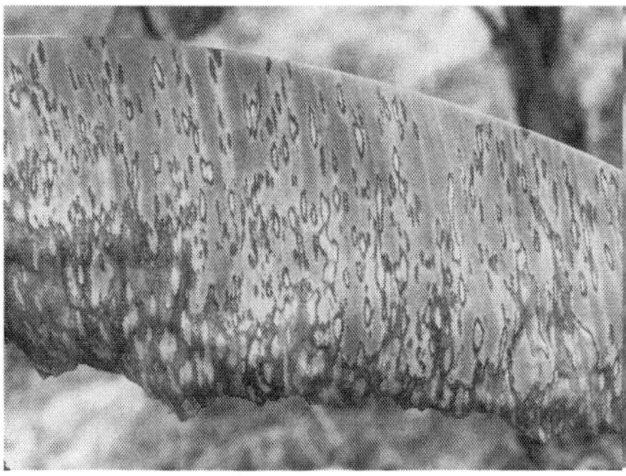

Photo 8. Stage 6 of the disease with perithecia (abaxial surface).

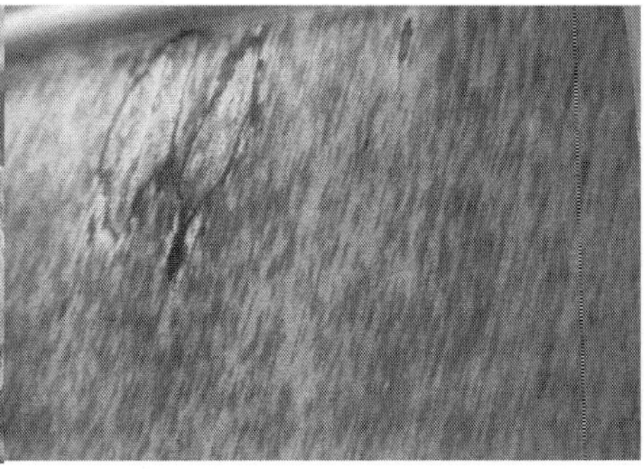

Photo 9. The density of infection is often very important resulting in necrotic patches by coalescence.

Plate 1. **Illustration of BLSD symptoms** (Photo credit: Cirad /E.Fouré & L. de Lapeyre de Bellaire)

The research therefore looked for reducing these constraints i) by studying more closely the biology of the pathogen to intervene only when necessary ii) limiting the volume of mixture applied per hectare and the number of fungicide applications, iii) focusing on the choice of more effective and persistent fungicides.

The first goal resulted in the development of disease observation and warning methods, with the early work conducted by Cuillé & Guyot in 1954.

To reach the second goal, the work achieved made it possible to replace the copper formulations applied through ground sprays at a rate of 2000 liters per hectare by a low volume of mixtures, initially applied with ground sprayers and later by plane or helicopter, to go down to 15 liters per hectare. This development was made possible by the discovery of the qualities of mineral oils which have proven to be excellent vehicles for fungicides, either alone or mixed with water. These mineral oils (i) showed a fungistatic activity on the pathogen, (ii) ensured a good repartition of droplets at leaf surface and a good penetration of fungicides in the leaf tissues, (iii)avoided the rainfall leaching occurring after treatment.

As a matter of fact, before 1972, aerial and ground sprays were currently done only with oil. With the appearance of systemic fungicides in the 1970', new formulations were introduced coupling the fungistatic action of the oil with the strong curative action of currently marketed systemic fungicides (Ganry & Laville, 1983). Fungicides used belong to different chemical families with different mode of actions such as benzimidazoles, triazoles, morpholins and more recently strobilurins (see Box 4), which have been used practically in this chronological order, taking into account the emergence of resistance of the fungus to benzimidazoles and triazoles..

Very soon appeared the need to abandon a strategy of systematic control and move towards less fungicide applications in order to (i) reduce production costs, (ii) reduce environmental impact and (iii) limit the occurrence of fungicide resistant strains. With such an objective of better management of the control strategy, CIRAD (formerly IRFA) has, since the seventies, focused its work on the development of a warning method in order to apply fungicides only when necessary.

The warning system used is based both on climatic data (Box 1) and on field observations of symptoms evolution in early stages (Box 2).

In Guadeloupe and Martinique, since the 70's, the information has been processed on an aggregate basis for each island and managed by a specialized team attached to the banana growers associations. Aerial sprays are made on a similar base and are generalized for the whole of each island.

As a result of the use of such a disease management strategy the control of the SLSD has been very satisfactory in the French Antilles since the seventies (Fig. 1), with a number of applications ranging from five to eight per year, while applications carried out systematically in similar situations would have required at least 10 to 15 treatments per year (Ganry, 2001).

This success story relies on some basic requirements:

- A generalized application over large geo-climatic areas including several growers,
- A very good spraying logistics in order to reduce as short as possible the period between the decision of the fungicide application according to the warning system and the completion of the spray.

- A strong curative effect of fungicide applications resulting from (i) a good quality of droplets coverage of aerial applications (application in good meteorological conditions, adaptated calibration of spraying equipments), (ii) the combination of systemic fungicides with mineral oil (oil compatible formulations)
- A targeted and strictly localized control of hotspots (through ground sprays or adapted crop practices)
- The alternation of systemic fungicides with different mode of action and the regular monitoring of fungicide resistance in order to adapt the fungicide use strategy.
- The centralization of decisions in a very specialized team

Box 1. The climatic forecast is based on two climatic parameters: Piche evaporation and temperature (Ganry & Meyer, 1972a).
Piche Evaporation is collected under a simplified meteorological shelter. This parameter integrates several factors affecting the leaf surface (wind, humidity, temperature). It gives a good expression of the moisture environment of the fungus, related to the development of the disease (Ganry & Meyer, 1972b). A scale of evaporation was established in relation with the development conditions of the disease:

Conditions for the development of the disease	Piche Evaporation in mm/week (PE)
Favorable	Less than 22 mm/week
Moderately favorable	Between 22 and 30 mm/week
Not very favorable	Between 30 and 40 mm/week
Unfavorable	More than 40 mm/week

The duration of efficacy of a fungicide application (oil + systemic fungicide) is estimated from PE
di= 0.5 PEi (Ganry & Laville, 1983) where di, according to a week "i" ,is the duration of efficacy that determines the next application if the current climatic conditions remain identical to the past conditions. This is generally not the case. So, each week, d is recalculated and, step by step, the next spray can be predicted a few weeks in advance.
Temperature is collected through standardized meteorological devices. It is used to estimate the weekly disease growth speed, derived from a fungus growth model in response to temperature changes (Ganry & Laville, 1983)

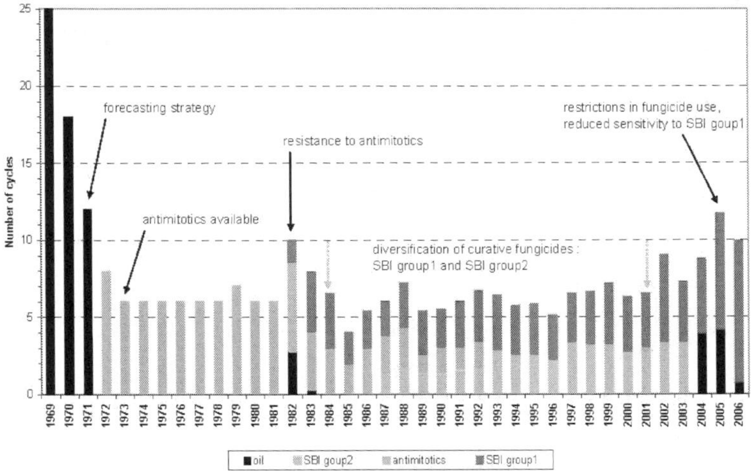

Fig. 2. Fungicide use for SLSD control in commercial banana farms of Guadeloupe (1969-2006). (from de Lapeyre et al., 2009)

This warning strategy developed by IRFA was very different from the systematic strategy used by U.S. companies using contact fungicides in water or systemic fungicides in oil emulsions or water.

The characteristics of these two control methods are summarized in the table 1.

Box 2. The Stage of Evolution of the SLSD, SEDs (Ganry & Meyer, 1972b; Ganry & Laville, 1983; Ganry et al, 2008)

Its purpose is to get a biological descriptor of the speed of evolution of the disease, according to climatic conditions and inoculum pressure.

It is adapted to the forecast as it is based on the earliest visible detection of infection and it is directly connected to climatic conditions

Observation plots are designed to weekly check the same 10 to 30 homogenous banana plants during their vegetative phase (between 5 to 18 leaves).

The observation is purely qualitative: the youngest leaves (range 1 to 5) are checked every week, and the most advanced stage of the disease according to Brun's scales (Brun, 1963) is scored.

In addition, the cigar development (score fom A to E) observation provides the evaluation of the weekly

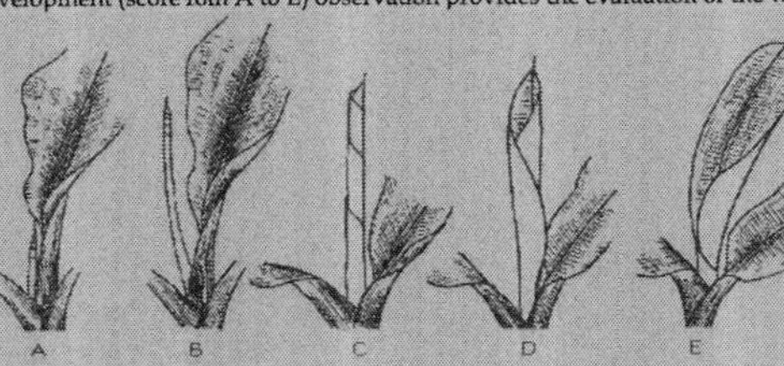

foliar emission rate.

The scoring of the disease is as follow from adaxial surface observations:

1. Stage 1: a spot is just visible to the naked eye as a minute yellowish-green speck less than 1mm.
2. Stage 2: the streak increases in size, notably in length, and remains yellowish-green, with a size superior to 1-2mm.
3. Stage 3: a streak begins to broaden slightly and increases in length as well, and begins to turn into rusty red.
4. Stage 4: this is the first spot stage. The streak turns dark-brown, is sunken and has reached its final size. A water-soaked yellow halo forms around the spot. Conidia are formed at this stage.
5. Stage 5: this is the final stage of the lesion. The central area of the spot turns grey and the margins of the spot are black. Some minute black points are visible, to the naked eye or with a magnifying glass, in the central area. They correspond to perithecia and spermogonia where ascospores and spermatia are produced

The couple leaf range/disease stage provides a coefficient that reflects the speed of evolution of the disease:

Stage of the disease	Leaf range				
	I	II	III	IV	V
1	100	80	60	40	20
2	120	100	80	60	40
3		120	100	80	60
4			120	100	80
5				120	100

In order to mitigate the scoring variations due to brutal change in leaf stage (equal at 20), a correction is done taking into account the cigar stage: for each leaf with symptoms (scoring >0), the following quantity is substracted: 4 for B, 8 for C, 12 for D and 16 for E.

As SEDs must be the expression of a growth speed, the couple "disease stage-leaf range" must be related to time through the leaf emission rate (LER), obtained by the current "leaf range-cigar stage" by the "leaf range-cigar stage" of the past week ($L_iC_i - L_{(i-1)}C_{(i-1)}$ with iC_i =,0 for A,.2 for B,.4 for C, 6 for D and 8 for E)

2.2 The basic principle of an integrated approach to control the banana BLSD

Inherited from this strong experience with the control of the SLSD, the integrated approach to control BLSD is based on the rational use of curative (systemic) fungicides applied in oil mixtures.

The sequence of fungicide sprays is driven by a forecasting system supported by biological indicators (Fouré & Ganry, 2008; Ganry et al., 2008). Since *M. fijiensis* is more aggressive than *M. musicola*, the system works efficiently for this pathogen when the disease pressure is not too high. It is the reason why it must be supplemented by sanitation practices consisting in necrotic and pre-necrotic leaf removal in order to eliminate new sources of inoculum. Other crop practices must be introduced to synergize with fungicide sprays and leaf removal dealing with crop management: plant nutrition, crop density, water management.

Depending on the type of fungicide used and on the selection pressure[1], there is always a risk of emergence of fungicide resistance in pathogen populations. It is one of the major reasons of the decrease of the chemical control efficiency which is often the starting point of a vicious circle.

Therefore it is highly important to monitor the pathogen sensitivity to the fungicides used. This monitoring must be considered as a key decision tool for the choice of appropriate fungicides (when it is possible) and their sequence in alternation, in order to minimize the selection pressure.

Such an approach was adopted for BLSD when it appeared in Gabon on plantains (see §2.2. BLSD control in Gabon in small scale plantain production for domestic markets) and in Cameroon (see §2.2. BLSD control in Cameroon in the banana industry for export), Ivory Coast and Ecuador on banana and plantain. Thanks to a strong experience on SLSD control and to the implementation of such disease management strategies, BLSD in these countries has been properly controlled with an average of 10 to 15 treatments per year[2].

In comparison, in some Latin American countries, the number of annual sprays can reach 30 to 60 (more than one per week) due to systematic fungicide applications and to the emergence of resistant strains to curative (systemic) fungicides that were used too frequently (Carlier et al., 2000; Romero & Sutton, 1997). As a consequence more contact, non curative, fungicides are used (Table 1.).

IRFA method	american method
Oil used with a rate of 12-15 liters/ha with curative (systemic) fungicide 10 to 15 sprays a year (depending on location and inoculums)	Water or oil-water emulsions (2 – 8 liters of oil + 10-20 liters of water) with 1. Contact fungicide 2. Curative fungicide 3. Contact and curative fungicide 30 to 60 systematic sprays depending on locations

Table 1. Compared characteristics of the IRFA and american methods for BLSD control

2.2.1 Component 1: Timing of decisions through a biological forecasting

The timing of decisions is based on a standardized set of criteria, taking into account the development of the fungus and the growth of the plant, both climate-driven. The first level

---

[1] Related to the duration of the pathogen exposure to the fungicide

[2] Cameroon: from 10 to 15 treatments / year have succeeded in the disease control for more than 10 years. However, the emergence of strains resistant to systemic fungicides, due to some failures in the implementation, is currently preventing the use of this disease management.

- Ivory Coast: control is achieved with 12-14 treatments / year in most production areas.
- Ecuador: from 10 to 12 treatments / year succeeded in disease control when this disease management strategy was experimented in this country in the late 80's.

of elaboration of reliable tools for strategic decisions is to implement a continuous monitoring of disease evolution in the field. It is also informative to fit these weekly data with weekly rainfall, in order to get a better appreciation of epidemiological conditions (de Lapeyre de Bellaire et al., 2010a).

- The Stage of Evolution of Disease of the BLSD, SEDb, is the main parameter that should be assessed. SEDb has been specifically designed for warning systems based on the early detection of new attacks, and has been specifically adapted from SLSD to BLSD (Fouré & Ganry, 2008) taking into account the important density of symptoms leading to necrotic development by coalescence (see plate 1 & box3). It is also a good indicator of the efficiency of the chemical control that is designed to reduce disease development in its early stages, that means on the youngest banana leaves (Fig. 3).

SEDb represents the speed of evolution of the disease, which reflects the climatic conditions and the intensity of the infection (Ganry & Meyer, 1972b). The early detection of new attacks on the youngest leaves will allow blocking their evolution towards necroses thanks to the application of a curative fungicide. In order to maintain the field inoculum as low as possible, it is then essential to stop the disease evolution before necrotic formation because sexual sporulation starts in necrotic stages of the disease.

A larger set of disease parameters should also be used and evaluated continuously every week on the same banana plot used for the assessment of SEDb in order to provide complementary information:

- The 'Youngest Leaf bearing Streaks' (YLSt) is the position of the youngest leaf bearing symptoms of the disease. It reflects the incubation period of the disease under a spraying program, which depends on both climatic conditions and fungicide inhibition of the pathogen. Its evolution gives information on the epidemiological conditions and on the performance of the chemical strategy as well. YLSt is scored jointly with SEDb. If none of the 4 leaves monitored have symptoms, the symptoms should be looked for on the older leaves.
- The 'Youngest Leaf Spotted' (YLS) should be scored according to Stover's method (Stover, 1971) where the YLS is the youngest leaf bearing at least 10 necrotic lesions. However, if none of the leaves bear any necrotic lesion, the value of YLS will arbitrary be N+1, where N is the total number of leaves of the banana tree. In addition, since the absolute value of the YLS is dependent of the phenological stage of the plant, all the banana plants monitored should be homogenous. Thus, better than the absolute value of the YLS we must consider its evolution from the establishment of a new plot of 10 homogenous young plants to the flowering of these 10 plants (all these plants should be changed at the same time just before flowering). The evolution of the YLS on a same banana plant is a balance between the rate of necrotic formation and the foliar emission rate. Thus, the evolution of this parameter is helpful to understand the efficiency of the fungicide control at the level of necrotic formation (*i.e.*, at a lower level than SED and YLSt) : (i) if the value of YLS decreases, the chemical control is failing because necrotic formation is faster than leaf emission; (ii) if the YLS is increasing, so we can conclude that the chemical strategy is efficient to slow down the necrotic formation (Fig. 4).

- The 'Number of Functional Leaves at Harvest' (NLH) is a final estimator of the effectiveness of the control strategy and a potential indicator of fruit exportability. It is often considered that a functional leaf should have less than 30% of necrotic surface, although no scientific data support this consideration. Banana companies generally use an empiric threshold of NLH (3–5 according to the companies) in order to decide which bunches are exportable or not. As no reliable experimental data is supporting this assumption the value of NLH should be regarded mostly as an indicator of the efficiency of the chemical strategy at the harvest stage.

Box 3. The Stage of Evolution of the BLSD - SEDb (Fouré &Ganry, 2008)
Its purpose is to get a biological descriptor of the speed of evolution of the disease, according to climatic conditions and inoculums pressure. It is adapted to the forecast as it is based on the earliest visible detection of infection. It is similar to SEDs for Sigatoka disease, but with some significant modifications due to the virulence of the disease in terms of evolution rate and streaks/spots density
Observation plots are designed to weekly check 10 to 30 homogenous banana plants in their vegetative phase between 5 to 18 leaves (always the same). The observation is qualitative and quantitative consisting of a checking of the youngest leaves (range 2 to 4) every week, and scoring the most advanced stage of the disease according both to Brun and Fouré's scales (Brun, 1963; Fouré, 1982c) The checking of the youngest leaves (from I to V) integrates the cigar development (score from A to E) and provides the weekly foliar emission rate

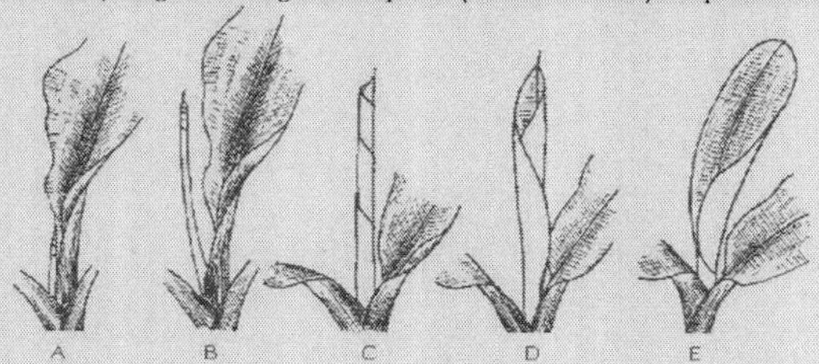

The scoring of the diseases is as follow:

1. Stage 1: a spot is just visible to the naked eye as a minute yellowish-green speck less than 1mm, only at the underside of the leaf.
2. Stage 2: red or brown streaks (rust colored) are visible firstly on the underside of the leaf, then on both sides. These streaks become progressively brown on the underside and black on the upper side of the leaf. The first outward sign of this stage 2 on the upper side of the leaf is a yellow streak which becomes progressively brown.
3. Stage 3: widening and extension of the stage 2.
4. Stage 4: this is the spot stage. The spot takes a circular or elliptical form.
5. Stage 5: this corresponds to the first of the two necrotic stages; a yellow halo firstly surrounds a black spot.
6. Stage 6: this corresponds to the second necrotic stage; drying out of the centre of the spot is observed, spots progressively take a grey color, but keep a black halo also surrounded by yellow.

The couple leaf range/ disease stage provides a coefficient that reflects the speed of evolution of the disease:

Stage of the disease[1]	Density of lesions	Leaf range		
		II	III	IV
1	-	60	40	20
	+	100	80	60
2	-	100	80	60
	+	140	120	100
3	-	140	120	100
	+	180	160	140
4	-	180	160	140
	+	220	200	180
5	-	220	200	180
	+	260	240	220
6	-	260	240	220
	+	300	280	260

1 For each stage of the disease, the mark attributed depends on density of lesions, which is estimated by "-" if there are less than 50 outward signs and by "+" if there are more than 50 outward signs.

(a) Observation of the disease on banana leaves.

Banana tree number (BTN)	Old leaf number (OLN)	Old cigar stage (OCS)	New leaf number (NLN)	New cigar stage (NCS)	FER	Leaf 2		Leaf 3		Leaf 4	
						Stage of disease	Lesion density	Stage of disease	Lesion density	Stage of disease	Lesion density
1	14	2	15	0	0.6	None	None	2	–	5	+
2	13	6	14	4	0.6	None	None	1	–	2	–
3	14	0	15	0	1	None	None	1	–	2	–
4	13	6	14	4	0.8	None	None	2	–	2	–
5	15	8	16	6	0.8	None	None	1	–	1	–
6	15	2	16	2	1	None	None	1	–	1	–
7	14	0	14	8	0.6	None	None	2	–	1	–
8	15	4	16	4	1	None	None	2	–	3	–
9	14	2	15	0	0.8	None	None	1	–	2	+
10	13	6	14	2	0.6	None	None	1	–	1	+

(b) Sum of coefficients attributed to leaves nos. 2, 3 and 4.

Stage of the disease	Number of stage observation coupled with lesion density in leaves						Coefficient attributed to the stage		
	Leaf 2		Leaf 3		Leaf 4		Leaf 2	Leaf 3	Leaf 4
	Frequency of disease stage	Lesion density	Frequency of disease stage	Lesion density	Frequency of disease stage	Lesion density			
1	None	None	6	–	3	–	None	240	60
					1	+			60
2	None	None	4	–	3	–	None	320	180
					1	+			100
3	None	None	None	None	1	–	None	None	100
4	None	None	None	None	None	None	None	None	None
5	None	None	None	None	1	+	None	None	220
6	None	None	None	None	None	None	None	None	None
Sum_{coef} per diseased leaf								560	720

(c) Calculation of weekly FER.

Sum_{FER}	$Number_{days}$	FER_{10d}	$FER_{past\ week}$	$FER_{new\ week}$	Sum_{coef}
8.4	7	1.2	1.3	1.25	1280

(d) Calculation of weekly SED: $Sum_{coef} \times FER_{new\ week} = 1600$.

Table 2. Example of a data sheet used for the calculation of the weekly SEDb in banana plants (FER = Foliar Emission Rate) *(from Fouré &Ganry, 2008)*

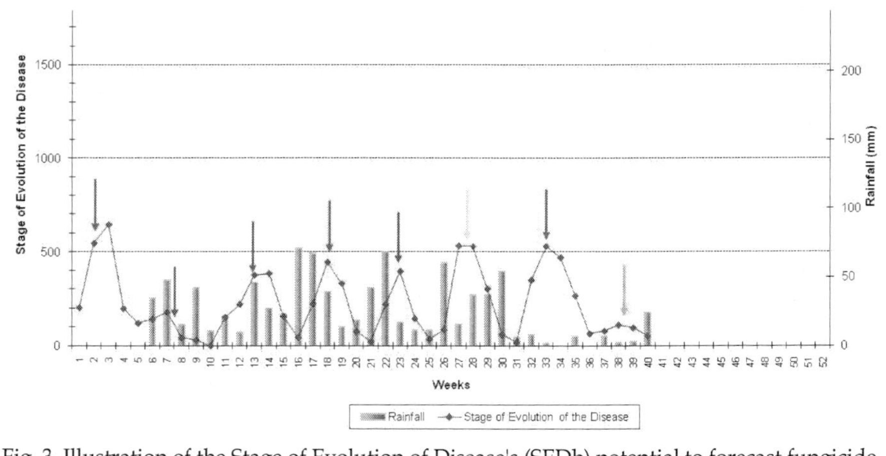

Fig. 3. Illustration of the Stage of Evolution of Disease's (SEDb) potential to forecast fungicide applications and to monitor the efficiency of the chemical control strategy in banana plantations. *All fungicide applications are indicated by an arrow above the SEDb curve (arrows with different colors correspond to different fungicides applied)* (from de Lapeyre de Bellaire et al., 2010a)

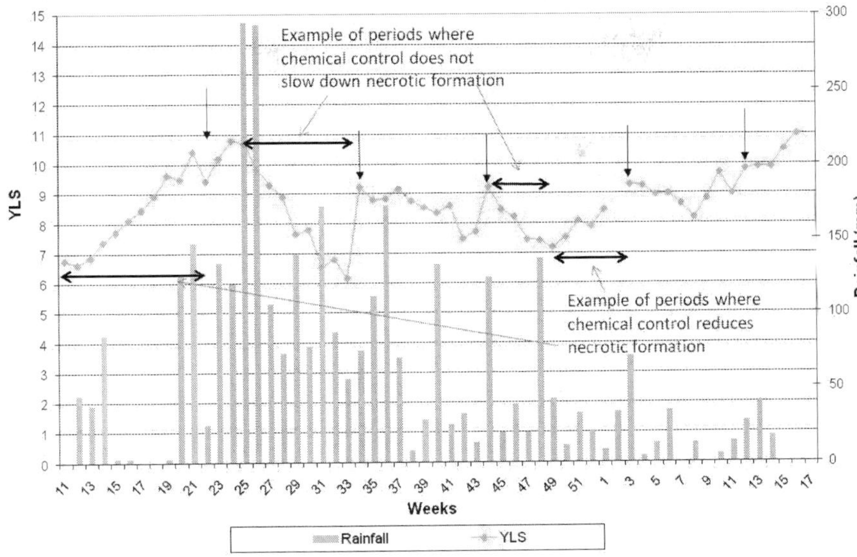

Fig. 4. Illustration of interpretation of the Youngest Leaf Spotted (YLS) from its weekly evolution in banana plantations. *Arrows indicate the date at which the whole 10 banana plants where changed At that date, the variation should not be considered as the value of YLS is artificially different. YLS is scored according to Stover's method* (from de Lapeyre de Bellaire et al., 2010a)

2.2.2 Component 2: Timing and quality of the fungicide applications

The timing of the fungicide applications is driven by SEDb and the analysis of other parameters. It is the reason why it must be implemented by highly skilled teams.

The time between decision and execution of one application should not exceed 2 days. Sprays are carried out by airplane, helicopter, or by backpack sprayer on small surfaces. However, care must be taken with the climatic conditions of aerial spraying: only a small window, early in the morning and late in the afternoon, is suitable; otherwise, thermal inversion and air turbulence do not allow correct spray deposition. Aerial application is not possible on rainy and windy days. Logistics available is therefore essential to optimize spraying during this small window.

The efficiency of sprays is dependent on the quality of the foliar application, and a good coverage is essential.

The weather conditions the day of application are critical: temperature must be rather cool (less than 28°C) and wind as low as possible (<1.3 $m.s^{-1}$) and regular. Sprayers and nozzles must be cleaned every day and calibrated frequently (calibration depends on the use of water or oil and on the rate of mixture used per ha). Significant technical improvement has been achieved when aircraft are guided precisely by GPS.

An irregular topography of the zone or the presence of obstacles might alter the uniformity of the treatment. The banana plots must be shaped to facilitate aerial sprays, taking into account the presence of obstacles and wind drift, and the crop density should not be too high to allow good penetration of the fungicide spray. A strong synergy between the aerial spraying companies and the banana growers is also important to ensure the quality of the aerial applications.

The use of mineral oils as a carrier considerably improves the quality of coverage through aerial spraying with low volumes (12–15 L ·ha–1). The quality of oil is essential and must fit with the norms defined by Cuillé and Blanchet in 1958 and always used. The quality of the mix with the fungicide is important as well. Fungicides must be used at full rate in order to limit the risk of fungicide resistance.

Characteristics	norms
Viscosity	4 à 7° Engler at 20°C
Density	0.83 to 0.90
Sulfonation index	85-90% maximum
Type	indifferently paraffinic or naphtenic
Acidity	less than 0.16
Addition ingredients	no sulfur compound

Table 3. Characteristics of oil used for sprays in the control of SLSD (Cuillé & Blanchet, 1958)

The efficiency of treatments relies on a strong curative effect. Thus, systemic fungicides (antimitotics, sterol inhibitors of group 1 and 2, strobilurines, pyrimidins) are preferred to contact fungicides (see Box 4).

Box 4. Various categories of fungicides used for BLSD control

a. Protectants

Contact fungicides (mancozeb, chlorothalonil) have only a preventive effect and do not penetrate in the banana leaf. They are generally applied every week in systematic frameworks because all new unfurling banana leaf must be protected. They have a multi-site effect on fungal biology and strains resistant to these fungicides have never been reported.

b. Systemic fungicides

Systemic fungicides penetrate in the banana leaf. The systemic properties of these fungicides are variable: some have only a translaminar transport and others can be distributed in the whole banana plant, and the biokinetics of these fungicides confer a more or less pronounced curative effect. The curative effect is more pronounced on young streaks (stages 1, 2), but lower on older lesions (stages 3, 4) and without any effect on necrotic stages (stages 5, 6), even if sporulation might be temporarily decreased. Different chemical families have been used:

Three groups have a strong curative effect:

- Antimitotic products that inhibit tubuline polymerisation (benzimidazole): very high risk of resistance

- Inhibitors of ergosterol biosynthesis (IBS) in group 1 that inhibit eburicol demethylation (DMI fungicides mostly belonging to the triazole family): high risk of resistance

- Mitochondrial respiration inhibitors that bind to the cytochrome b complex (QoI inhibitors of the strobilurin family): very high risk of resistance

Two groups have a moderate curative effect:

- Sterol inhibitors in group 2 that inhibit several enzymes like reductases and isomerases (amines)/ moderate risk of resistance

- Pyrimethanil, a supposed inhibitor of methionine biosynthesis whose mode of action is still not clearly established / moderate risk of resistance

The use of these fungicides in pure oil strengthens their curative effect, because mineral oils are fungistatic. Attention should be paid to the formulation of the fungicide used. Only formulations compatible with pure oil should be used. Phytotoxic effects of the oil have to be preliminarily determined. Fungicide concentration must also be determined according to preliminary field experiments.

2.2.3 Component 3: The control of the inoculum pressure in the field

It is achieved through prophylactic leaf removal coupled with relevant crop management, such as plant density and water and fertilizer management.

The experience of BLSD control in Gabon (see §2.2) has clearly shown that when plantain fields were isolated into a forest environment and far from infected plots, preventing them from external contamination, the disease control was strongly facilitated.

Thus, keeping the sources of inoculum at a very low level is very important to ensure the success of the chemical control strategy. Where extensive spotting is present, new infections will develop quickly because chemical sprays do not remove the disease from spotted leaves and the only solution is to remove them mechanically from the banana tree. Therefore, such necrotic leaf removal must be a regular practice in banana growing countries where BLSD is present. It should be based on a risk management approach, taking into account that it could affect the production beyond certain level.

Experimental leaf removal has shown that bunch weight, the main yield component, is poorly affected as long as 5 to 7 leaves remain on the plant from flowering to harvest (Ramsey et al., 1990; Daniells et al., 1994; Vargas et al., 2009). A stronger leaf removal could affect the final yield. In addition a lack of resources during this phase, due to severe spotting could compromise the yield of the next ratoon (Eckstein et al., 1995; Dens et al., 2008).

Leaf removal is particularly needed in critical areas where regular sprays are not well performed due to natural obstacles or to regulatory measures such as in buffer areas where aerial spraying is prohibited (e.g. 50 m from houses, gardens, rivers and roads in the FWI). A strong communication effort should be done towards the farm managers in order to improve the situation, because the chemical control strategy will not be successful if the leaf removal is not done properly. Such leaf removal might be mandatory on BLSD susceptible banana plants that cannot be part of the spraying agenda: in home gardens, near houses, along roads or rivers. In such situations the best solution would be to replace the susceptible bananas by resistant varieties. It requires to take into account social and political components, such as the acceptability of these varieties by the consumers in order to encourage voluntary behavior. In some cases like in Australia, more coercive measures were used with an objective of complete eradication which was successfully achieved (Henderson et al., 2006).

In addition it is essential to destroy systematically all abandoned plantations and isolated roadside banana plants that are potential reservoir of disease inoculums.

In areas where fungicide sprays are done, it is essential to realize necrotic leaf removal before any fungicide spray in order to lower the inoculum pressure and thus enhance the efficiency of the fungicide and to minimize the exposure of the pathogen to the fungicide and thus the risk of appearance of resistant strains to the fungicide.

Box 5a - Technical Sheet used in the FWI for the mechanical sanitation of banana for the Leaf Spot Diseases control
(part 1)

iT² INSTITUT TECHNIQUE TROPICAL	cirad

Technical Sheet
Banana Leaf Spot Diseases : Mechanical Sanitation

Code : NT_IT² 2011/03	15/05/2011	Page(s) : 2
Rédacteurs : David DURAL et Eric FOURE		

SANITARY LEAF REMOVAL = removal of leaves showing necrosis and symptoms that may move rapidly into necrosis

• SANITARY LEAF REMOVAL is a priority cultural practice that is mandatory accompanying chemical control.

• SANITARY LEAF REMOVAL is fully part of the disease control as it can significantly reduce the amount of inoculum present in plantations.

• SANITARY LEAF REMOVAL should be performed regularly to prevent further necrosis appearance on the leaves. In this respect spots that cannot be blocked by the fungicide sprays must be withdrawn.

It is therefore essential to be familiar with different symptoms and stages of the disease to achieve the proper leaf removal.

Stages of the disease	Sigatoka Leaf Spot Disease (SLSD) or Yellow Sigatoka	Black Leaf Streak Disease (BLSD) or Black Sigatoka
Stage 1	Yellow spot < 1 mm	Yellowish depigmentation spot
Stage 2	Yellow streak 1 to 5 mm	Extension of the lesion- Brown streaks
Stage 3 Contamination (conidia)	Enlargement of the streak takes a rusty color	Elongation and enlargement of the lesion visible on the under and upper sides
Stage 4 Contamination (conidia)	Brown lesion with a yello	Brown oval lesion
Stage 5 : necrosis Contamination (ascospores)	Necrotic lesion with gray	Black lesion
Stage 6 : necrosis Contamination (ascospores)	No stage 6	Necrotic lesion with gray color - yellow halo for isolated lesion

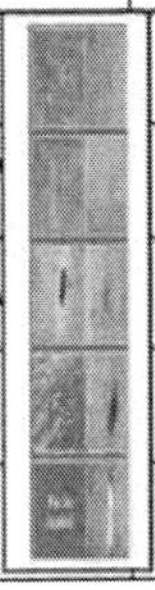

Box 5b - Technical Sheet used in the FWI for the mechanical sanitation of banana for the Leaf Spot Diseases control
(part 2)

SANITARY LEAF REMOVAL must be completed before the chemical spray (no fungicide spray on necrosis / risk of resistance emergence).

SANITARY LEAF REMOVAL increases the efficiency and sustainability of the current chemical control strategies.

Black Sigatoka: the presence on the leaves of stages 3 and 4 may result in a few days in the emergence of new necrotic stages 5 and 6. These symptoms of stage 3 and 4 cannot be blocked by fungicide sprays.
This requires that SANITARY LEAF REMOVAL is performed as soon as stages 3 and 4 are detected.

TERMS OF THE LEAF REMOVAL:
50% of the leaf area must be removed if stages 3-4 and / or 5-6 (BLSD) and / or 4-5 (SLSD) are present only on this half leaf. The leaf will be cut (total removal) if the symptoms are present throughout the leaf area

In some cases and in some producing plots, a severe sanitation can lead to the removal of most of the leaves present on the banana when it is the only way to rapidly and permanently clean the related plots

The cutted leaves with necrosis will be returned with the upper side against the ground to limit the spread of ascospores.

In the first cycle, it is recommended to put the cutted leaves in a pile at the end of the row(in this case, only the sheet placed on top is likely to emit spores).

A spotted leaf left hanging along the pseudostem produces spores for about 4 to 5 months against 2 to 3 weeks if it is on the ground.

A SANITARY LEAF REMOVAL correctly and regularly done reduces the production of spores of more than 80%; thus it must be considered as a priority control practice

The achievement of this sanitary leaf removal by ALL THE GROWERS is a key element of the success in the control of black Sigatoka ; IT IS OUR PRIORITY!

Other measures for the control of Black Sigatoka:

• Elimination of all isolated banana plants that may be on the edge of the plot; they are sources of contamination.

• Regular desuckering to avoid the ratoon bunches in the field as they are also sources of contamination.

• Rapid switch to fallow by injecting glyphosate and rapid burial (2 weeks after injection) in order to avoid necrosis of the leaves on the plants that will remain a source of contamination for a long time and to avoid regrowth.

2.2.4 Component 4: The fungicide sequence management

The forecasting strategies enable BLSD control with a limited use of fungicides. In such strategies, the cost of control is lower but, above all, the environmental impact is limited.

Such strategies should be preferred to systematic strategies that are conducive to significant environmental impact and they should be combined with other measures such as sanitation (leaf removal), drainage, under-cover irrigation and good fertilization practices. Forecasting strategies have faced some drawbacks in some countries, especially in Cameroon, because of the emergence of fungicide resistance to systemic fungicides. It is the reason why such strategies should be implemented under specific conditions in the related banana industry: (i) banana areas with no or few fungicide resistance to available fungicide families and particularly those where adverse climatic conditions exist during a part of the year (dry season). This is still the case in some parts of Ivory Coast, Ghana and the Caribbean; (ii) availability of curative fungicides with different mode of actions that can be used in alternation and offering alternatives in case of fungicide resistance is detected in one family; (iii) logistic capacity to operate when necessary without any delay and in the best conditions (window); (iv) strong involvement of the plantation managers and cooperation of the citizens and administration to contribute to the sanitation of banana plants (see §1.4).

When new plantations are established in new areas of a country where fungicide resistance is present, special attention should be paid to germplasm movement in order to prevent the introduction of resistant strains through planting material. Nurseries for planting material must be established in areas where the presence of resistant strains has not been demonstrated. This is, for instance, the case in Cameroon and Ivory Coast. In such situations, the sustainability of the forecasting strategy would be a significant challenge.

It is clear that a key component of a good fungicide sequence management relies on a regular monitoring of fungal populations (see box 6).

A special attention should be paid to the management of fungicide resistance that might develop following the repetitive use of curative fungicides that have a monosite action. Alternation between different groups of fungicides, or mixtures with contact fungicides, is essential to reduce selection pressure and delay the emergence of such resistance.

In several countries where such attention was not so high, the regular and intensive use of systemic fungicides has resulted in an important development of resistant strains in *Mycosphaerella fijiensis* populations. These data are well documented inside the banana working group of FRAC (Fungicide Resistance Action Committee). The resistance risk is considered as severe for strobilurins and antimitotics, and important for IBS of group 1.

The current status of fungicide resistance to these chemicals is a key information to monitor control strategies for BLSD based on a forecasting strategy because these strategies rely on the use of fungicides having a strong curative effect, i.e. systemic fungicides. It is also important to note that the cost of systemic fungicides is higher than the protectants, and that their use is then justified if they really enable a good control of BLSD with a lower number of applications.

It is the reason why a regular resistance monitoring is of prime importance in an integrated disease management system for BLSD (box 6)

Box 6. Resistance monitoring: The basic methodology relies on the comparison of the sensitivity to the different fungicides in fungal populations (50-100 spores) sampled in commercial farms (treated with the fungicides) and fungal populations sampled in untreated locations. The monitoring of sensitivity is based on germination tests: the germination of spores grown on agar media added with different concentrations of fungicides is compared with the germination of spores grown on agar (de Lapeyre de Bellaire et al., 2010b).

- For benzimidazoles, susceptible strains do not germinate on agar added with the fungicide or have distorted germtubes. Resistant strains have a normal or short germtube as compared with the control on ager medium.
- For sterol inhibitors, the germ tube length is measured and the % of growth inhibition (GI) is calculated for each strain at a specific concentration. The distribution of the GI in the population sampled in the treated farm is then compared to the distribution of the population sampled in the untreated farm. Another possibility is to evaluate the EC50 concentration (concentration for 50% GI) for each strain from GI assessed over a large range of concentration of the target fungicide.
- For strobilurines, the germ tube length is measured and the % of growth inhibition (GI) is calculated for each strain at a specific concentration. A strain is considered as resistant if germ tube length or GI is over a threshold value.

In the methodology approved by the FRAC (see FRAC website, monitoring methods (MYCOFI)), fungal populations consist in ascospores obtained from necrotic leaf samples (20-25 plants are sampled in each location). At laboratory, these necrotic leaf samples are bulked and incubated in a moist chamber for 48h and then leaf pieces are used for ascospore discharge on petri dishes enriched with the targeted fungicide.

This method has several drawbacks : (i) ascospore production is very fluctuable and the population analyzed in the Petri dishes might be very different from the population initially sampled in the field; (ii) in certain cases sporulation does not occur and the test cannot be carried, especially in the dry season; (iii) this method does not allow to use a predetermined sampling design; (iv) the populations analysed on each fungicide concentration and the control are always different; (v) other ascospores belonging to the genus *Mycosphaerella* might be confused with the ascospores of *M. fijiensis*, especially on fungicide amended media.

Then new methods are presently developed from conidia to overcome these drawbacks.

2.2.5 Component 5: Organization of the control

The success story of SLSD control in Guadeloupe and Martinique is particularly due to the mode of organization.

Centralization of decisions and operations is essential and the banana growers should be grouped in an association that would perform the control strategy.

Since ascospores are transported by wind over long distances, the control strategy should be the same in all banana plantations to prevent any disruption. The organization of the treatments is more efficient when a centralization of the decision is performed by a unique technical service operating according to rational guidelines rather than if each grower implements his own strategy.

2.3 BLSD control in Cameroon in the banana industry for export

In Cameroon, *M. fijiensis* was first reported in 1981. In the late 80s, a warning method using biological descriptors has been developed based on the strong experience gained with the control of SLSD in the FWI (Ganry & Laville, 1983; Bureau & Ganry, 1987; Bureau et al., 1982) and on results obtained in Gabon with BLSD on plantain (Fouré 1982a, 1982b, 1983, 1984, 1985; Fouré & Grisoni, 1984, Fouré et al., 1984). It was successfully applied, thus limiting the number of applications to 12-14 per year. This rational control by warning relied heavily on the use of systemic fungicides with a high curative effect (Fouré, 1988a, 1988b, 1988c; Fouré & Mouliom Pefoura, 1988; Fouré & Moreau, 1992).

This situation was sustainable for 10 years, but disease control then became unsuccessful due to logistic failures (shortage of airplanes), which led to the more intensive use of systemic fungicides.

As a consequence, since 1996 the emergence of strains resistant to systemic fungicides resulted in the progressive replacement of this rational control strategy by more frequent applications of systemic fungicides. As a result, resistance to site-specific strobilurins has developed particularly swiftly. Systemic fungicides have been progressively abandoned and replaced by contact fungicides, with chlorothalonil the most frequently used (Fig. 4). The contact fungicides do not cause the emergence of resistant strains, but have not the curative effect required for a prolonged action, as in the case of systemic's.

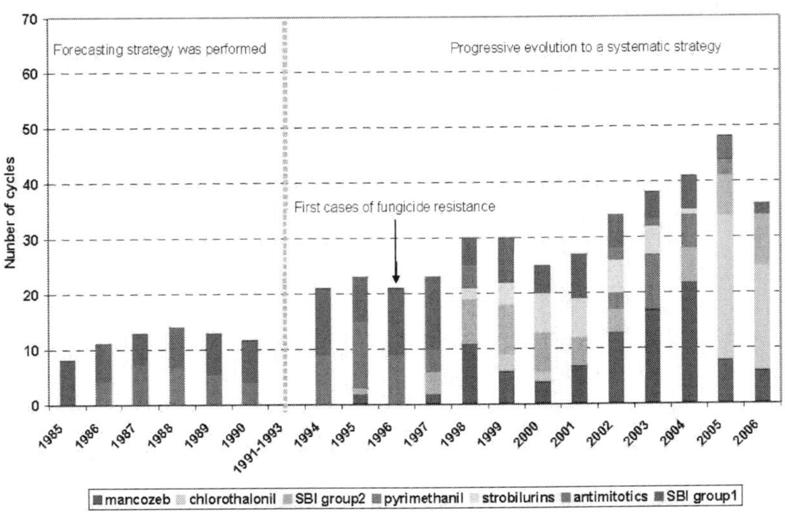

Fig. 5. History of fungicide use for BLSD control in a representative commercial banana farm in Cameroon from 1985 to 2006 (de Lapeyre de Bellaire et al., 2009).

Thus, in 2006, despite a continuing effort to drive the chemical control from the observation of biological descriptors, about 40 treatments were performed on most plantations. This increase in the number of treatments resulted in an increase in the cost of the control, but also in environmental risks, as contact fungicides are applied at higher rates than systemic fungicides (de Lapeyre de Bellaire et al., 2009). So, this evolution has led to an important increase of negative environmental effects since 30-40 kg/a.i/ha/year are now applied (vs. 2-4 kg/ha/year in the former forecasting system)

Recent observations of the latest monitoring show there is a decrease in resistance levels in some commercial plantations, especially since the systemic fungicides are no longer or rarely used (de Lapeyre de Bellaire et al., 2010b)

This trend suggests that the phenomena of resistance to fungicides may be reversible (see Box 7 and 8) and thus that new treatment strategies can be redefined in a more sustainable - both economic and environmental - way.

In the horizon of 2 or 3 years, it seems possible to recover one or two fungicides and reintegrate them with newly available chemical family as part of an integrated strategy that would reduce the number of fungicide applications and the amount of active ingredient spread.

Box 7.
Resistance to benzimidazoles and strobilurins has actually dropped since they are no longer used. But we still need a background of resistance that can even advise their use. However there is every reason to be optimistic about the possibility of reuse, at the horizon of a few years in some areas under warning strategies that would also benefit the registration of systemic fungicides with new mode of action, like succinate deshydrogenase inhibitors (SDHI) that will be soon released. With regard to the triazoles, the sensitivity deteriorates steadily mainly because they still are used in most plantations. Their use should be postponed in order to recover sensitivity.

Box 8.
Several mechanisms may be behind this recent drop of resistance levels observed in Cameroon:
- Gene flow from untreated areas (because of the absence of chemical treatments, populations have high size and are susceptible to all fungicides in untreated neighboring plantains), to commercial plantations (because of fungicide applications populations have lower size in commercial treated farms) that could cause a progressive "dilution" of fungicide resistance
- A loss of competitiveness of resistant strains which would be phased out when the fungicide selection pressure is off. Effectively, in some cases and particularly for DMI fungicides, fungicide resistance has a fitness cost (Karaoglanidis *et al*, 2001)

2.4 BLSD control in Gabon in small scale plantain production for domestic markets

Here are briefly presented the activities conducted on the agro-industrial plantation of plantains at N'toum (100 ha, 60 km from Libreville, Gabon) created in the late 1980s to supply the urban markets of Libreville

For the first time in Central Africa, the BLSD was detected in this area of Ntoum in 1980 (Frossard, 1980). Given the pathogenic activity of the fungus, a research program was implemented in this country. It was based on the study of certain aspects of the biology and the epidemiology of the causal agent of the BLSD, *Mycosphaerella fijiensis* and the development of a warning method for a rational chemical control of the disease (Fouré, 1983; 1984, Fouré et al., 1984). Continuous analysis of biological descriptors (observation of various stages and progression of the disease on the foliage) and the use of systemic fungicides has produced very satisfactory results in this intensive plantation of plantain and effectively controlled the BLSD on 100 ha of plantain in Ntoum (ten fungicide sprays / year; alternation of systemic fungicides). One of the reasons why it was possible to reach such good results was related to the situation of these plantain fields which were very isolated into a forest environment, preventing from external contamination.

This program was an important contribution in the adaptation of a biological warning method for BLSD control.

2.5 BLSD control in the French West Indies

Up to recently, the French West Indies islands were still free of the BLSD.

The SLSD has been controlled effectively and at lower economic and environmental costs through the implementation of a pest management strategy based on a warning system over more than 30 years (5-7 treatments / year). It has been already described in the §1.

Recent developments in the phytosanitary regulations in France (withdrawing of most fungicides and difficulty for registration of new active ingredients) resulted in a sharp decrease of the 'in vitro' susceptibility of the fungicides that are still registered for the control against the SLSD. Thus, the number of treatments performed each year has increased very recently (10-12 treatments per year). In addition, there was also a change in French legislation on aerial spraying and the establishment of untreated buffer zones in a distance of 50 m from houses, gardens, rivers, roads.... Ultimately, it is possible that the aerial treatments will be prohibited.

Recent developments in the spread of the BLSD in the Greater Antilles and more recently in the Lesser Antilles (Fig. 5) suggested that its arrival in Martinique (and probably a later deadline in Guadeloupe) was inevitable in a more or less short term.

Inexorably, BLSD was detected in Martinique in September 2010 and since then it has spread very fast, making it unfeasible any eradication attempt.

Thus, the unique solution is to apply an integrated disease management approach based on the key principles inherited from the lessons learned with BLSD in other areas, and taking into account the excellent know-how of the banana industry in control of the SLSD through a forecasting system.

The effectiveness of pest management is also based on the common management and centralized mode of organization that is working out in Martinique and Guadeloupe for over 30 years. This type of organization is undeniably a major asset for the successful implementation of this strategy. Nevertheless, the implementation of this strategy might be hampered by the limited number of fungicides registered in the FWI for BLSD control and by the effective regulation of BLSD in the 50 m buffer areas.

Fig. 6. Geographic expansion of BLSD in the Americas

2.6 BLSD control in Latin America in the banana industry for export: The case of Belize

The forecasting system described in previous chapters, has been adapted and implemented in several countries in Latin America (Marin, 2003). In some situations in Central America, the commercial adaptation of the early warning system resulted in a significant reduction in the number of fungicide applications. However, due to various factors related to weather patterns and high resistance levels, there was a "come-back" to systematic sprays with contact fungicides. Some commercial programs still use the system only to help management decisions (Marin, 2003)

In Belize, BLSD is controlled on an area of 2600 Ha through aerial spraying of fungicides. The control of the disease is centrally managed by the Banana Growers Association (BGA). The Sigatoka service of BGA is accountable for the number of cycles, timing of applications and for the type of fungicides used during the campaign for BLSD control on each farm belonging to a same pedo-climatic area. At the beginning of 2006, the Sigatoka service of BGA was in charge of weekly disease assessment in the different farms, based on the evaluation of the youngest leaf with visible streaks from the ground (YLS) and of the total number of functional leaves (see §1.1). Some farms had started to implement disease assessments for a biological forecasting (de Lapeyre de Bellaire, 2006 & 2007).

The centralization of BLSD control was very suitable for disease management because it guarantees that a same technical guideline is used over the whole banana area. Nevertheless, an indirect negative consequence of this centralization was that banana growers were less involved directly in BLSD, especially for leaf removal.

Three airplanes and two pilots were available for spraying, which is the minimum to control the disease over 2600 ha in order to optimize spraying during the best conditions (small window in the morning).

Fungicides used for BLSD control fall in the 2 categories described earlier: protectants, and systemic fungicides (Box 3).

The spraying program was generally based on the use of contact fungicides during the dry season (February to May) and systemic fungicides being mainly used in the rainy season.

Many fungicides are registered for BLSD control in Belize (Table 3), but despite the intensive use of systemic fungicides in recent years (Fig. 6), control of the disease was not successful at that time, and significant losses were registered (de Lapeyre de Bellaire, 2006). High disease outbreaks resulted in bunch reject, bunch weight losses and strong quality problems linked with early ripening and heterogeneity of ripening in ripening rooms.

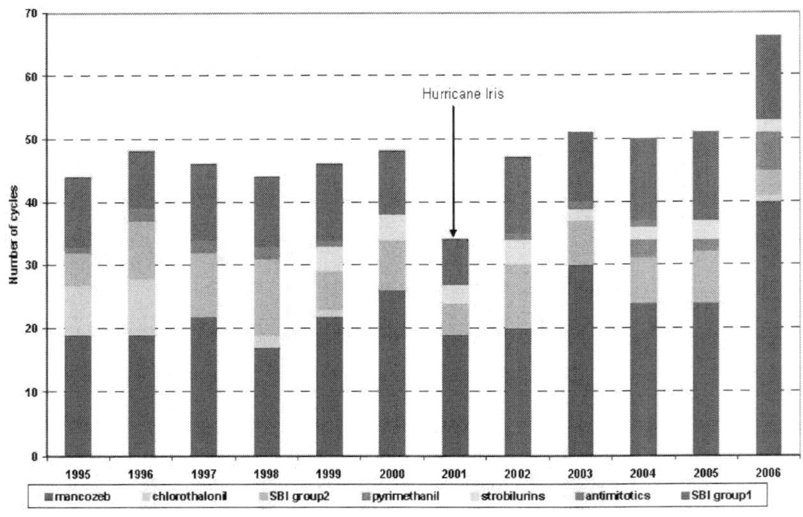

Fig. 7. History of fungicide use for BLSD control in a representative commercial banana farm in Belize (1995-2006) (from de Lapeyre de Bellaire, 2006)

In 2006, a fungicide resistance monitoring campaign was achieved thanks to the contribution of CIRAD's laboratory in Montpellier. The results of this monitoring analysis showed a very worrying situation in terms of fungicide resistance in all the commercial banana farms. For the three chemical groups evaluated (strobilurins, antimitotics and IBS group 1) high levels of resistance have been observed (de Lapeyre de Bellaire, 2006), which explained the poor control obtained with systemic fungicides in this country.

In such conditions, satisfactory control could be achieved through a systematic use of contact fungicides, such as mancozeb and chlorothalonil. However, this strategy does not reduce the environmental impact since the number of fungicide applications/year remained high (de Lapeyre de Bellaire, 2007).

Nevertheless, as observed in Cameroon, this strategy could create the opportunity for a possible reversion of resistance, especially for IBS of group 1 and give the possibility to implement in the future the biological forecasting system developed by CIRAD in this country.

2.7 BLSD control in Latin America in plantain production for export: The case of Panama

In Panama, as in other Central American countries, a very significant decrease in the production of plantains was observed after the emergence of BLSD. The production in this country was divided by three in five years, from 100,910 tons in 1979 to 31,134 tons in 1984 (Diaz, 1986). The effects were very sensitive on the supply of domestic markets as well as on the prices that have almost doubled over this period. In addition all exports, while expanding (661 tons in 1980, 2338 tons in 1982) were stopped due to a very high quality deterioration caused by the disease.

It was therefore considered interesting to use the experience gained with the control of the Sigatoka Disease in banana plantations of the French West Indies (Bureau, 1990).

The warning system has been implemented on two types of production: small farms of 4-5 ha on average, with a low technicality and family labor and medium farms of 10 to 30 hectare with a good technical level, hired labor and more productive.

In the first case the control was done through ground sprays and in the second case by aircraft.

The results were very promising with a very good control of BLSD in both situations with nine applications a year in family farms and 6.5 applications in medium-sized farms. Such promising results have been achieved through a rigorous execution giving its full preventive character to the system. It is interesting to notice that under these conditions, a very good relationship was found between the evaporation Piche and the duration of treatment efficacy. It appears that the aerial sprays are much more efficient than ground sprays

Apart from Panama, the forecasting system was also used on plantain in other countries such as Costa-Rica. In this latter country the forecasting system has been simplified (Marin, 1992) as well as combined with climatic factors to develop a bioclimatic forecasting system for plantain (Jimenez et al., 1995; Lescot et al., 1998)

3. Lessons drawn from various experiences and perspectives

Beyond the biological and technical components there are economic, political, logistical, environmental and social issues that are key points to consider in a holistic approach, otherwise leading to unpredictable failures.

Chemical control of BLSD would not appear sustainable in the long term. In several countries, fungicide resistance to systemic fungicides is increasing, and chemical control using these fungicides is becoming no longer efficient. In such situations, a systematic use of contact fungicides, as shown in the case studies of Belize and Cameroon, needs to be implemented. As a consequence, warning strategies, which could help reduce costs and environmental impact of chemical control, are becoming useless, because of systematic and frequent sprays. As a consequence, fungicides for BLSD control are the most important contribution to the annual amount of pesticide used in all countries were BLSD is present and where favorable conditions prevail (Risède et al., 2010). In the FWI, fungicide resistance is particularly worrying since the SBI of group 1 are the only fungicides approved. The introduction of very restrictive legislation significantly affects the sustainability of chemical control (table 4). More restrictive legislation aimed to further protect human and environmental health may be passed in the future. New solutions are then necessary to guarantee the sustainability of banana cropping systems (de Lapeyre de Bellaire et al., 2009).

Active ingredient group	Belize	Cameroon	Guadeloupe
mancozeb	8	7	0
chlorothalonil	2	5	0
SBI group 2	1	3	0
pyrimidins	1	2	0
strobilurins	2	2	0
antimitotics	1	1	0
SBI group1	5	7	2

Table 4. Number of fungicide products registered in various countries for BLSD or SLSD control in 2006. (from de Lapeyre de Bellaire et al., 2009)

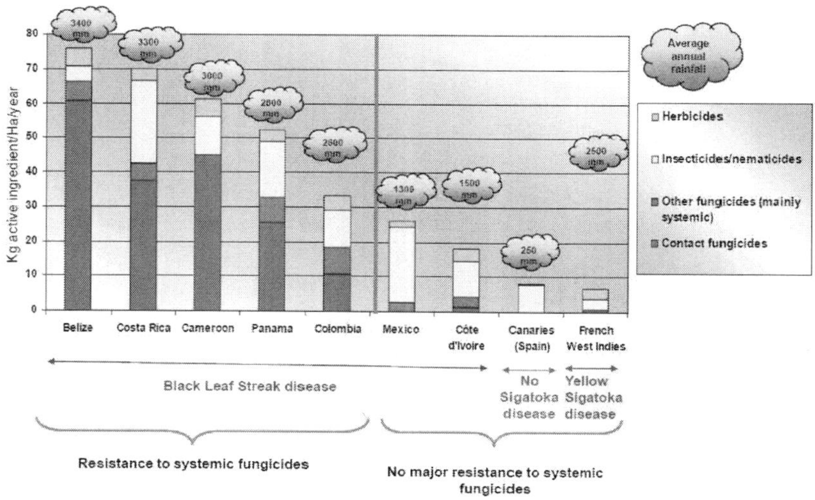

Fig. 8. Estimated total pesticide quantities used by the dessert banana industry in some countries, including European Community areas (2006-2007).© *Thierry Lescot, CIRAD, France*

3.1 Short-term solutions

The introduction of more eco-friendly fungicides would be beneficial to address environmental and health impact of contacts fungicides. In the past 5 years, organic fungicides or bio-fungicides, such as essential oils, alimentary additives, organic acids, potassium carbonates, leachates of decomposed banana material (bunch stems, fruits), and bio-control agents have been experimented in Cameroon. None of these fungicides gave good control of BLSD under high inoculum pressure. However, recent experimental data suggest that the combination of some bio-control agents (*Bacillus subtillis* and *B. pumilis*) applied in mixtures with contact fungicides could enable the reduction of the amount of fungicide applied (de Lapeyre de Bellaire et al., 2009)

In addition, forecasting strategies should be devoted and implemented in areas where specific conditions are fulfilled: (i) areas free of fungicide resistance, (ii) new banana areas, (iii) low disease pressure areas. Where fungicide resistance is established, the reintroduction of forecasting strategies relies on possible fungicide resistance reversion and incoming of new mode of action fungicides with a high curative effect. For instance, in situations where the current fungicide resistance is reversible as shown in Cameroon (see § 2.1), it is possible to carefully reintroduce adequate curative fungicides and thus implement warning strategies.

3.2 Long-term solutions

3.2.1 At the cross-road between genetics and landscape management

As already mentioned at the beginning, only Cavendish bananas, highly susceptible to BLSD, are grown in the banana industry, which is a high risk for the sustainability of the industry.

For this reason and although the current market organization of the banana industry that is an obstacle to the diversification of banana cultivars in the commodity chain, recourse to resistant varieties in an integrated strategy is certainly part of the future of an integrated BLSD control.

Resistant banana, edible or wild, already exist and two types of resistance have been described (Fouré et al., 1990,Beveraggi et al., 1995, , Jones, 2000).

The first one is a high resistance due to a hypersensitive reaction of the host and characterized by the blockage of symptoms at early stages. It is found in cultivars such as Yangambi km 5 (AAA, Ibota) and in various diploids already used in breeding programs as a source of resistance (Paka, AA and some genotypes from the Mlali group and originated from the Comoros archipelago).

The second one is a partial resistance characterized by a slower evolution of disease symptoms as compared with susceptible varieties. This type of resistance is characteristic of cultivars belonging, for example, to the subgroups Pisang Awak (ABB, *i.e.*, Fougamou) and Mysore (AAB).

Since *M. fijiensis* has significant adaptation capacities, already observed in some situations[3], the type of resistance used should be polygenic instead of monogenic and thus the breeding

---

[3] Virulent strains of *M. fijiensis* were observed on Paka in the Cook Islands (Fullerton & Olsen, 1995)[3]. These authors consider that these strains are widespread in the Pacific Islands and also mention that they are virulent on Yangambi km 5 which is also considered as highly resistant to BLSD.

strategy must aim at producing partially resistant rather than highly resistant cultivars. As genetically modified bananas are facing consumer resistance and legislation constraints in most importing countries, the more promising way is to look at innovative ways of conventional breeding.

Few programs are currently focusing on the creation of resistant cultivars through such an approach (Abadie et al., 2009).

Some partially resistant hybrids issued from these programs are already tested but their adaptation to an industry exclusively based on Cavendish cultivars is not an easy task. Consumer and market requirements are major constraints, and selection of suitable export cultivars, if at all possible, is a very long process.

The introduction of resistant cultivars into the agro-system could contribute to a decrease in epidemic development of BLSD on spatial scales that remain to be determined, from field to landscape (Ganry, 2004).

3.2.2 Definition of acceptable disease thresholds

Instead of only targeting a perfect control of the disease through chemical control, the banana industry would probably have better question the economically acceptable level of disease, and define the disease management accordingly. Acceptable disease thresholds should be determined through the modeling of the effects of BLSD on bunch mass. Such models should rely on a better understanding of disease effect on dry matter accumulation at different phenological stages and on the differential mobilization of resources by the different organs at these different stages. Since banana is a semi-perennial crop, such models should integrate successive crop cycles.

This global approach for BLSD on the various components of yield will enable the optimization of bunch weight according to a fixed stage of harvest (through agronomic practices), and *vice versa*.

3.2.3 A more integrated approach based on strategic decision tools

While data on the cost of BLSD control are generally only focusing on direct costs for fungicides and fungicide applications, it is necessary to take into account other components including direct costs (spraying operations, leaf removal, etc.), but also indirect costs such as (i) disease monitoring, (ii) losses (bunches rejected, weight reduction, quality reduction), and (iii) the cost of environmental measures (de Lapeyre de Bellaire et al., 2009).

It is the reason why it is important to think about reliable tools for strategic decisions is the evaluation of the global economic incidence of BLSD, as there are no existing tools for this purpose. Such information should be collected in databases and specific models should be defined in order to simulate the potential benefit of changes in the industry. Only this global approach should justify changes in the industry.

4. Conclusion

The experience gained in the control of the Sigatoka leaf spot diseases, compared to situations of concern observed in Latin America, shows that an integrated approach applied on a sufficiently long period, can master the agricultural risk related to the disease and minimize economic impact and environmental damages as well.

The limited use of fungicides through this system is more sustainable than the systematic strategy, and more than ever it is important to adopt such an approach in the control of the BLSD. Particularly this strategy should be used wherever it is possible, and particularly where fungicide resistance does not prevent it. This assumption should be especially considered in the frame of new banana projects either in traditional banana growing countries (Cameroon, Ivory Coast,….) or in new banana growing countries (Mozambique, Angola, China.…)

It is clear that any failure in the implementation of one component of the system is condemning all the system, jeopardizing any effort done in the implementation of any other components.

For instance, if the inoculum density is becoming too high, due to some failure in leaf removal, in the control of "hotspots", or in the timing of the sprays, the effectiveness of the control will be strongly reduced. Thus, the control strategy would not be able to reduce the inoculum density which will continue to increase, if nothing is done to cut such an amplifying effect. In such a situation, the priority is to reduce drastically the inoculum density through appropriate measures (strong leaf removal, systematic applications with contact fungicides to prevent the emergence of resistant strains to systemic fungicides…) and to consider that it is an essential punctual investment necessary for the sustainability of this strategy. The potential benefit of a temporary shift to a systematic use of protectants is probably the most important lesson that should be learnt from the experience in Cameroon, and specific adaptations of the strategy must be adopted in order to manage efficiently fungicide resistance and thus the sustainability of this system that relies on systemic fungicides.

Taking into account that such a control strategy is dealing with a pathogen with a high evolutionary potential, there could not be any "routine" behavior. On the contrary a permanent "alert" behavior is needed to anticipate any new evolution. The control strategy must be shaped to be prepared to evolve and adapt to new constraints.

It is the case with the emergence of resistant strains to curative fungicides, which requires permanent efforts for fungicide resistance management and also to bring curative fungicides with new mode of action. It is also the case with the increasing social pressure related to environment and health in production areas. In this context, it will be probably essential, in the future, to consider the use of resistant varieties to BLSD, at least in certain critical situations (residential areas, hotspots,...). A permanent innovation is the corollary of an integrated and sustainable approach to control the Black Leaf Streak Disease of bananas. It is the reason why a strong interaction must be kept between the technical services in charge of its implementation and research teams able to provide support for anticipation and adaptation.

5. Acronyms

AA: diploid Acuminata
AAA: triploid Acuminata
BGA: Banana Growers Association

BLSD: Black Leaf Streak Disease
CARBAP : Centre Africain de Recherches sur Bananiers et Plantains
CIRAD: Centre de Cooperation Internationale en Recherche Agronomique pour le Développement.
FRAC: Fungicide Resistance Action Committee
FWI: French West Indies
GPS: Global Positioning System
IBS: Inhibitors of ergosterol biosynthesis
IFAC: Institut Français de Recherches Fruitières Outre-Mer
IRFA: Institut de Recherche sur les Fruits et Agrumes
LER : leaf emission rate
Mt : Million tons
NLH: number of functional leaves at harvest
PE: Piche evaporation
SEDb: The Stage of Evolution of the Black Leaf Streak Disease
SEDs : the Stage of Evolution of the Sigatoka Disease
SLSD: Sigatoka Leaf Spot Disease
YLS: the youngest leaf spotted
YLSt: the youngest leaf bearing streaks

6. References

Abadie, C., Hubert, O., Ngando Essoh, J., Ngoh, G., Mbéguié-A-Mbéguié, D., de Lapeyre de Bellaire, L., Chillet, M. (2008). Evidence of the effects of *Mycosphaerella* leaf spot diseases on fruit quality. In: C. N. J.S. Borja, C. Orrantia, R. Paladines, V. Quimí, L. Tazán, (Ed.), *XVIIIth ACORBAT meeting, 10-14 November 2008, Guayaquil, Ecuador.*

Abadie, C., Chilin-Charles, Y., Huat, J., Salmon, F., Pignolet, L., Carlier, J., Lescot, T., Côte, F., Jenny, C. (2009). New approaches to select cultivars of banana with durable resistance to Mycosphaerella leaf spot diseases. *Acta Horticulturae*, Vol.828, pp. 171-178

Beveraggi, A., Mourichon, X., Salle, G. (1995). Comparative study of the first stages of infection in sensitive and resistant banana plants with *Cercospora fijiensis* (*Mycosphaerella fijiensis*), responsible for Black Leaf Streak

Disease, *Can. J. Bot.*, Vol. 73, pp. 1328–1337

Brun, J. (1963).La Cercosporiose du bananier. Thèse Doctorat d'Etat, Université de Paris

Bureau, E., Ganry, J., Zapater, M. F., Laville, E. (1982). Les cercosporioses du bananier et leurs traitements. Evolution des populations pathogènes. Distribution géographique et évolution des populations de *Mycosphaerella musicola* résistantes aux benzimidazoles dans les zones bananières de Guadeloupe. *Fruits*, Vol.37, pp. 665-672

Bureau, E., Ganry, J. (1987). A climatic forecasting system to control banana Sigatoka (*Mycosphaerella fijiensis*), using sterol-biosynthesis inhibiting fungicides. *Fruits*, Vol.42, pp. 199-205

Bureau, E. (1990). Adaptation d'un système d'avertissement à la lutte contre la cercosporiose noire (*Mycosphaerella fijiensis* Morelet) en plantation de banane plantain au Panama. *Fruits*, Vol.45, pp. 329-338

Carlier, J., Fouré, E., Gauhl,F., Jones, D. R., Lepoivre, P., Mourichon, X., Pasberg-Gauhl, C., Romero, R.A. (2000). Fungal diseases of the foliage - Sigatoka Leaf Spots, in *"Diseases of banana,abaca and Ensete"*, D.R. Jones, cabi Publishing, U.K., 37-91

Chillet, M., Abadie, C., Hubert, O., Chilin-Charles, Y., De Lapeyre de Bellaire, L. (2009). Sigatoka disease reduces the greenlife of bananas. *Crop Protection*, Vol.28, pp. 41-45

Cuillé, J., Guyot, H. (1956). Les traitements aériens en bananeraie contre *Cercospora musae*. *Fruits*, Vol. 11, pp. 435-441

Cuillé, J., Blanchet, B. (1958). Les huiles de traitements, leur phytotoxicité. *Fruits*, Vol.13, pp. 53-65

Daniells, J.W., Lisle, A.T., Bryde, N.J. (1994). Effect of bunch trimming and leaf removal at flowering on maturity bronzing, yield, and other aspects of fruit-quality of bananas in North Queensland, *Aust. J. Exp. Agric.*, Vol. 34, pp. 259–265

de Lapeyre de Bellaire, L. (2006). Belize, disease monitoring and management exper FWC Lot 1/2005/111696. Report for mission 1: 22 January to 5 February 2006. Cirad, Montpellier, France. 25 p.

de Lapeyre de Bellaire, L. (2007). Belize, disease monitoring and management exper FWC Lot 1/2005/111696. Report for mission 5 : 16/04 to 28/4/2007. Cirad, Montpellier, France. 120 p.

de Lapeyre de Bellaire, L., Essoh Ngando, J., Abadie, C., Chabrier, C., Blanco, R., Lescot, T., Carlier, J., Côte, F.(2009). Is chemical control of *Mycosphaerella* foliar diseases of bananas sustainable? *Acta Hortic.,Vol.* 828, pp. 161–170

de Lapeyre de Bellaire, L., Fouré, E., Abadie, C., Carlier, J. (2010a). Black leaf streak disease is challenging the banana industry, *Fruits*, Vol.65, pp. 327-342

de Lapeyre de Bellaire, L., Ngando, J. E., Rieux, A., Carlier, J., Ravigné, V., Zapater, M.F., Careel, F. (2010b). Etude sur les conditions de réintroduction des fongicides systémiques dans les programmes de lutte contre la maladie des raies noires au Cameroun dans la zone de production de la banane dessert d'exportation : Rapport narratif final N°9 décembre 2010 . Douala : CARBAP, 43 p.

Dens, K.R., Romero, R.A., Swennen, R., Turner, D.W. (2008). Removal of bunch, leaves, or pseudostem alone, or in combination, influences growth and bunch weight of ratoon crops in two banana cultivars, *J. Hortic. Sci. & Biotechnol.* Vol. 83, pp. 113–119

Diaz, E. (1986). Una alternativa de solucion al problema de la Sigatoka negra en la produccion de platanos en Panama, Carta Bananera, No.2-3, pp. 29-32

Eckstein, K., Robinson J.C., Davie S.J. (1995). Physiological responses of banana (*Musa*-AAA, Cavendish subgroup) in the subtropics. 3. Gas-exchange, growth analysis and source-sink interaction over a complete crop cycle, *J. Hortic. Sci.,* Vol.70, pp. 169–180

Fouré, E. (1982a). Les cercosporioses du bananier et leurs traitements. Comportement des variétés. Études de la sensibilité variétale des bananiers et plantains à *Mycosphaerella fijiensis* Morelet au Gabon (maladie des raies noires). I. Incubation et évolution de la maladie. *Fruits*, Vol.37, pp. 749-759

Fouré, E. (1982b). Les cercosporioses du bananier et leurs traitements. Comportement des variétés. Etude de la sensibilité variétale des bananiers et plantains à *Mycosphaerella fijiensis* Morelet au Gabon (maladie des raies noires). II- Etude de quelques paramètres. *Fruits*, Vol.37, pp. 760-771

Fouré,E.(1982c).La maladie des raies noires des bananiers et plantains M.fijiensis (Morelet) Etude comparée des différents symptômes et stades de la maladie au Gabon IRFA-CIRAD-France et SONADECI-Gabon

Fouré, E. (1983). Les cercosporioses du bananier et leurs traitements. Sélection de molécules fongicides nouvelles. Activités comparées de différentes molécules fongicides sur *Mycosphaerella fijiensis* Morelet, agent de la 'maladie des raies noires' des bananiers et plantains au Gabon. *Fruits*, Vol.38, pp. 21-34

Fouré, E. (1984). Les cercosporioses du bananier et leurs traitements. Sélection de molécules fongicides nouvelles. Activités comparées de différentes molécules fongicides sur *Mycosphaerella fijiensis* Morelet, agent de la maladie des raies noires des bananiers et plantains du Gabon. (suite IV). *Fruits*, Vol.39, pp. 753-765

Fouré, E., Grisoni, M. (1984). Les cercosporioses du bananier et leurs traitements. Sélection de molécules fongicides nouvelles. Activités comparées de différentes molécules fongicides sur *Mycosphaerella fijiensis* Morelet, agent de la maladie des raies noires des bananiers et plantains du Gabon. (suite III). *Fruits*, Vol.39, pp. 427-440

Fouré, E., Grisoni, M., Zurfluh, R. (1984). Les cercosporioses du bananier et leurs traitements. Comportement des variétés. Etude de la sensibilité variétale des bananiers et plantains à *Mycosphaerella fijiensis* Morelet et de quelques caractéristiques biologiques de la Maladie des raies noires au Gabon. *Fruits*, Vol.39, pp. 365-378

Fouré, E. (1985). Les cercosporioses du bananier et leurs traitements. Comportement des variétés. Etude de la sensibilité variétale des bananiers et plantains à *Mycosphaerella fijiensis* Morelet au Gabon (maladie des raies noires). suite III. *Fruits*, Vol.40, pp. 393-399

Fouré, E. (1988a). Activités comparées de différentes molécules fongicides sur *Mycosphaerella fijiensis*, agent de la maladie des raies noires des bananiers et plantains au Cameroun. *Fruits*, Vol.43, pp. 15-19

Fouré, E. (1988b). Les cercosporioses du bananier et leurs traitements. Efficacités comparées du pyrazophos et du triadiménol sur *Mycosphaerella fijiensis* Morelet (agent de la cercosporiose noire des bananiers et des plantains au Cameroun) lors de traitements sur grandes surfaces. *Fruits*, Vol.43, pp. 143-147

Fouré, E. (1988c). Stratégies de lutte contre la cercosporiose noire des bananiers et des plantains provoquée par *Mycosphaerella fijiensis* Morelet. L'avertissement biologique au Cameroun. Evaluation des possibilités d'amélioration. *Fruits*, Vol.43, pp. 269-274

Fouré,E.,Mouliom Pefoura,A.(1988).Les cercosporioses des bananiers et leurs traitements.Efficacités comparées de différentes molécules fongicides sur *Mycosphaerella fijiensis* Morelet, agent de la Maladie des Raies Noires des bananiers et des plantains au Cameroun. *Fruits*, Vol.43, N°3, pp 67-72

Fouré, E, Lescot, T. (1988) Variabilité génétique des *Mycosphaerella* inféodés au genre *Musa*. Mise en évidence de la présence au Cameroun sur bananiers et plantains d'une cercosporiose (*Mycosphaerella musicola*) au comportement pathogène atypique, *Fruits*, Vol.43, No7, 8, pp. 407-415

Fouré, E, Mouliom, Pefoura A, Mourichon, X. (1990). Etude de la sensibilité des bananiers et des plantains à *M.fijiensis* Morelet au Cameroun. Caractérisation de la résistance au champ de bananiers appartenant à divers groups génétiques. *Fruits*, Vol.45, No.4, pp. 339-345

Fouré,E.,Moreau,A.(1992).Contribution à l'étude épidémiologique de la cercosporiose noire dans la zone bananière du Mungo de 1987 à 1989. *Fruits,* Vol.47, N°1, pp.3-16

Fouré, E., Ganry, J. (2008). A biological forecasting system to control Black Leaf Streak disease of bananas and plantains. *Fruits,* Vol.63, pp. 311-317

FRAC. 2004. Recommendations of the FRAC Banana Working Group: *www.frac.info*

Frossard, P. (1980). Apparition d'une nouvelle et grave maladie foliaire des bananiers et plantains au Gabon : la maladie des raies noires : *Mycosphaerella fijiensis* Morelet. *Fruits,* Vol.35, pp. 519-527

Fullerton, R., Olsen, T. (1995). Pathogenic variability in Mycosphaerella fijiensis Morelet cause of black Sigatoka in banana and plantain. *New Zealand Journal of Crop and Horticultural Science,* Vol.23, pp. 39-48

Ganry, J., Meyer, J. P. (1972). La lutte contrôlée contre le cercospora aux Antilles. 1-Bases climatiques de l'avertissement. *Fruits,* Vol.27, pp. 665-676

Ganry, J., Meyer, J-P. (1972). La lutte contrôlée contre le Cercospora aux Antilles. Technique d'observation et de numérotation de la maladie, *Fruits,* Vol.27, pp. 767–774

Ganry, J., Laville, E. (1983). Les cercosporioses du bananier et leurs traitements. 1- Traitements fongicides. 2- Avertissement. *Fruits,* Vol.38, pp. 3-20.

Ganry,J. (2001). Maitrise de la culture du bananier pour une production raisonnée face aux nouveaux défis, *C.R. Acad. Agric. Fr.,* Vol.87, No.6, pp. 119-127

Ganry, J. (2004). Specific diversity of plant populations at rainfed scale and crop protection: the example of banana production in the French West Indies. *Comptes rendus Biologies,* Vol.327, No.7, pp. 621-627

Ganry J., Lapeyre de Bellaire L., Mourichon X., 2008. A biological forecasting system to control Sigatoka leaf spot of bananas and plantains, *Fruits,* Vol.63, pp. 381-387

Guyot, H., Cuillé,J. (1954).Les formules fongicides huileuses pour le traitement des bananeraies, *Fruits,* Vol.9, No.7, pp. 289-292

Guyot, H., Cuillé,J. (1955). Les traitements fongicides des bananeraies - Efficacité des différents modes de traitement - Rôle de l'huile, *Fruits,* Vol.10, No.3, pp. 101-107

Henderson, J., Pattemore, J. A., Porchun, S. C., Hayden, H. L., Van Brunschot, S., Grice, K. R. E., Peterson, R. A., Thomas-Hall, S. R., Aitken, E. A. B. (2006). Black Sigatoka disease: new technologies to strengthen eradication strategies in Australia. *Australasian Plant Pathology,* Vol.35, pp. 181-193

Jones, D.R. (2000). Diseases of banana, abaca and ensete, *CAB Int.,* Wallingf., UK

Jimenez, F., Tapia, A. C., Gribius, N., Escalant, J. V. (1995). Relation entre la durée de pluie et le développement de la cercosporiose noire sur le bananier plantain. Proposition d'un système d'avertissement biométéorologique. *Fruits,* Vol.50, pp. 87-99

Karaoglanidis G.S., Thanassoulopoulos C.C., Ioannidis P.M. (2001). Fitness of Cercospora beticola field isolates - resistant and - sensitive to demethylation inhibitor fungicides. *European Journal of Plant Pathology,* Vol.107, pp.337-347

Lescot, T., Simonot, H., Fages,O., Escalant, J.V. (1998). Avertissement biométéorologique pour lutter contre la cercosporiose noire en plantations de banane plantain au Costa Rica. *Fruits,* Vol.53, pp. 3-16.

Lescot, T. (2011). La diversité génétique des bananiers, *FruiTrop* Vol.89, pp. 58-62

Loeillet, D. (2005). Le commerce international de la banane : entre évolution et révolution, Frui-Trop, Vol.129, pp. 2–19

Marin, D.H., Corrales, O. (1992). Sistema de preaviso simplificado para el combate de la Sigatoka Negra en platano, *in Corporacion Bananera Nacional, Informe Anual 1991*, san José, Costa-Rica, pp. 90-91

Marin, D.H.,Romero, R.A.,Guzmàn, M., Sutton, T.B. (2003). Black Sigatoka : An Increasing Threat to Banana Cultivation, Plant Disease, Vol.87, No. 3, pp. 208-222

Mobambo, K.N., Gauhl, F., Pasberg-Gauhl, C., Zuofa, K. (1996). Season and plant age effect evaluation of plantain for response to black sigatoka disease, *Crop Prot.* Vol.15

Pennisi ,E. (2010). Armed and dangerous, *Science, Vol.*327, pp. 804–805

Ramsey, M.D., Daniells, J.W., Anderson, D.J. (1990). Effects of Sigatoka leaf spot (*Mycosphaerella musicola* Leach) on fruit yields, field ripening and greenlife of bananas in North Queensland, *Sci. Hortic.* Vol.41, pp. 305–313

Risède, J.M., Lescot, T., Cabrera Cabrera, J., Guillon, M., Tomekpé, K., Kema, G.H.J., Côte, F. (2010). Challenging short and mid-term strategies to reduce pesticides in bananas. [Online]. [S.l.] : ENDURE, [8] p. (Banana Case Study - Guide, 1).

Romero, R. A., Sutton, T. B. (1997). Sensitivity of Mycosphaerella fijiensis, causal agent of Black Sigatoka of Banana, to Propiconazole. *Phytopathology*, Vol.87, pp. 96-100

Stover, R.H. (1971). A proposed international scale for estimating intensity of banana leaf spot (*Mycosphaerella musicola* Leach),*Trop. Agric.*, Vol.48, pp. 185–196

Stover, R.H. (1972). Banana, plantain and abaca diseases, *Commonw. Mycol. Inst.*, Kew, Surrey, UK

Stover, R.H. (1983). Effet du cercospora noir sur les plantains en Amérique centrale, *Fruits* Vol.38, pp. 326–329

Stover, R.H. (1986). Disease management strategies and the survival of the banana industry. *Annual Review of Phytopathology*, Vol.24, pp. 83-91

Stover, R.H., Simmonds, N.H. (1987). Bananas, 3rd edit. *Longman Scientific and Technical*, Harlow, Essex. UK, 468 pp

Vargas, A., Araya M., Guzman M., Murillo G. (2009). Effect of leaf pruning at flower emergence of banana plants (*Musa* AAA) on fruit yield and black Sigatoka (*Mycosphaerella fijiensis*) disease, *Int. J. Pest Manag.*, Vol.55, pp. 19–25

11

Yield Response to Foliar Fungicide Application in Winter Wheat

Stephen Wegulo, Julie Stevens,
Michael Zwingman and P. Stephen Baenziger
University of Nebraska-Lincoln
USA

1. Introduction

Fungicides are routinely applied to control fungal diseases of wheat and other cereal crops, with the main goal of preventing yield loss (or increasing yield) and hence maximizing economic returns. In North America, the fungicides used to control foliar fungal diseases of wheat belong to two major classes with a broad spectrum of activity against fungal pathogens. These are the strobilurins and triazoles. Fungicides in both classes are used as foliar fungicides and seed treatments. The strobilurins are named in recognition of a mushroom, *Strobilurus tenacellus,* the original source of the chemical compound that formed the basis of the chemistry of this fungicide class. They are quinone outside inhibitors (QoI) and work by interfering with energy production in fungi (Vincelli, 2002). They act as local systemics by inhibiting fungal spore germination and early infection, and are highly effective when applied preventively. The strobilurins have a single-site mode of action. Examples of strobilurin fungicides used in cereal crop production in North America are azoxystrobin, pyraclostrobin and trifloxystrobin.

The triazoles are characterized by having a five-membered ring of two carbon atoms and three nitrogen atoms. They are curative and move systemically through the plant xylem. Triazoles slow fungal growth through the inhibition of sterol biosynthesis (Buchenauer, 1987). Sterols are essential building blocks of fungal cell membranes and are inhibited at a single site by triazoles. Because of their curative activity against early fungal infections and their ability to redistribute in the crop, triazoles are highly effective and reliable (Hewitt, 1998). Examples of triazoles used in cereal crop production in North America are metconazole, propiconazole, prothioconazole, and tebuconazole.

In the Great Plains of the United States, the most common foliar diseases of winter wheat are leaf rust (*Puccinia triticina*), powdery mildew (*Blumeria graminis* f. sp. *graminis*), tan spot (*Pyrenophora tritici-repentis*) (anamorph: *Drechslera tritici-repentis*), Septoria tritici blotch (*Mycosphaerella graminicola*) (anamorph: *Septoria tritici*), spot blotch (*Cochliobolus sativus*) (anamorph: *Bipolaris sorokiniana*), and Stagonospora nodorum blotch (*Phaeosphaeria nodorum*) (anamorph: *Stagonospora nodorum*). Stripe rust (*Puccinia striiformis* f. sp. *tritici*) and stem rust (*Puccinia graminis* f. sp. *tritici*) also occur, but less commonly.

The magnitude of yield loss caused by these diseases in winter wheat is variable and depends on several factors including environmental conditions during the growing season, cultural practices, and cultivar resistance. Leaf rust occurs every year in the wheat-producing regions of the U.S. In 2007, severe epidemics of leaf rust occurred in the Great Plains region of North America, causing yield losses of up to 14% (Kolmer et al., 2009). Stripe rust is more frequent in the western U.S., especially the Pacific Northwest (Sharma-Poudyal & Chen, 2011). However, it can be widespread in certain years, as in 2010 when severe epidemics occurred throughout the wheat-producing regions of North America. Yield losses of up to 74% due to stripe rust have been documented in experimental fields (Sharma-Poudyal & Chen, 2011). Stem rust has been effectively controlled in the U.S. through genetic resistance and eradication of barberries (*Berberis vulgaris* and *B. Canadensis*), which act as alternate hosts. Stem rust has the potential to cause 100% yield loss (Murray et al., 1998). Powdery mildew occurs wherever wheat is grown and is common where high humidity prevails during the growing season. Yield losses of up to 25% due to powdery mildew have been reported (Murray et al., 1998).

Spot blotch occurs commonly in the Great Plains of the United States (Murray et al. 1998). The causal agent, *C. sativus*, also causes common root rot and seedling blights in wheat. Spot blotch often occurs together with tan spot (Duveiller et al., 2005). In wet growing seasons, Septoria tritici blotch also can occur as part of this foliar disease complex. This leaf spot disease complex is favored by cultural practices that leave crop residue on the soil surface (Watkins & Boosalis, 1994). Yield losses of up to 50% have been documented to be caused by these leaf spot diseases in winter wheat (Murray et al. 1998; Villareal et al., 1995; Wegulo et al., 2009).

2. The use of fungicides to control foliar fungal diseases of wheat

Fungicides have been used routinely in cereal production since the development of systemics in the late 1960s (Hewitt, 1998). New fungicide chemistries have been developed steadily over the last several decades, in part to increase efficacy and overcome resistance to older chemistries in pathogen populations. The benefits of fungicide use in crop production have long been acknowledged. Ordish and Dufour (1969) noted the popularity of spraying fungicides to control crop diseases; returns of up to three times the cost involved often were realized from fungicide application. In the United Kingdom, experiments conducted from 1978 to 1982 showed that applying fungicides to winter wheat resulted in a yield response of up to 89%, and the value of the increased yield from fungicide application to cereals in 1982 was nearly double the fungicide costs (Cook and King, 1984). In Denmark, fungicide application to control powdery mildew and *Septoria* diseases resulted in yield increases of 400-2700 kg ha^{-1} with margin over cost varying from -500 kg ha^{-1} to 2000 kg ha^{-1} (Jørgensen et al., 2000). An economic evaluation of fungicide use in winter wheat in Sweden showed a mean net return of US$28 ha^{-1} during the period 1995-2007 and $16 ha^{-1}during the period 1983-2007 (Wiik and Rosenqvist, 2010).

In the U.S., various studies have demonstrated yield increases in winter wheat due to fungicide application. Wegulo et al. (2009) showed that up to 42% yield loss was prevented by applying foliar fungicides to winter wheat. Kelley (2001) found that over a period of six years, the fungicide propiconazole significantly increased winter wheat yield 77% of the time. Vamshidhar et al. (1998) demonstrated significant yield increases from fungicide application to control the disease complex of leaf rust, tan spot, and Septoria tritici blotch in winter wheat. They found that cultivar specific economic benefits were associated with improved wheat quality from fungicide treatment. Ransom and McMullen (2008) showed that within an environment and averaged across winter wheat cultivars, fungicides improved yields by 5.5 to 44.0%. Tebuconazole applied at Zadoks growth stage (GS) 37 (Zadoks, 1974) and propiconazole applied at GS 37 followed by triadimefon + mancozeb at GS 55 to control leaf rust and Septoria tritici blotch consistently resulted in the lowest disease severities and highest winter wheat yields (Milus, 1994).

In the Great Plains region of the U.S., the prevalence, incidence, and severity of tan spot and other residue-borne diseases such as spot blotch and Septoria tritici blotch have increased over the last several decades due to a shift toward conservation tillage practices that leave crop debris on the soil surface (Watkins and Boosalis, 1994). The damage caused by these and other foliar fungal diseases has promoted the use fungicides in winter wheat production in the region.

3. Timing of foliar fungicide application in winter wheat

Fungicides are generally applied to winter wheat 1-2 times per season. Some farmers apply a fungicide early in the growing season during the stem elongation growth stage to control early season diseases such as tan spot. Often these early fungicide applications are done in combination with herbicide or fertilizer application. A second fungicide application is usually timed to protect the flag leaf. A high risk of Fusarium head blight may necessitate a third fungicide application at early flowering. Results from previous studies on the effect of fungicide application timing on yield in winter wheat have been inconsistent. Some studies have demonstrated yield loss from early season infections and a benefit from early fungicide application in winter wheat. Shabeer and Bockus (1988) found that about 17% of total yield loss from tan spot occurred from early season infections. Marroni et al. (2006) found that the lowest area under the disease progress curve (AUDPC) and the best level of protection against early season Septoria tritici blotch were achieved with azoxystrobin applied at the pre-stem extension stage of crop growth. They also found good control of the disease when a mixture of azoxystrobin and epoxiconazole was applied at the pre-stem extension stage or at the stem extension stage. Cromey et al. (2004) found no consistent effects of crop growth stage when the fungicides azoxystrobin and tebuconazole were applied at three alternative growth stages between flag leaf emergence and flowering to control *Didymella exitialis* (anamorph: *Ascochyta spp.*). Bockus et al. (1997) found the optimum timing to be between the boot and the fully headed growth stages. Duczek and Jones-Flory (1994) found the optimum timing to be between extension of the flag leaf and the medium milk growth stages. Wiersma and Motteberg (2005) found that across cultivars, the optimum timing for foliar fungicide application was GS 60 rather than GS 39. Because of the inconsistent results from previous studies, experiments were conducted in Nebraska, USA to investigate the effects of fungicides and fungicide application timing on disease severity, yield and economic returns in winter wheat.

4. Methods

4.1 Field experiments

The methods used in field experiments have been described previously (Wegulo et al., 2009; Wegulo et al., 2011).

4.1.1 2006 field experiments

In autumn 2005, seed of winter wheat cv. Millennium was planted with a small plot drill at the University of Nebraska's Agricultural Research and Development Center (ARDC) near Mead (9 Oct), the South Central Agricultural Laboratory (SCAL) near Clay Center (22 Sep), the West Central Research and Extension Center (WCREC) near North Platte (21 Sep), and the High Plains Agricultural Laboratory (HPAL) near Sidney (6 Sep) (Fig. 1).

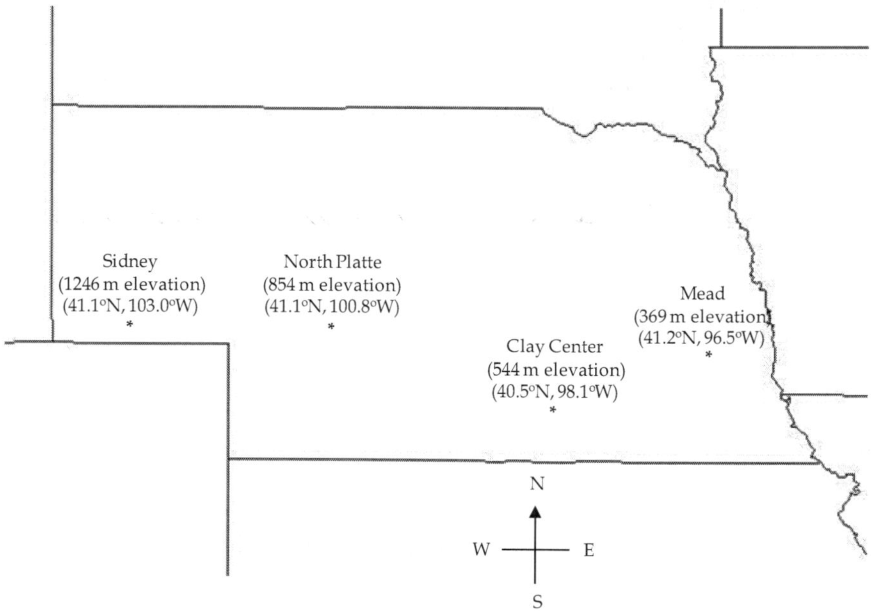

Fig. 1. Map of Nebraka, USA (not to scale) showing the locations where field experiments were conducted in 2006 and 2007 to determine the effects of fungicides and fungicide application timing on foliar fungal disease severity, yield increase and net return in winter wheat cv. Millennium.

Standard agronomic practices for wheat production were followed at each location. Seeding rate was 98, 84, 72, and 50 kg ha^{-1} at Mead, Clay Center, North Platte, and Sidney, respectively. Row spacing was 25.4 cm and plot size was 1.8 m x 4.6 m at Mead, 1.2 m x 8.2 m at Clay Center and Sidney, and 2.1 m x 4.6 m at North Platte. Four fungicides were each applied once at GS 31 (first node on the stem detectable) or GS 37 (flag leaf just visible) (Table 1). The fungicides were azoxystrobin (7.0% of marketed product) + propiconazole (11.7%) (Quilt, Syngenta Crop Protection, Greensboro, NC), pyraclostrobin (23.6%) (Headline, BASF Ag Products, Research Triangle Park, NC), azoxystrobin (22.9%) (Quadris, Syngenta Crop Protection, Greensboro, NC), and trifloxystrobin (11.4%) + propiconazole (11.4%) (Stratego, Bayer CropScience, Research Triangle Park, NC). Treatments were arranged in randomized complete blocks with four replications.

Due to minimum or no-tillage practices and inclusion of winter wheat in crop rotation schemes, primary inoculum of *P. tritici-repentis* was provided naturally at all locations from pseudothecia on wheat straw from previous wheat crops. Inoculum of other fungal foliar pathogens such as *M. graminicola*, *B. graminis* f.sp. *tritici*, and *P. triticina* also occurred naturally. At GS 31, plots were inoculated with conidia of *B. sorokiniana* on 2 May at Mead and Clay Center and on 5 May at Sidney and North Platte. Conidia were obtained by culturing mycelia from a single spore isolate of *B. sorokiniana* on V8 agar media in 9-cm-diameter petri plates at 20°C for 7 to 14 days in continuous darkness. Sterile distilled water was added to each petri plate and conidia were dislodged with a rubber policeman. The conidial/mycelial suspension that resulted was filtered through several layers of cheese cloth to obtain the conidial suspension. Conidial concentration was determined with a haemacytometer.

Thirty millilitres of inoculum containing 70 000 conidia ml^{-1} m^{-2} were sprayed onto wheat leaves in each plot with a hand-pumped back pack sprayer. Fungicide treatments were applied 24 h after inoculation at each location. Fungicides were applied with a CO_2-powered back pack sprayer set at 276 kPa, with a 1.2-m-wide boom and 4 Teejet # 800-1VS nozzles spaced 0.3 m apart. Tan spot and spot blotch severity (%) was visually estimated together on the flag leaf of thirty randomly selected plants per plot at growth stage GS 55 (50% of inflorescence emerged) at Sidney and GS 60 (beginning of anthesis) at Mead, Clay Center, and North Platte. At maturity, plots were harvested with a small plot combine and grain yield was determined.

4.1.2 2007 field experiments

Seed of winter wheat cv. Millennium was planted with a small plot drill in autumn 2006 at the University of Nebrasaka's Agricultural Research and Development Center near Mead (26 Sep), the South Central Agricultural Laboratory near Clay Center (27 Sep), the West Central Research and Extension Center near North Platte (17 Sep), and the High Plains Agricultural Laboratory near Sidney (13 Sep) (Fig. 1). Standard agronomic practices for wheat production were followed at each location. Seeding rate was 72, 84, 72, and 50 kg ha^{-1} at Mead, Clay Center, North Platte, and Sidney, respectively. Row spacing was 25.4 cm and plot size was 2.4 m x 2.4 m at Mead, Clay Center, and North Platte and 1.2 m by 6.7 m at Sidney.

Primary inoculum of *P. tritici-repentis* was provided naturally at all locations from pseudothecia on wheat straw from previous wheat crops. To ensure development of spot blotch, plots were inoculated with conidia of *B. sorokiniana* at GS 30 (pseudostem erection) on 24 Apr, 25 Apr, 26 Apr, and 27 Apr at Sidney, North Platte, Clay Center, and Mead, respectively. A second inoculation was similarly done at GS 31 (first node of stem detectable) on 6 May, 7 May, 8 May, and 9 May at Sidney, North Platte, Clay Center, and Mead, respectively. Inoculum was obtained, prepared, and applied as in 2006.

Five fungicides were each applied once at GS 31 (first node on the stem detectable) or GS 39 (ligule/collar of flag leaf just visible) (Table 1). The fungicides were azoxystrobin (7.0% of marketed product) + propiconazole (11.7%) (Quilt, Syngenta Crop Protection, Greensboro, NC), pyraclostrobin (23.6%) (Headline, BASF Ag Products, Research Triangle Park, NC), propiconazole (41.8%) (Tilt, Syngenta Crop Protection, Greensboro, NC), azoxystrobin (22.9%) (Quadris, Syngenta Crop Protection, Greensboro, NC), and trifloxystrobin (11.4%) + propiconazole (11.4%) (Stratego, Bayer CropScience, Research Triangle Park, NC). Fungicides were applied with a CO_2-powered back pack sprayer set at 276 kPa, with a 1.2-m-wide boom and four Teejet # 800-1VS nozzles spaced 0.3 m apart. Treatments were arranged in randomized complete blocks with four replications. Tan spot and spot blotch severity (%) was visually estimated together on the foliage of thirty plants at each of three arbitrarily selected sites per plot at GS 60 (beginning of anthesis). At maturity, plots were harvested with a small plot combine and grain yield was determined.

4.2 Economic analysis

Table 1 shows the fungicide costs, fungicide application cost, and wheat prices used in economic analysis. Average wheat prices were calculated from data provided by the USDA Agricultural Marketing Service. The average local prices during thirteen months were used. The months used were August prior to crop planting to August following crop harvest in 2005-2006 and 2006-2007. Fungicide prices ($ ha^{-1}) were obtained by surveying local retailers and chemical manufacturers and averaged. Adjuvant and surfactant costs were omitted because of the wide variation in their uses and costs. Fungicide application costs were obtained by surveying commercial applicators in Nebraska. All surveys were conducted in 2009 by telephone. Information provided by those surveyed was obtained from 2006 and 2007 records. Aerial application cost was used. Because aerial fungicide application is by contract between the grower and the commercial applicator, machinery and machinery maintenance costs were omitted.

2006			2007		
Fungicide treatment	Fungicide Cost ($ ha^{-1})	Fungicide application cost ($ ha^{-1})	Fungicide treatment	Fungicide cost ($ ha^{-1})	Fungicide Application cost ($ ha^{-1})
Quilt GS 31, 0.58 l ha^{-1}	16.83	18.19	Quilt GS 31, 1.02 l ha^{-1}	32.76	18.19
Quilt GS 37, 0.58 l ha^{-1}	16.83	18.19	Quilt GS 39, 1.02 l ha^{-1}	32.76	18.19
Headline GS 31, 0.58 l ha^{-1}	27.48	18.19	Headline GS 31, 0.66 l ha^{-1}	34.22	18.19
Headline GS 37, 0.58 l ha^{-1}	27.48	18.19	Headline GS 39, 0.66 l ha^{-1}	34.22	18.19
Quadris GS 31, 0.58 l ha^{-1}	35.21	18.19	Tilt GS 31, 0.29 l ha^{-1}	27.08	18.19
Quadris GS 37, 0.58 l ha^{-1}	35.21	18.19	Tilt GS 39, 0.29 l ha^{-1}	27.08	18.19
Stratego GS 31, 0.73 l ha^{-1}	25.38	18.19	Quadris GS 31, 0.88 l ha^{-1}	58.73	18.19
Stratego GS 37, 0.73 l ha^{-1}	25.38	18.19	Quadris GS 39, 0.88 l ha^{-1}	58.73	18.19
...	...	...	Stratego GS 31, 0.73 l ha^{-1}	28.17	18.19
...	...	...	Stratego GS 39, 0.73 l ha^{-1}	28.17	18.19

Table 1. Fungicide treatments and fungicide and fungicide application costs used to calculate net return from applying fungicides to winter wheat cv. Millennium to control foliar fungal diseases at four locations in Nebraska, USA in 2006 and 2007.

Net return from fungicide application was calculated as

$$R_n = Y_i P - (F_c + A_c) \quad (1)$$

where R_n is the net return from fungicide application (\$ ha^{-1}); Y_i is yield increase from fungicide application (kg ha^{-1}), obtained by subtracting the yield in the check treatment from the yield in the fungicide treatments; P is the wheat price (\$ kg^{-1}); F_c is the fungicide cost (\$ ha^{-1}); and A_c is the fungicide application cost (\$ ha^{-1}).

4.3 Data analysis

Data from each of the four locations were subjected to analysis of variance using the the GLM procedure of SAS (SAS Institute, Cary, NC). These data from individual locations have been published previously (Wegulo et al., 2009; Wegulo et al., 2011). To determine the overall effect of fungicides and fungicide application timing on disease severity, yield increase, and net return (the data reported in this chapter), the means from each location were further subjected to analysis of variance using the GLM procedure of SAS. In this latter analysis, each of the four locations was considered a replication. Fisher's least significant difference test at $P = 0.05$ (Gomez and Gomez, 1984) was used to compare pairs of treatment means. Linear regression analysis (Gomez and Gomez, 1984) was used to model the relationships between disease severity and yield increase, between disease severity and net return, and between yield increase and net return with disease severity and yield increase as independent variables and yield increase and net return as dependent variables.

4.4 Results and discussion

4.4.1 Effect of weather on disease severity

Average total rainfall across the four locations for the months of May, June, and July (the period of active vegetative growth and grain filling in the winter wheat crop in Nebraska) was 15.6 cm in 2006 and 30.6 cm in 2007 (Table 2). Therefore, although average temperature was similar in both years, the growing season in 2006 was unusually dry whereas it was excessively wet in 2007. Consequently, disease severity was very low in 2006 compared to 2007 (Table 2).

Environment has a major influence on the development of plant disease epidemics (Campbell and Madden, 1990). Temperature and moisture are especially critical to the development, reproduction, and survival of plant pathogens. In this study, since temperature was similar in both years, the difference in disease severity between 2006 and 2007 was attributable to moisture. In 2006, dry conditions considerably slowed down disease development, resulting in very low disease severity. Excessive moisture in 2007 favored the development of severe epidemics, resulting in considerable disease severity even in sprayed plots (Tables 2 and 3).

This study demonstrates that variation in weather from year to year can significantly impact not only disease development but overall yield. Average yield in sprayed plots in 2006 was 36.4% less than average yield in sprayed plots in 2007 (Table 2). This was likely because of lack of adequate moisture during the grain filling period in 2006. Due to the low disease severity in 2006, the average yield from unsprayed plots was only 12.6% lower than the average yield in sprayed plots compared to 2007 when the average yield from unsprayed plots was 29.4% lower than the average yield from sprayed plots.

This variable effect of weather on disease and yield has been demonstrated in other studies. Using results from fungicide field trials conducted from 1983 to 2007 and disease surveys conducted from 1988 to 2007 in winter wheat in southern Sweden, Wiik and Elwadz (2009) showed through regression analysis that air temperature and precipitation explained more than 50% of the variation in yield increase between years. They found May precipitation to be the factor most consistently related to Septoria tritici blotch, Stagonospora nodorum blotch, and tan spot. In the UK, Gladders et al. (2001) showed that year to year variation in the severity of Septoria tritici blotch was greater than spatial variation.

	2006	2007
Average total rainfall (cm)	15.6	30.6
Average temperature (°C)	21.3	20.7
Average disease severity (%)		
Sprayed plots	2.4	21.8
Unsprayed plots	4.3	58.6
Average yield (kg ha^{-1})		
Sprayed plots	2963	4658
Unsprayed plots	2589	3288
Average yield increase (kg ha^{-1})	394	1369
Average net return ($ ha^{-1})	12	189
Probability of a positive net return	0.63	1.00

Table 2. Average total rainfall and temperature (May, June, and July), disease severity, yield increase, and net return; and the probability of a positive net return from experiments conducted to determine the effects of fungicides and fungicide application timing on disease severity, yield increase and net return in winter wheat cv. Millennium in Nebraska, USA in 2006 and 2007.

4.4.2 Effects of fungicides and fungicide application timing on disease severity

There were no significant differences among fungicides in their efficacy in controlling disease in both application timings in both years (Table 3). In 2006, disease severity in unsprayed plots (4.3%) was significantly higher than that in plots sprayed with all fungicides except Stratego in the GS 31 application timing. In 2007, disease severity in unsprayed plots was significantly higher than in all sprayed plots. Although fungicides did not significantly differ in the level of disease control in each year, Headline was the most efficacious, especially in 2007 when disease severity was high. Disease severity did not significantly differ between the two fungicide application timings for any of the fungicides (Table 3) or when averaged across fungicides (Fig. 2), but generally was higher in the GS 31 than in the GS 37/GS 39 application timing. This was expected since the time between fungicide application and disease assessment was longer in the earlier (GS 31) application timing. This resulted in a greater reduction in fungicide residual activity in the earlier application timing, leading to higher disease severity in this timing compared to the later (GS 37/GS 39) timing.

In a previous study in the UK, Cook et al. (1999) showed that the effect of fungicide application timing on disease intensity varied with the fungicide applied and the disease controlled. Reduction in Septoria tritici blotch (STB) area under the disease progress curve (AUDPC) was greater when chlorothalonil was applied at GS 35-39 than at GS <35. Propiconazole reduced STB AUDPC only slightly, reduced powdery mildew and leaf rust AUDPC significantly, and reduced stripe rust AUDPC equally well when applied at GS 35-39 compared to GS <35. In general, efficacy of fungicides in controlling disease declined as the growth stage at which the fungicides were applied increased (Cook et al., 1999). In our study, fungicides were not applied beyond GS 39 and only leaf spot disease severity was assessed.

2006			2007		
Fungicide treatment	Disease severity (%) GS 31 timing	Disease severity (%) GS 37 timing	Fungicide treatment	Disease severity (%) GS 31 timing	Disease severity (%) GS 39 timing
Headline 0.58 l ha^{-1}	2.0 b	2.2 b	Headline 0.66 l ha^{-1}	16.5 b	16.7 b
Quadris 0.58 l ha^{-1}	2.4 b	2.0 b	Quadris 0.88 l ha^{-1}	21.1 b	20.2 b
Quilt 0.58 l ha^{-1}	2.7 b	2.5 b	Quilt 1.02 l ha^{-1}	25.0 b	21.7 b
Stratego 0.73 l ha^{-1}	3.2 ab	2.2 b	Stratego 0.73 l ha^{-1}	28.2 b	22.3 b
...	...	...	Tilt 0.29 l ha^{-1}	26.4 b	20.3 b
Check	4.3 a	4.3 a	Check	58.6 a	58.6 a

Table 3. Effects of fungicides and fungicide application timing (Zadoks growth stage GS 31 versus GS 37 or GS 39) on foliar disease severity in winter wheat cv. Millennium in field experiments conducted in Nebraska, USA in 2006 and 2007. Means followed by the same letter within a column are not significantly different according to Fisher's least significant difference test at P = 0.05. Means with an asterisk within a row in a year are significantly different according to Fisher's least significant difference test at P = 0.05.

4.4.3 Effects of fungicides and fungicide application timing on yield increase

In 2006, yield increase due to fungicide application was low and did not significantly differ among fungicides in both application timings. In 2007, yield increase due to fungicide application was much higher than in 2006, but also did not significantly differ among fungicides in both application timings (Table 4). Yield increase did not significantly differ between fungicide application timings in 2006. However, in 2007, fungicide application timing had a significant effect on yield increase for the fungicides Headline, Quilt, and Tilt (Table 4), and when averaged across fungicides (Fig. 3), with the GS 39 timing resulting in a higher yield increase than the GS 31 timing. The reason for the higher yield increase in the GS 39 timing compared to the GS 31 timing may be explained by the protection provided to the flag leaf which contributes significantly to yield (Ali et al., 2010; CiuHua et al., 2010; Rawson et al., 1983). In a GS 31 application, the residual fungicide activity would have waned by GS 39 and therefore would not provide the same level of protection to the flag leaf as a GS 39 application. Cook et al. (1999) showed that fungicide application to winter wheat at the GS 37 growth stage to control powdery mildew resulted in significantly higher yield than application at GS 33.

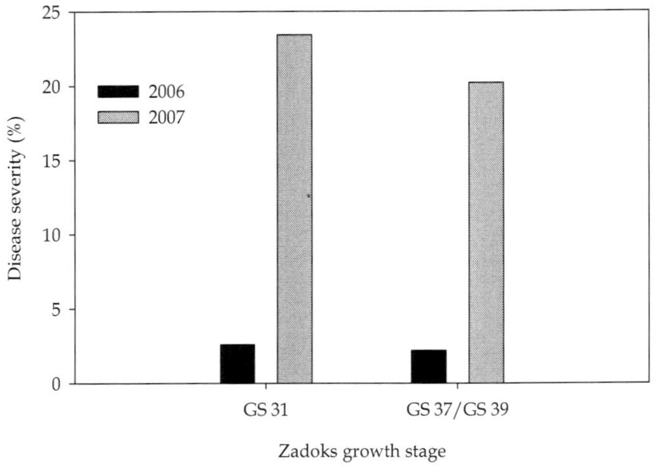

Fig. 2. Effect of fungicide application timing (Zadoks growth stage GS 31 versus GS 37 or GS 39), averaged across fungicides, on disease severity in winter wheat cv. Millennium in experiments conducted in Nebraska, USA in 2006 and 2007. Bars with an asterisk within a year are significantly different according to Fisher's least significant difference test at P = 0.05.

2006			2007		
Fungicide treatment	Yield increase (kg ha^{-1}) GS31 timing	Yield increase (kg ha^{-1}) GS37 timing	Fungicide treatment	Yield increase (kg ha^{-1}) GS31 timing	Yield increase (kg ha^{-1}) GS39 timing
Headline 0.58 l ha^{-1}	290 a	451 a	Headline 0.66 l ha^{-1}	1370 a*	1585 a*
Quadris 0.58 l ha^{-1}	375 a	279 ab	Quadris 0.88 l ha^{-1}	1332 a	1430 a
Quilt 0.58 l ha^{-1}	518 a	424 a	Quilt 1.02 l ha^{-1}	1123 a*	1518 a*
Stratego 0.73 l ha^{-1}	474 a	339 a	Stratego 0.73 l ha^{-1}	1156 a	1350 a
...	...	...	Tilt 0.29 l ha^{-1}	1229 a*	1600 a*
Check	0 a	0 b	Check	0 b	0 b

Table 4. Effects of fungicides and fungicide application timing (Zadoks growth stage GS 31 versus GS 37 or GS 39) on yield increase in winter wheat cv. Millennium in field experiments conducted in Nebraska, USA in 2006 and 2007. Means followed by the same letter within a column are not significantly different according to Fisher's least significant difference test at P = 0.05. Means with an asterisk within a row in a year are significantly different according to Fisher's least significant difference test at P = 0.05.

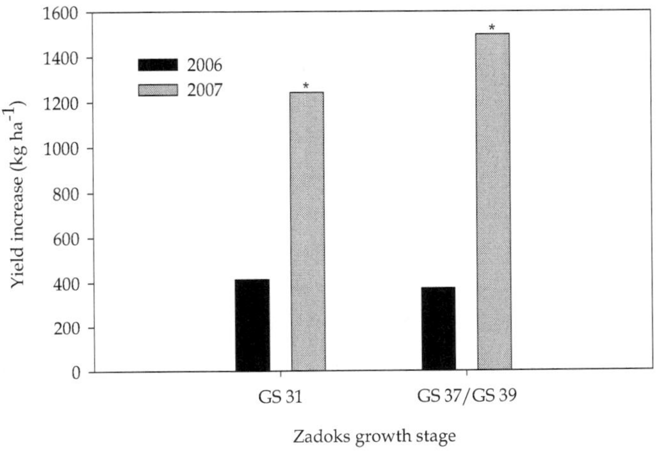

Fig. 3. Effect of fungicide application timing (Zadoks growth stage GS 31 versus GS 37 or GS 39), averaged across fungicides, on yield increase in winter wheat cv. Millennium in experiments conducted in Nebraska, USA in 2006 and 2007. Bars with an asterisk within a year are significantly different according to Fisher's least significant difference test at P = 0.05.

4.4.4 Effects of fungicides and fungicide application timing on net return

In 2006, net return from fungicide application was very low. It ranged from \$-5 ha^{-1} to \$39 ha^{-1} in the GS 31 timing and from \$-14 ha^{-1} to \$25 ha^{-1} in the GS 39 timing and did not significantly differ from zero or among fungicides (Table 5). In 2007, net return from fungicide application was significantly higher than zero and ranged from \$148 ha^{-1} to \$191 ha^{-1} in the GS 31 timing and from \$177 ha^{-1} to \$239 ha^{-1} in the GS 39 timing (Table 5). Although net return did not significantly differ among fungicides, Headline resulted in the highest return in the GS 31 application timing whereas Tilt resulted in the highest return in the GS 39 application timing. The effect of application timing on net return was not significant for any fungicide in 2006. However, in 2007 it was significant for Headline, Quilt, and Tilt (Table 5), and when averaged across fungicides (Fig. 4), with the GS 39 application timing resulting in a higher net return than the GS 31 application timing. As explained above for yield increase, the higher net return in the GS 39 timing compared to the GS 31 timing is attributable to protection provided to the flag leaf by a GS 39 fungicide application. The probability of a positive net return was 0.63 and 1.00 in 2006 and 2007, respectively. It should be noted, however, that positive net returns in 2006 were very small compared to 2007.

2006			2007		
Fungicide treatment	Net return ($\$ ha^{-1}$) GS31 timing	Net return ($\$ ha^{-1}$) GS37 timing	Fungicide treatment	Net return ($\$ ha^{-1}$) GS31 timing	Net return ($\$ ha^{-1}$) GS39 timing
Headline 0.58 l ha^{-1}	-5 a	19 a	Headline 0.66 l ha^{-1}	191 a*	229 a*
Quadris 0.58 l ha^{-1}	0 a	-14 a	Quadris 0.88 l ha^{-1}	160 a	177 a
Quilt 0.58 l ha^{-1}	39 a	25 a	Quilt 1.02 l ha^{-1}	148 a*	218 a*
Stratego 0.73 l ha^{-1}	24 a	5 a	Stratego 0.73 l ha^{-1}	159 a	193 a
...	...	...	Tilt 0.29 l ha^{-1}	173 a*	239 a*
Check	0 a	0 a	Check	0 b	0 b

Table 5. Effects of fungicides and fungicide application timing (Zadoks growth stage GS 31 versus GS 37 or GS 39) on net return in winter wheat cv. Millennium in field experiments conducted in Nebraska, USA in 2006 and 2007. Means followed by the same letter within a column are not significantly different according to Fisher's least significant difference test at P = 0.05. Means with an asterisk within a row in a year are significantly different according to Fisher's least significant difference test at P = 0.05.

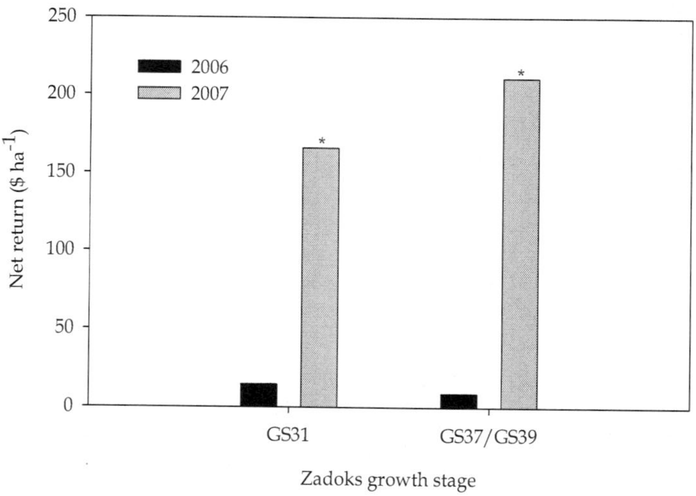

Fig. 4. Effect of fungicide application timing (Zadoks growth stage GS 31 versus GS 37 or GS 39), averaged across fungicides, on net return in winter wheat cv. Millennium in experiments conducted in Nebraska, USA in 2006 and 2007. Bars with an asterisk within a year are significantly different according to Fisher's least significant difference test at P = 0.05.

4.4.5 Relationship between disease severity and yield increase

By applying different fungicides at two growth stages, different levels of disease were generated which resulted in corresponding yield increases. Thus, in 2007 when environmental conditions favored disease development, data were generated and used to model the relationship between disease severity and yield increase using linear regression analysis. This relationship can be used to estimate the yield increase to expect from a certain level of disease control from fungicide application. The results showed a significant, linear inverse relationship ($r^2 = 0.56$, $p = 0.0122$) between disease severity and yield increase (Fig. 5), implying that higher yield increases were realized in plots with lower disease severity and vice versa. Disease severity explained 56% of the variation in yield increase. Every unit of disease severity reduction resulted in a yield increase of 32.9 kg ha^{-1}. This result demonstrates the potential for a yield benefit if a fungicide is applied to winter wheat when environmental conditions favor disease development.

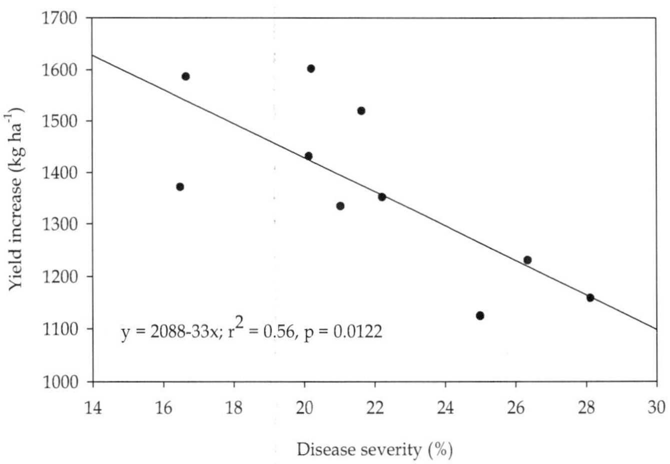

Fig. 5. Relationship between disease severity and yield increase due to fungicide application in winter wheat cv. Millennium. Data were obtained from field experiments conducted in Nebraska, USA in 2007.

4.4.6 Relationship between disease severity and net return

Net return from fungicide application was linearly and inversely related to disease severity (Fig. 6), which mirrored the relationship between yield increase and disease severity. However, the relationship between net return and disease severity was weaker ($r^2 = 0.32$, $p = 0.0658$) than the relationship between yield increase and disease severity. This was because factors that do not directly affect yield, such as fungicide and fungicide application costs, were used to calculate net return. Nevertheless, this result shows that when environmental conditions favor disease development, a higher level of disease control will result in a higher net return.

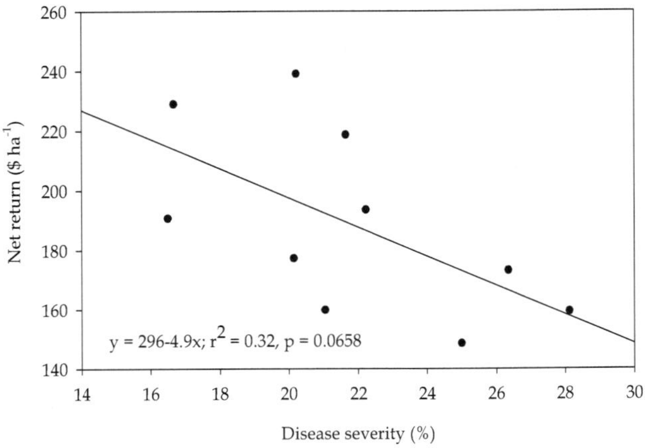

Fig. 6. Relationship between disease severity and net return due to fungicide application in winter wheat cv. Millennium. Data were obtained from field experiments conducted in 2007 and economic analyses conducted in 2009-2010 in Nebraska, USA.

4.4.7 Relationship between yield increase and net return

In 2007, there was a strong, positive linear relationship between yield increase and net return (Fig. 7). Eighty five percent of the variation in net return was explained by yield increase. Every unit (kg ha^{-1}) of yield increase resulted in a net return of \$0.17 ha^{-1}.

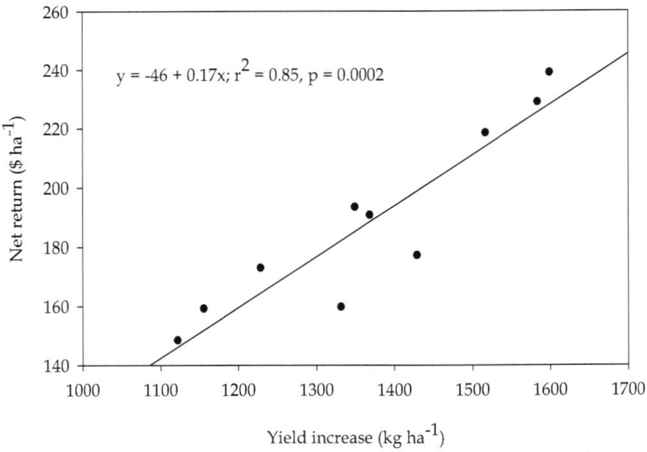

Fig. 7. Relationship between yield increase and net return due to fungicide application in winter wheat cv. Millennium. Data were obtained from field experiments conducted in 2007 and economic analyses conducted in 2009-2010 in Nebraska, USA.

It should be noted, however, that profitability from fungicide application is dependent on many factors, including weather conditions favorable to disease development, the level of disease intensity during the growing season, the price of wheat, fungicide and fungicide application costs, fungicide application rates and timing, cultivar resistance, and cultural practices.

5. Conclusions

We conclude that the fungicides Quilt, Headline, Tilt, Quadris, and Stratego effectively controlled foliar fungal diseases in winter wheat, resulting in yield increase and a profitable net return in 2007. Environment had a significant effect on yield increase and net return. In 2006 when dry conditions led to development of low levels of disease, yield increase and net return were very low. However, in 2007 when excessively wet weather favored development of high levels of disease, yield increase and net return were high. These results suggest that fungicide application to winter wheat can be profitable when environmental conditions favor development of damaging levels of disease. Timing of fungicide application at GS 39 generally resulted in a higher yield increase and a higher net return than a GS 31 timing. Therefore, under Nebraska conditions, when a farmer can afford only one spray in a growing season, spraying at GS 39 or later would likely be more beneficial than spraying earlier. Regression analysis of 2007 data showed an inverse linear relationship between yield increase and disease severity and between net return and disease severity, and a positive linear relationship between yield increase and net return, confirming the negative effect of disease on yield and suggesting a potential benefit from fungicide application to control foliar fungal diseases in winter wheat when environmental conditions favor the development of damaging levels of disease.

6. References

Ali, M. A., Hussain, M., Khan, M. I., Ali, Z., Zulkiffal, M., Anwar, J., Sbir, W. & Zeeshan, M. (2010). Source-sink relationship between photosynthetic organs and grain yield attributes during grain filling stage in spring wheat (*Triticum aestivum*), *International Journal of Agriculture and Biology,* Vol. 12 No. 4, pp. 509-515, ISSN 1560-8530

Buchenauer, H. (1987). Mechanism of action of triazolyl fungicides and related compounds, In: *Modern selective fungicides: properties, applications, mechanisms of action,* H. Lyr, (Ed.), 205–231 Wiley, ISBN: 047020799X ,New York, New York, USA

Bockus, W. W., Bowden, R. L., Claasen, M. M., Gordon, W. B., Heer, W. F. & Shroyer, J. P. 1997. Time of application and winter wheat genotype affect production of large seed after fungicide application. *Canadian Journal of Plant Science,* Vol. 77, pp. 567-572, ISSN 0008-4220

Campbell, C. L., and Madden, L. V. (1990). *Introduction to Plant Disease Epidemiology,* John Wiley & Sons, ISBN 0-471-83236-7, New York, New York, USA.

CiuHua, G., ZhiQiang, G. & GuoYuan, M. (2010). Effect of shading at post flowering on photosynthetic characteristics of flag leaf and response of grain yield and quality to shading in wheat. *Acta Agronomica Sinica,* Vol. 36, No. 4, pp. 673-679, ISSN 0496-3490

Cook, R. J., Hims, M. J. & Vaughan, T. B. (1999). Effects of spray timing on winter wheat disease control. Plant Pathology, Vol. 48, No. 1, pp. 33-50, ISSN 0032-0862.

Cook, R. J. & King, J. E. 1984. Loss caused by cereal diseases and the economics of fungicidal control, *Plant Diseases: Infection, Damage and Loss*, Blackwell, pp. 237-245, ISBN 0632011262, Oxford, UK

Cromey, M. G., Buttler, R. C., Mace, M. A., & Cole, A. L. J. 2004. Effects of the fungicides azoxystrobin and tebuconazole on *Didymella exitialis*, leaf senescence and grain yield in wheat. *Crop Protection*, Vol. 23, pp. 1019-1030, ISSN 0261-2194

Duczek, L. J., and Jones-Flory, L. L. 1994. Effect of timing of propiconazole application on foliar disease and yield of irrigated spring wheat in Saskatchewan from 1990 to 1992. *Canadian Journal of Plant Science*, Vol. 74 pp. 205-207, ISSN 0008-4220

Duveiller, E., Kandel, Y. R., Sharma, R. C., & Shrestha, S. M. 2005. Epidemiology of foliar blights (spot blotch and tan spot) of wheat in the plains bordering the Himalayas. *Phytopathology*, Vol.95, No. 3, (March 2005), pp. 248-256, ISSN 0031-949X

Gladders, P., Paveley, N. D., Barrie, I. A., Hardwick, N. V., Hims, M. J., Langton, S., S. & Taylor, M. J. (2001). Agronomic and meteorological factors affecting the severity of leaf blotch caused by *Mycosphaerella graminicola* in commercial wheat crops in England. *Annals of Applied Biology*, Vol. 38, No. 3, pp. 301-311, ISSN 00034746

Gomez, K.A., Gomez, A.A. (1984). *Statistical Procedures for Agricultural Research*, Second Edition, John Wiley & Sons, ISBN 0-471-87092-7, New York, New York, USA

Hewitt, H. G. (1998). *Fungicides in Crop Protection*, CAB International, ISBN 0-85199-164-5,Wallingford, UK

Jørgensen, L.N., Henriksen, K. E. & Nielsen, G.C. 2000. Margin over cost in disease management in winter wheat and spring barley in Denmark, *Brighton Crop Protection Conference: Pests & Diseases – 2000: Volume 2: Proceedings of an International Conference*, pp. 655-662, ISBN 1-901396-59-2, Brighton, UK, November 13-16, 2000

Kelley, K. W. 2001. Planting date and foliar fungicide effects on yield components and grain traits of winter wheat. *Agronomy Journal*, Vol. 93, No. 2, (March 2001) pp. 380-389, ISSN 0002-1962

Kolmer, J. A., Long, D. & Hughes, M. (2009). Physiologic specialization of *Puccinia triticina* on wheat in the United States. *Plant Disease*, Vol. 93, No. 5, (May 2009), pp. 538-544, ISSN 0191-2917

Marroni, M. V., Viljanen-Rollinson, S. L. H., Butler, R. C., & Deng, Y. (2006). Fungicide timing for the control of *Septoria tritici* blotch of wheat. *New Zealand Plant Protection*, Vol.59, pp. 160-165 ISSN 1175-9003

Milus, E.A. 1994. Effect of foliar fungicides on disease control, yield and test weight of soft red winter wheat. *Crop Protection*, Vol. 13, No. 4, (February 1994), pp. 291-295 ISSN 0261-2194

Murray, T. D., Parry, D. W., & Cattlin, N. D. (1998). *A Color Handbook of Diseases of Small Grain Cereal Crops*, Iowa State University Press, ISBN 0-8138-2529-6, Ames, Iowa

Ordish, G., Dufour, D. (1969). Economic bases for protection against plant diseases. *Annual Review of Phytopathology*, Vol. 7, No. 1, (January 1969), pp. 31-50, ISSN 0066-4286

Ransom, J. K. & McMullen, M. P. 2008. Yield and disease control on hard winter wheat cultivars with foliar fungicides. *Agronomy Journal,* Vol. 100, No. 4, (October 2008) pp. 1130-1137, ISSN 0002-1962

Rawson, H. M., Hindmarsh, J. H., Fischer, R. A. & Stockman, Y. M. (1983). Changes in leaf photosynthesis with plant ontogeny and relationships with yield per ear in wheat cultivars and 120 progeny. *Australian Journal of Plant Physiology,* Vol. 10, No. 6, pp. 503-514, ISSN 0310-7841.

Shabeer, A. and Bockus, W. W. (1988). Tan spot effects on yield and yield components relative to growth stage in winter wheat. *Plant Disease,* Vol. 72, No. 7, pp. 599-602, ISSN 0191-2917

Sharma-Poudyal, D. & Chen, X. M. (2011). Models for predicting yield loss caused by stripe rust in the U.S. Pacific Northwest. *Phytopathology,* Vol. 101, No. 5, (May 2011), pp. 544-554, ISSN 0031-949X

Vamshidhar, P., Herrman, T.J., Bockus, W.W. & Loughin, T.M. (1998). Quality response of twelve hard red winter wheat cultivars to foliar disease across four locations in central Kansas. *Cereal Chemistry,* Vol. 75 No. 1 (January 1998), pp. 94-99, ISSN 0009-0352

Villareal, R. L., Mujeeb-Kazi, A., Gilchrist, L. I. & Del Toro, E. (1995). Yield loss to spot blotch in spring bread wheat in warm nontraditional wheat production areas. *Plant Disease,* Vol. 79, No. 9, (September 1995), pp. 893-897, ISSN 0191-2917

Vincelli, P. 2002. Q_oI (Strobilurin) Fungicides: Benefits and Risks. *The Plant Health Instructor.* DOI: 10.1094/PHI-I-2002-0809-02. Available from http://www.apsnet.org/edcenter/advanced/topics/Pages/StrobilurinFungicides.aspx

Watkins, J. E. & Boosalis, M. G. (1994). Plant disease incidence as influenced by conservation tillage systems, In: Unger, P.W. (Ed.), *Managing Agricultural Residues,* P.W. Unger, (Ed.), 261-283, Lewis Publishers, ISBN 0-8737-1730-9, Boca Raton, Florida, USA

Wegulo, S. N., Breathnach, J. A. & Baenziger, P. S. 2009. Effect of growth stage on the relationship between tan spot and spot blotch severity and yield in winter wheat. *Crop Protection,* Vol. 28, No. 8, (May 2009), pp. 696-702, ISSN 0261-2194

Wegulo, S. N., Zwingman, M. V., Breathnach, J. A. & Baenziger, P. S. 2011. Economic returns from fungicide application to control foliar fungal diseases in winter wheat. *Crop Protection,* Vol. 30, No. 6, (April 2011), pp. 685-692, ISSN 0261-2194

Wiersma, J. J. & Motteberg, C. D. 2005. Evaluation of five fungicide application timings for control of leaf-spot diseases and Fusarium head blight in hard red spring wheat. *Canadian Journal of Plant Pathology,* Vol.27, pp. 25-37 ISSN 0706-0661

Wiik, L. & Ewaldz, T. (2009). Impact of temperature and precipitation on yield and plant diseases of winter wheat in southern Sweden 1983-2007. *Crop Protection,* Vol. 28, No. 11, pp. 952-962 ISSN 0261-2194

Wiik, L. & Rosenqvist, H. (2010). The economics of fungicide use in winter wheat in southern Sweden. *Crop Protection,* Vol. 29, No. 1, (January 2010), pp. 11-19, ISSN 0261-2194

Zadoks, J. C., Chang, T. T. & Konzak, C. F. 1974. A decimal code for the growth stages of cereals. *Weed Research,* Vol. 14, No. 6, (June 1974), pp. 415-421, ISSN 0043-1737

12

Efficacy and Utilization of Fungicides and Other Antibiotics for Aseptic Plant Cultures

Simon A. Mng'omba[1], Gudeta Sileshi[1],
Elsa S. du Toit[2] and Festus K. Akinnifesi[1]
*[1]World Agroforestry Centre, SADC - ICRAF, Agroforestry Programme,
Chitedze Agricultural Research Station, Lilongwe,
[2]Department of Plant Production and Soil Science,
Faculty of Natural and Agricultural Sciences, University of Pretoria, Pretoria,
[1]Malawi
[2]South Africa*

1. Introduction

1.1 *In vitro* culture

In vitro plant culture which encompasses cell, tissue, organ and also embryo culture has been a vital technique for mass multiplication of plants (propagules), elimination of plant diseases thorough meristematic tissue culture technique, plant conservation (through cryopreservation) and crop improvement through gene transfer (Pierik, 1987; George, 1993; Singh & Chand, 2003; Sarasan et al., 2011). *In vitro* culture techniques have been used to shorten breeding cycles of plants and to achieve genetic transformation (Singh & Chand, 2003). Despite the several merits of *in vitro* culture techniques to the modern world where plant diseases, flower/fruit abortions and low plant regeneration capacity are major challenges, many developing countries, especially in Africa do not utilize them to address the numerous challenges in crop production and improvements. This is mainly due to the high investment costs in equipment, chemicals and intensive and skilled labor demands (Sarasan et al., 2011).

The high costs for equipment, water, chemicals (such as plant growth hormones or regulators, surface sterilants, disinfectants etc.) and culture losses due to *in vitro* culture contamination among many others can cripple several *in vitro* plant culture techniques in many developing countries, especially in Africa. Possibly, this is the main reason we have limited or no interest to invest in tissue culture laboratories in Africa, except for the academic institutions for the sole purpose of learning the technique.

1.2 *In vitro* contaminants

The most common challenge for *in vitro* plant culture protocols in the tropics, especially stock plants derived from mature ortet has been elimination of culture contaminants. These culture contaminants have been problematic to the effect that there are instances where propagation and mass multiplication of useful plants have failed. To a great extent, asepsis has always been a key factor towards successful *in vitro* plant culturing and mass multiplication. However, there have been limited research studies undertaken to document the success or failure of asepsis in many *in vitro* cultures of tropical plants.

There are several types of *in vitro* culture contaminants. They include various types of fungi, bacteria, viruses and other micro-organisms. Among these *in vitro* contaminants, fungi have been the most common and conspicuous microbes that cause many *in vitro* plant culture contaminations. In this chapter, we mainly outline a few common fungicides and other antibiotics (such as sterilants and/or disinfectants) and procedures employed to eliminate or reduce fungal *in vitro* culture contaminations to a manageable level as proliferation of fungi can cripple the whole plant culture process. We further highlight some common surface sterilants and disinfectants frequently utilized in plant culture laboratories to reduce or destroy fungal contaminants. In this respect, disinfectants and surface sterilants are chemicals used in plant culture laboratories to control or kill pathogens which may not necessarily be fungi. These surface sterilants or disinfectants may not be purely fungicides although they act against fungi. This chapter also presents the efficacy, handling, storage and the possible effects of fungicides on plant cultures and the roles of these fungicides in stock plant preconditioning to improve the efficacy of fungicides or other disinfectants. However, details of other forms of *in vitro* contaminants (such as bacteria, viruses and other micro-organisms) are outside the scope of this chapter.

2. *In vitro* culture pathogens

There are several pathogens (microbial contaminants) which have been a major threat to *in vitro* cultures due to their rapid proliferation characteristics (Enjalric et al., 1998). Axenic cultures are often mandatory throughout the entire *in vitro* culture process. The contaminated cultures are usually discarded at any stage of sub-culturing because many *in vitro* culture contaminants eventually cause plant death by exuding their phytotoxins (George, 1993).

Many fungal contaminants are visible right from the primary *in vitro* plant culture initiation except for the cryptic contaminants. Generally, these contaminants overgrow plant cultures and often outcompete them for resources in the nutrient media (Pierik, 1987). Consequently, *in vitro* contaminants often adversely interfere with culture growth and survival. Therefore, freedom of pathogens (asepsis) in plant cultures has been the ultimate goal in many *in vitro* protocols.

2.1 Fungal contaminants

Fungal contaminants have posed a major threat at every stage of the *in vitro* plant culture process. These fungal contaminants are characterized by the presence of hyphae (thread-like) appearing on the explants and culture media. Colonization of fungi progresses with time and the plant cultures are eventually covered in fungal mycelia. Unlike other contaminants, fungal contaminants are visible in the culture media. Furthermore, the development and growth of fungal contaminants are faster than the growth of the plant cultures. However, the rapid proliferation of fungi in the plant cultures depends on the type of fungi colonizing the culture media.

The fungi colonizing the plants could be endogenous, cryptic or endophytic. Herman (1990) reported that endogenous or endophytic microbes are often difficult to decontaminate. It is known that many endophytes are beneficial for the growth of ortet plants in the wild, but they also become contaminants in culture media. Helander et al. (1996) reported that mutualism between endophytes and the host plant depends on the prevailing plant condition. Endophytes can be harmful to plant cultures although they are useful outside the plant culture laboratories. Examples of some trees that live and thrive with symbiotic microbes (mycorrhizae) include *Uapaca kirkiana* (Mwamba, 1995), but this mutualistic association also breaks once the host plant is stressed.

In the tropics, fungal contaminants are very common due to the prevailing favorable environmental conditions (especially the high temperatures and humidity) for the proliferation of fungal pathogens on the ortet plants. Furthermore, the conditions in the plant culture incubation chambers are favorable for the rapid growth of fungi and other *in vitro* contaminants. For instance, the range of temperatures (usually between 23 - 26 °C), high humidity and plant nutrients (available inside the culture vials or test-tubes) cannot deter or inhibit the proliferation of many fungi in the tubes placed in the culture incubation chambers. Moreover, some plants live and survive by mutual association with fungi in the wild. However, such fungi are likely to become '*vitropathogenic*' in the plant cultures.

2.1.1 Exogenous fungal contaminants

Generally, exogenous fungal pathogens are easy to eliminate from the stock plants using many fungicides. This is because such pathogens are found on the outer surface of the stock plants, and hence adequate contact between the fungicides and the pathogens and even the spores. However, insufficient surface disinfection might not dislodge exogenous pathogens and spores from the stock plants, and hence endogenous contaminants still pose a threat in plant culture protocols.

It is always difficult to detect the presence of either exogenous, endogenous or both pathogens on the stock plants. So it might be advisable to use systemic fungicides (e.g. Benomyl) before other disinfectants such as sodium hypochlorite are used. It is also a good practice to ensure that all possible culture contaminants are dealt with at the initial decontamination stage instead of re-disinfecting the explants at later stages (sub-culturing stages). Re-decontamination of explants might weaken the ensuing explants due to cell injury. This can easily cause death of explants.

2.1.2 Endogenous fungal contaminants

Symbiotic microbes which are beneficial to many plants may become contaminants in the culture media (Herman, 1990). It has been reported that endogenous or endophytic fungi become pathogenic to the host plants when the plants are stressed, for example, when the cell walls are weakened or under other unfavorable *in vitro* conditions (Darworth & Callan, 1996). The endogenous or endophytic microbes are often hard to decontaminate. According to Cassells (1991), culture asepsis is important in all plant culture protocols. However, many trees in the tropics live and survive in association with endogenous or cryptic microbes (fungi) in the wild (Darworth & Callan, 1996). It remains unclear whether this association may continue to be mutual or lethal to the growth of plant cultures. This is because some fungi may not be '*vitropathogenic*', despite being prolific in the cultures (Herman, 1990).

Fungal contaminants increase plant culture mortality as they compete for the same resources (nutrients and oxygen) in the culture media or produce phytotoxins to harm the plant cultures. Obuekwe & Osagie (1989) reported that fungi such as *Aspergillus niger* and *Aspergillus flavus* produce oxalate and aflatoxin poisons respectively that can cause death to plant cultures. To overcome the fungal contamination problem, commercial fungicides have been used to control *in vitro* fungal contaminations (Table 1).

Common chemicals	Type and some functions
Benomyl and Captain	fungicides - act as multiplication inhibitors
Formaldehyde (aldehyde)	fungicide - wide microbial activity (kills spores)
Sodium & calcium hypochlorite	oxidizing agents – kills wide range of pathogens
Teepol (detergent)	sterilant
Mercuric chloride	increases culture mortality
Copper oxychloride	both a fungicide and sterilant
Nystatin	antibiotic toxic to fungi and yeast

Table 1. Common fungicides and disinfectants used in plant culture laboratories

Establishing complete axenic cultures has not always been easy since some potential culture contaminants, especially endophytes mutually co-exist with plants. It is accepted that the success of achieving culture asepsis depends on the nature of stock plants used. This is because mature stock plants are highly loaded with fungal pathogens and spores, but mature stock plants are selected due to their fruiting precocity, especially for fruit trees. To achieve plant culture asepsis for such mature stock plants, there is need for pre-conditioning of ortet plants before collection of stock plants.

2.1.3 Antibiotics

Although the main focus of this chapter is on fungicides, antibiotics also play a vital role in achieving aseptic *in vitro* cultures. It has been observed that the use of fungicides alone can lead to proliferation of other *in vitro* contaminants such as yeast, bacteria, protozoa and other microbes (www.phytotechlab.com) hence broad-spectrum antibiotics are used in plant culture to control many *in vitro* contaminants. Other disinfectants and surface sterilants may not be necessarily fungicides but broad-spectrum antibiotics which are often used.

3. Fungicides for stock plant pre-conditioning

Fungicides are often used to reduce the pathogen loads from the stock plants to achieve culture asepsis in the culture laboratories. However, attention must be paid to the source of stock plants as many fungal contaminants are derived from the ortet plants. A few are then introduced during the culture initiation or at later stages of *in vitro* plant culturing. Leifert (1990) reported that sub-culturing processes are potential sources of contaminants and it is estimated that 5 - 15% of contaminants are introduced per sub-culture. This possibly suggests that a large number of fungal contaminants originate from the plants themselves. Other possible sources of fungal culture contamination have been due to poor handling of plant cultures during excision, initiation and sub-culturing and insufficient test tube sterilization and poor media preparation. These potential sources of fungal culture contaminations can be avoided with proper care when handling plant cultures.

Stock plants, especially from the wild are often difficult to decontaminate. For instance, *Uapaca kirkiana*, one of the wild trees of southern Africa, has been a difficult plant to eliminate *in vitro* fungal contaminants, especially when derived from mature ortet plants (Mng'omba et al., 2007). Success in culture asepsis of *U. kirkiana* has been achieved using seedling stock plants and 3.5% sodium hypochlorite (NaOCl) as a surface disinfectant (Maliro, 1997). However, they did not achieve any success in decontaminating adult *U. kirkiana* plant materials due to high fungal contamination at the initial stage.

Plants that live in association with endogenous fungi require preconditioning before collection. Application of Benomyl (Benlate), a systemic fungicide has been found effective on ortet plants before stock plant collection for *in vitro* culture. For instance, this method has been an effective way in reducing fungal load on *U. kirkiana* (Mng'omba et al., 2007). However, this process was time consuming as it involved isolation of grafted plants and placing in a screen or greenhouse where Benlate solution was regularly applied to eliminate fungi and the spores. The stock plants collected from such isolated plants are often free from heavy fungal infestation. It is advisable to avoid watering the whole ortet plants, but watering should be done to the polyethylene bags through drip irrigation or a hose pipe as most fungal pathogens and spores proliferate on wet plant surfaces.

To increase the chances of eradicating fungal pathogens and spores on the mother plants, pruning might be necessary to induce proliferation of new and rejuvenated lateral shoots which may not be highly loaded with fungal pathogens, and hence ease decontamination unlike the old plant shoots. Furthermore, the prevailing weather conditions during collection of stock plants play a vital role. Collecting stock plants from the forests or open field on a rainy day increases fungal load on the stock plants, and hence makes it difficult to eliminate the pathogens as this would require a rigorous decontamination process. It is, therefore, clear that fungicides and other disinfectants do play a vital role in the entire *in vitro* plant culture.

4. Surface sterilants and disinfectants

Surface sterilants can be described as chemicals rendering plants free from any pathogens including fungal spores. Generally, surface sterilants act on the outside of the explants. There are several chemicals which have been used to free plants from pathogens before culture initiation onto the media (chemotherapy). They deter growth and proliferation of pathogens (fungi, bacteria and other types of microbes). In many cases, a combination of several surface sterilants, disinfectants and/or fungicides is used to improve the efficiency of subsequent fungicides in decontaminating stock plants. For instance, a few drops of Teepol (0.05%) are often applied to the water and this increases wettability of the plant surfaces. Also, penetration of fungicides into the outer plant cells might be enhanced by rinsing the stock plants in the water. Generally, Teepol (soapy water or detergent) has been one of those disinfectants or surface sterilants commonly used in many plant culture laboratories to enhance the removal of pathogens and/or fungal spores from the stock plants.

Another common disinfectant in plant culture laboratories has been ethanol (an alcohol derivative). Ethanol has been one of the commonest disinfectants used in plant tissue culture to eliminate pathogens and spores. It is used from the preparation room up to the laminar air floor to kill pathogens and spores. This alcohol derivative has often been used to reduce culture contaminants as it kills fungal pathogens and spores on the stock plants.

The concentrations of sterilants and disinfectants widely used vary depending upon the nature of stock plants (soft vs. lignified plant surfaces) to be sterilized. A wide range of ethanol concentrations (20 - 100%) has been used, but high concentrations of ethanol (100% concentration) are rarely used, especially on soft stock plants because they can easily damage or injure the plant tissues. Generally, many epithytic contaminants (fungi) are eliminated, but not endophytes, and hence high concentrations of disinfectants are required to destroy these endophytic culture contaminants. Apart from the concentration of the disinfectants, exposure time must also be considered. Generally, the exposure time should be short (5 - 20 seconds), especially with the tender stock plants which are highly infested. Low concentrations of ethanol (≤50%) have often been used for tender and soft stock plants to avoid injuring the plant cells or tissues.

4.1 Improving decontamination process

For adequate plant sterilization (decontamination), stock plants are stirred in a beaker to ensure sufficient surface contact between stock plant and the fungicide. In this case, the exposure time, disinfectant concentration, and active ingredient (a.i.) are important factors determining the efficacy of the entire process of *in vitro* decontamination of stock plants. Rinsing the stock plants in the water assists in stopping the reaction between the chemicals (fungicide) and stock plants. The disinfection by fungicides and other surface sterilants in a sealed beaker (with aluminum foils) is followed by rinsing stock plants under running tap water for some time to remove the contaminants and also stop any reaction between the stock plants and the disinfectants.

5. Selection and efficacy of fungicides

Some fungicides commonly used to control plant culture contaminations during the *in vitro* culture include Benomyl, Captain and several others (Table 1). Utilization of these fungicides and other sterilants and disinfectants largely depends on their availability and the costs, especially in some Africa countries where these fungicides, sterilants and/or disinfectants are imported. Generally, they are not locally available in many African countries for use in plant culture laboratories.

The fungicides and disinfectants can be in the form of powder or liquid. Generally, the quantity of fungicides normally used per decontamination is very little. For instance, a low amount of Benomyl such as 0.14g per litre of water could be used to decontaminate some stock plants. Fungicides and disinfectants are harmful to human beings when swallowed or inhaled. Therefore, proper handling of all fungicides must be a priority.

5.1 Selecting fungicides

In many plant culture laboratories, systematic fungicides (such as Benomyl) are preferred in case of endogenous fungi. Selection of a particular fungicide depends on the nature of fungal contaminants as some are hard to eliminate (e.g. endogenous fungi). The use of systemic fungicides (such as Benomyl) has often been recommended. In some cases, stronger disinfectants or fungicides have been used to remove culture contaminants, especially those which are hard to decontaminate. For instance, mercuric chloride ($HgCl_2$) has been effective in decontaminating pre-conditioned mature *U. kirkiana* stock plants (Mng'omba et al., 2007) where sodium hypochlorite (NaOCl) and Calcium hypochlorite (Ca $(OCl_2)_2$) have not been effective in decontaminating the above stock plants. There have been no aseptic cultures obtained regardless of the type and concentration of such disinfectants used on *U. kirkiana* stock plants. These fungi were suspected to be endogenous.

5.2 Efficacy of fungicides

Efficacy of fungicides depends on many factors including their active ingredient (a.i.), concentration (dosage), the type of stock plants (mature vs. new shoots), type of fungi (exogenous vs. endogenous), and the exposure time. Generally, rigorous decontamination is required for stock plants derived from mature ortet, especially when weaker disinfectants are used. This will also require longer exposure time. Generally, the use of concentrated (stronger) fungicides or disinfectants to eliminate contaminants from the stock plants requires a shorter exposure time.

Mercuric chloride as a culture disinfectant is stronger than sodium or calcium hypochlorite solutions for disinfecting stock plants and this could be the reason for its efficacy in decontaminating endogenous fungal contaminants. Danso et al. (2011) reported successful *in vitro* decontamination of sugarcane explants with mercuric chloride where NaOCl and Ca $(OCl_2)_2$ were less effective. Therefore, $HgCl_2$ is ideal for the control of endophytic fungal culture contaminants.

6. Effects of fungicides on plant cultures

6.1 Cell integrity

The use of stronger disinfectants and fungicides could weaken cell membrane and cell wall of the plants. This can lead to the discharge of cell sap (nutrients) which stimulates an outgrowth of endogenous fungi once placed onto the culture media. These endogenous fungi might become pathogenic to the explants under *in vitro* conditions (Darworth & Callan, 1996). For example, $HgCl_2$ might be strong to some plants and hence can easily damage their cells, especially after a long exposure (Mng'omba et al., 2007). Danso et al. (2011) reported a low survival of sugarcane plantlets when decontaminated with $HgCl_2$. Therefore, the application of fungicides should consider the right concentration (dosage) and exposure time in order to reduce injury to the cells of the explants.

The mortality of plant cultures could be high due to cell damage by the use of concentrated (strong) fungicides or disinfectants. Where possible, it is important to maintain cell integrity of the stock plants to avoid any undesirable effects on the plant cultures. However, this may be difficult with heavily fungal loaded stock plants. For instance, the use of disinfectants or surface sterilants such as NaOCl, $Ca(OCl_2)_2$ and many others for a heavy fungal loaded stock plants could result in the resurgence of endogenous contaminants at any stage after the culture initiation. Therefore, the utilization of strong disinfectants or surface sterilants is warranted for the highly fungal loaded stock plants such as those derived from the old plants.

6.2 Preconditioning mother plants

Application of fungicides to the mother plants in the field before collection of explants (preconditioning of the ortet plants) greatly contributes to aseptic cultures. This could reduce the need to use stronger disinfectants such as $HgCl_2$. Furthermore, it may also improve decontamination efficacy. For instance, the use of $HgCl_2$ alone was less effective on field collected *U. kirkiana* (Mng'omba et al., 2007). Thus, preconditioning of grafted *U. kirkiana* stock plants using Benomyl contributed to achieving culture asepsis. Therefore, preconditioning reduces the fungal loads on the stock plants.

6.3 Fungicide toxicity

The problem with the use of concentrated disinfectants and fungicides is that there is always a lengthy washing of stock plants in running tap water and distilled water to remove pathogens and, especially excess residues that may be lethal to the explants in the culture media. Our experience has indicated a possibility that there are some reactions between $HgCl_2$ and aluminum foil. This suggests a possible affinity for chloride (Cl^-) in that Al^{3+} could be replacing Hg^{2+}. We observed that the foil was completely eaten up as a result of this reaction. Since stock plants were covered with aluminum foil throughout the disinfection, rinsing and culture excision processes which also took time, this could mean accumulation of aluminum deposits in the cultures, and hence possible aluminum toxicity to plantlets.

There are several studies undertaken and illustrate the deleterious effect of heavy metals such as aluminum and mercury on plant growth in the fields. However, such studies have not been accomplished in plant culture laboratories. It is possible that the poor performance of some plant cultures or explants might be attributed to the presence of residues of the strong disinfectants and fungicides used during decontamination process and toxicity of the heavy metals. However, research studies are needed to confirm or establish whether the use of such disinfectants and heavy metals *in vitro* culture contributes to plant culture toxicity or mortality.

7. Storage and disposal of fungicides

7.1 Storage of fungicides

Fungicides, surface sterilants and disinfectants are obviously harmful to human beings, and hence proper handing and storage are always good and important practices in plant tissue culture laboratories. Poor storage of fungicides/disinfectants could lead to their fast deterioration. This can make them less efficient in decontaminating plant cultures. Poor storage can also accelerate the expiry of the fungicides and disinfectants as they may be exposed to moisture. Long exposure to moisture due to the poor storage could alter the chemical composition of the fungicides or disinfectants, and hence making them less efficient. Proper storage of fungicides should include observing its ideal storage temperature and good containers to avoid any possible damage or leakage.

7.2 Disposal of used fungicides

Proper disposal of fungicides and other disinfectants must be adhered to. In many plant culture laboratories, disinfectants, surface sterilants and fungicides are decanted through water drainage system after rinsing the stock plants. Disinfectants such as $HgCl_2$ are known to be corrosive, and hence can damage the water drainage system. Proper disposal of such disinfectants must be taken into consideration. For instance, it may be advisable to dig a pit where $HgCl_2$ solutions must be disposed off and buried. This is because heavy metals are also not friendly to the environment, and hence proper disposal is needed.

8. Future research areas on fungicides

The effects of most fungicides and many other disinfectants on different stock plants and explants remain unanswered, especially for the strong fungicides or disinfectants used in decontaminating stock plants in the plant culture laboratories. To our knowledge there has been limited research in this area. Therefore, there is the need for more research studies in this area. For instance, the effects of the fungicide and disinfectant residuals on the growth and survival of plantlets in *in vivo* have not been thoroughly researched. Our experience has shown that the use of $HgCl_2$ may require several rinsing of stock plants. This is to get rid of the residues and we hypothesize that the residues could have deleterious effects on explants and later on the plantlets.

9. Conclusion

This chapter has outlined the significant roles that fungicides and other disinfectants play in plant culture laboratories in achieving aseptic *in vitro* plant cultures. This is a vital process for the success of *in vitro* plant culture protocols. However, selection of an appropriate fungicide or disinfectant is important since endogenous or endophytic pathogens require the use of systemic and strong fungicides to effectively disinfect stock plants. Proper handling of fungicides and disinfectants and also proper disposal of the used disinfectants in the plant culture laboratory are equally important as they may become corrosive to the water drainage systems, and hence dangerous to the environment. Many research studies are needed in order to increase our understanding with respect to various disinfectants used in the plant culture laboratories.

10. Acknowledgement

We would like to acknowledge the financial support from the World Agroforestry Centre (ICRAF), southern Africa region without which this chapter could have not been published.

11. References

Cassells, AC. (1991). Problems in Tissue Culture: Culture Contamination. In: *Micropropagation: Technology and Application*, P.C. Derbergh & R.H. Zimmerman (Eds), 31- 44, Kluwer, Dordrecht.

Danso, K.E., Azu, E., Elegba, W., Asumeng, A., Amoatey, H.M. & Klu, G.Y.P. (2011). Effective decontamination and subsequent plantlet regeneration of sugarcane (*Sacchrum officinarum* L.) *in vitro*. *International Journal of Integrative Biology*, Vol. 11 No.2, pp 90 - 96 (May 2011)

Darworth, C.E. & Callan, B.E. (1996). Manipulation of endophytic fungi to promote their utility as vegetation biocontrol agents. In: *Endophytic Fungi in Grasses and Woody*

Plants, Systematics, Ecology and Evolution. S.C. Redlin & L.M. Carris (Eds), pp 209–216

Enjalric, F.; Carron, M.P. & Lardet, L. (1998). Contamination of Primary cultures in tropical areas. The case of *Hevea brasiliensis*. *Acta Horticulturae*, Vol. 223, pp 57- 65

George, E.F. (1993). Plant Propagation by Tissue Culture, Part 1. The Technology, Second Edition. Exergetics Ltd. Edington, Wilts, England.

Helander, M.L.; Neuvonen, S. & Ranta, H. (1996). Natural variation and effects of anthropogenic environmental changes on endophytic fungi in trees. In: S.C. Redlin & L.M Carris (Eds) *Endophytic Fungi in Grasses and Woody Plants. Systematics, Ecology and Evolution* pp 197- 207

Herman, E.B. (1990). Non-axenic plant tissue culture: possibility and opportunities. *Acta Horticulturae* Vol. 280 pp 233 - 248

Maliro, M. (1997). Propagation of *Uapaca kirkiana* using tissue culture techniques. MSc. Thesis, Bunda College of Agriculture, Lilongwe, Malawi, 98 pp.

Leifert, C. (1990). Contaminants of plant tissue cultures, Ph. D. Thesis, Nottingham University, School of Agriculture.

Mng'omba, S.A.; du Toit, E.S. & Akinnifesi, F.K. (2007). Effective preconditioning methods for *in vitro* propagation of *Uapaca kirkiana* Müell Arg. tree species. *African Journal of Biotechnology*, Vol. 6, No. 14, pp 1670 - 1676

Mwamba, C.K. (1995). Effect of root - inhabiting fungi on root growth potential of *Uapaca kirkiana* Muell Arg. seedlings. Applied Soil Ecology, Vol. 2, pp 217 - 226

Obuekwe, C.O. & Osagie, I.J. (1989). Morphological changes in infected wilt resistant and wilt-susceptible oil palm progenies and hydrolytic enzyme activities associated with *Fursarium oxysporum* f sp elaeidis pathogens, *Oeagureux*,Vol. 44, No. 11, pp 8 - 9

Pierik, R.L.M. (1987). *In Vitro* Culture of Higher Plants. Dordecht, Netherlands, p 344

PhytoTechnology Laboratories. Antibiotic preparation and storage. www.phytotechlab.com. Cited on 9th August 2011.

Sarasan, V., Kite, G.C., Sileshi, G.W., Stevenson, P.C. (2011) Applications of phytochemical and *in vitro* techniques for reducing over-harvesting of medicinal and pesticidal plants and generating income for the rural poor. *Plant Cell Reports* Vol. 30, pp 1163 - 1172

Singh, A.K. & Chand, S, (2003). Somatic embryogenesis and plantlet regeneration from cotyledon explants of a timber-yielding leguminous tree *Dalbergia sissoo* Roxb. J. Plant Physiology, Vol. 160, pp 415 - 421

Part 3

Chemical Fungicidal Agents

13

Copper in Horticulture

Denis Rusjan
Biotechnical Faculty, University of Ljubljana
Slovenia

1. Introduction

Cu as a solid, conductive chemical (Cu) material has the ability to deform under tensile stress (ductile metal); its freshly exposed surface has a reddish-orange colour, it is easily recyclable and with an annual production of more than 5 million tons of copper nowadays, due to the increasingly vital part of the metal's use in many branches of modern technology, especially in architectural structures and electronic devices (Table 1), the material is gaining importance. Copper is an essential material of the future regarding solar heating, desalination of water, environmentally-sound cultivation practices in agriculture and its contribution to the production of linear motors. Its unique chemical characteristics provide for more than 200 year's use of Cu in phytochemistry especially in the role of fungicides.

Cu^{2+} and Cu^{+} ions are soluble in water and provide antifungal and antibacterial effects (biostatic elements) at low concentration levels, wherefore their contribution to the production of fungicides is, up to the present day, irreplaceable. On the other hand high concentrations of copper salts affect physiological and biochemical processes in higher organisms. Copper takes part in numerous physiological processes and is an essential cofactor for many metalloproteins. However, copper excess leads to problems in cell function and metabolism, as copper surplus inhibits plant growth and impairs important cellular processes (*i.e.,* photosynthetic electron transport).

Copper compounds are commonly encountered as salts of Cu^{2+}, which often impart blue or green colors, in the past widely used as pigments, therefore. As a native metal it is one of the few metals to occur naturally as an un-compounded mineral and it seems to be the first metal used by man dating back at least 10,000 years in history. A wider usage of copper started in the Copper Age (5th millennium BC) and throughout the Antiquity and Middle Age to Nowadays being considered as one of the most important metals.

Copper plays an important function in human diet, because of shared similarities with iron it is crucial for the reddish coloration of hemoglobin in blood. It is essential and indispensable at all higher organisms (plants, animals) especially at cycles and functions of growth and reproduction. The average daily recommended uptake of Cu should be 0.9 mg, falling short of which frequently leads to an increase of cholesterol level and coronary diseases. Copper toxicity in terms of human health can be observed at concentrations higher than 11.0 mg kg^{-1}, affecting functions of main vital organs. A protracted exposition to toxic concentration of Cu leads to irreparable damages of stomach, kidneys, liver and brain, therefore the daily food should be under permanent sanitary (chemical) control.

Characteristic	Attribute / Value
Category	transition metal
Phase	solid
Atomic number	29
Atomic weight	63.546 g $\cdot$mol^{-1}
Density (liquid density at m.p.)	8.94 g $\cdot$cm^{-3} (8.02 g $\cdot$cm^{-3})
Boiling	2562 °C (2835 °K)
Heat of fusion	13.26 kJ $\cdot$mol^{-1}
Specific heat capacity	(25 °C) 24.440 J $\cdot$mol^{-1} $\cdot$K^{-1}
Oxidation states	+1, +2, +3, +4 (mildly basic oxide)
Crystal structure	face-centred cubic

Table 1. Chemical and physical properties of copper (Cu)

2. Copper in past

Copper has been part of the human civilisations since ancient times up to present days, its first records date back 5,000 and 6,000 years in the past. More specified utilization of copper as metal was achieved in the time of Sumerians and Chaldeans in Mesopotamia, who developed considerable skills in copper handling. Copper fabricated in Mesopotamia was quickly introduced to the the Egyptian Empire where its use flourished thousands of years. These peoples used copper for the fabrication of different jewels, ornaments, but also in the fields of armament and tools. They soon realized that pure copper, because of its softness, is not suitable for shaping tools used in agriculture, cultivation. Later, in the era of the Romans other metals, especially iron and bronze as the main metals in weaponry, gained in importance. In that time copper was first used also for architectural intentions, what can be witnessed on the roof sheathing of the Pantheon.

In Medieval times the use of copper additionally decreased because of the turbulent events in Europe involving the French, Germans, Burgundians, Anglo-Saxons and many more; when bronze had given off its place largely to iron in context of paramount agriculture, but also wars were common in that time. In the context of currency, silver took over the main role, although some copper coins were occasionally hammered in Mercia during the 8th Century.

The copper boom could be linked to immense changes in the time of industrial revolution, when the demand for better raw materials rapidly increased. In the 17th and 18th Centuries massive copper mines and smelting furnaces were exploited and erected to produce several hundreds of tons of pure copper per week.

Nowadays copper presents an important metal, what is undoubtly affirmed by its worldwide annual production of around 5 million tons. Many branches of modern technology use and include copper in their products, especially in the electrical, architectural and chemical industry as well as medicine. The biggest copper sources are located in Chile, USA, Indonesia, Australia and China, where it became the most exported metal.

3. Copper in horticultural environment

In the context of metal and environment emphasis has to be laid on a complex system which embraces relations among soil characteristics, plant necessities and human expectations. Copper and its different substances constantly traverse from one to the other stage of this system where they are ascribed many roles and have different functions, which is rendered possible with permanent transformation.

Metal	Sediment soils			
	Limestone	Sandstone	Slate	Average
Chrome (Cr)	10-11	35	90-100	70
Manganese (Mn)	620-1100	4-60	850	650
Cobalt (Co)	0.1-4	0.3	19-20	7
Nickel (Ni)	7-12	2-9	68-70	40
Copper (Cu)	5.5-15	30	39-50	30
Zinc (Zn)	20-25	16-30	100-120	50

Table 2. Average content of metals (mg kg^{-1}) in different soil and stoneware types (Ross, 1994; Šajn et al., 1998)

Copper as a microelement effects the characteristics of the soils, that are also its preliminary source, while its content depends on parental matter (Table 2). Copper reaches the soil almost exclusively in divalent form (Cu^{2+} ions) incorporated in minerals but also bound with soil organic matter. Reed & Martens (1996) reported the specifically adsorption of Cu to carbonates, soil organic matter, phyllosilicates, and hydrous oxides of Al, Fe, and Mn. Usually it is dissolved in soil solutions as Cu^{2+} or as an organic complex adsorbed to inorganic and organic negatively charged groups. The presence of Cu^{2+} ions in soil solutions decreases with increasing pH, whereas Cu complexes are less dependent on soil pH. Cu^{2+}, compared to other divalent cations (Ni, Co, Zn, Mn, Mg), has a strong affinity to soil organic matter, its concentration in soil ranges from 1×10^{-5} to 6×10^{-4} mol m^{-3}, what is controlled by Cu adsorption to organic and inorganic soil particles. Because of this the transition of Cu in deeper soil layers rarely occurs, however it can be promote by soil cultivation (deep ploughing) and by decreasing soil pH to 5.5-6.5, at which values the mobility of Cu increases (Fig. 1).

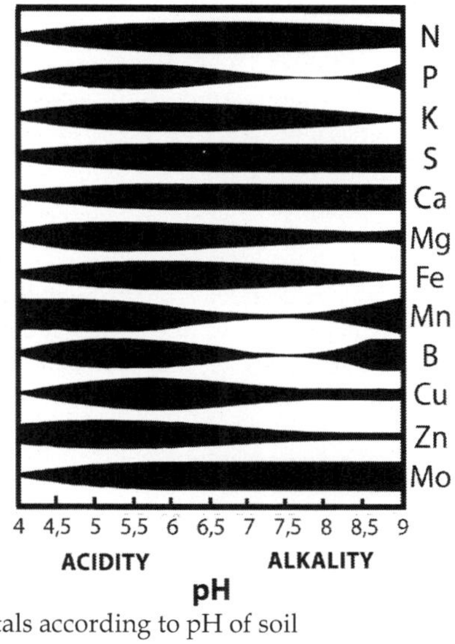

Fig. 1. The mobility of metals according to pH of soil

The bounding and bioavailability of Cu depends also on the concentration of other elements in soil, where an antagonistic effect is observed between N, P and Mo : Cu and between Cu : Zn, Mn, while synergism has not been reported yet (Fig. 2).

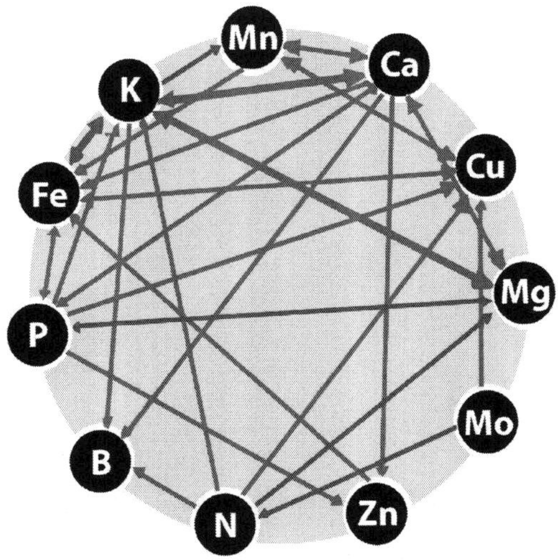

Fig. 2. Synergism (blue) and antagonism (red) among elements in soil

Beside the parental matter of soils the significant "sources" of copper are agriculture (spraying, fertilisation), industry (metallurgy, pharmacy, mining etc.) and cities and roads, which together can be denoted as pollutants. Intensive agricultural practices based on substantial use of copper compounds in fungicides have in the last 200 years lead to copper accumulation in soils, especially in traditional horticultural areas (Fig. 3), like some winegrowing regions in France, Northern Italy, Germany an many more, where copper contents in soils are likely to reach many hundreds ppm. Long time use of Cu preparations leads to the accumulation of copper, where many data from viticultural area has been already reported (France: Bordeaux 800 mg kg^{-1}, Alsace, Burgundy and Champagne 400 - 500 mg kg^{-1}; Italy: Valle d'Aosta 300 mg kg^{-1}, Lombardia 260 mg kg^{-1}, Trentino Alto Adige 161 mg kg^{-1}, Piemonte 90 mg kg^{-1}, Tuscany 34 mg kg^{-1} etc. (Brun et al., 1998) and Slovenia 5-150 mg kg^{-1} (Rusjan et al., 2007)). Rusjan et al. (2007) reported that Cu concentration in soils in horticultural environments with intensive cultivating practices increases with years, especially in the upper soil layers (< 20 cm) reaching more than 150 mg kg^{-1} after 30 years of permanent Cu usage. The accumulation of exceeded concentrations of Cu in soils leads to the perish of flora and fauna in the soils; however the greatest concerns are residuals of Cu in food (vegetable and fruit).

In the sense of a greater awareness of environmental issues European Union member states and some countries in particular are adopting many regulations to decrease the use of copper compounds in agriculture, where its permitted annual quantity does not exceed 5 kg of pure Cu per hectare. An increase of environmentally-sound cultivation practices is frequently mentioned as the most significant reason of excessive Cu pollution, leading to a paradox, as Cu is in this cultivation practices beside sulphur the only permitted element in pathogen controlling.

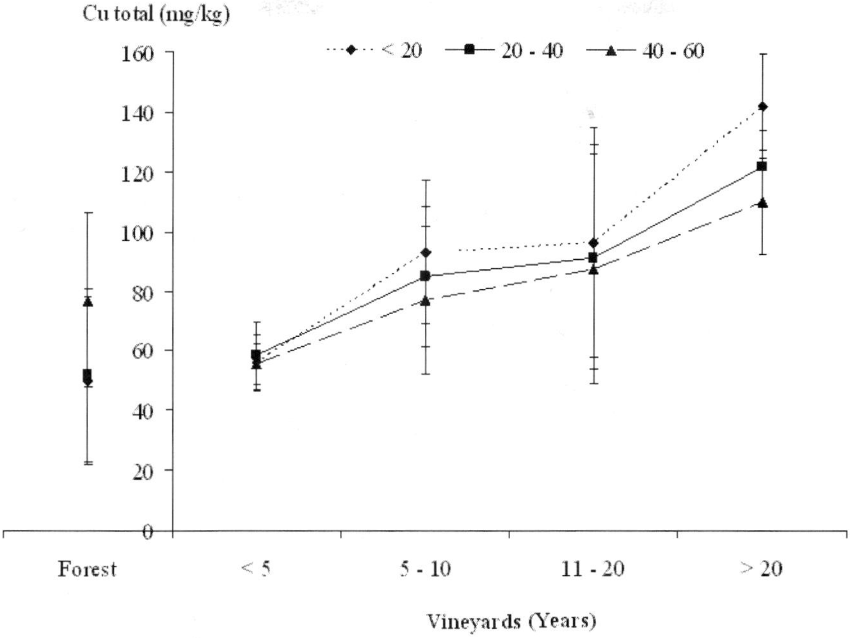

Fig. 3. Differences in total copper accumulation (mg kg^{-1}) according to soil depth (cm) and to vineyards age. Means and standard errors are presented (Rusjan et al., 2007).

3.1 Copper and horticultural plants

The essentiality of copper for optimal growth and reproduction in higher plants is largely due to its occurrence as a constituent of several proteins (copper protein/enzyme), which regulate many essential biochemical reactions in plants. Sandmann & Böger (1983) report three different forms of proteins where copper plays an important component (Cu-proteins): blue proteins (plastocyanin; without oxidase activity) with function in one-electron transfer, non-blue proteins which represents peroxidises and oxidize monophenols to diphenols, and multicopper proteins with at least four Cu-atoms per molecule, which acts as oxidases (ascorbate oxidase and diphenol oxidase). Consequently Cu plays an essential role at the following physiological and other functions:

a. photosynthesis and respiration: both can be inhibited by Cu deficiency, as Cu-enzymes catalyse or activate various steps in these processes. Cu is required for the formation of chlorophyll and other thylakoid constituents - with chlorosis being one of the symptoms of copper deficiency - Cu is particularly involved in photosynthesis as a component of ribulose-disphosphate carboxylase, while at respiration it is present as the copper enzyme cytochrome-c-oxidase, where its deficiency may reduce respiration rates.

b. lignifications and phenol metabolism: copper deficiency can significantly reduce the activity of the phenol oxidases which are involved in the lignin synthesis, leading to formations of rather weak tissues and the distortion of leaves and stems. Cu presence affects the synthesis of phenol compounds which inhibit cell elongation. Cu intensifies oxidation of phenol compounds to chinons, wound healing and pigment formation.
c. regulation of auxins (growth): shortage of auxin is frequently caused by copper deficiency, resulting in a lack of germination. The monophenol compounds control the activity of enzyme auxin oxidase, but diphenols inhibit it.
d. reproduction: deficiency of copper leads to the production of higher polyphenols contents (diphenols) which inhibit the auxin oxidase and consequently the auxin accumulation in anthers. The shortage of auxin causes the formation of a large mass of cellular material, which eventually fills the cavity normally occupied by the pollen grains and consequently renders the pollen grains sterile (Alloway et al., 1985).

The uptake of copper in the form of Cu^{2+} is done after the metal ion dissociation from the organically complex molecules in soil and it can be enhanced by fungi associated with roots, also known as vesicular-arbuscular mycorrhizae. Research on the interaction between phosphorus and copper contents in soil has already been done; phosphorus improves root growth and thus increases copper absorption, therefore the fertilisation should be adapted at presence or absence of mycorrhizae.

Not only the deficiency of copper contents but also its sufficiency (excess) causes irreversible consequences on plant growth, its cultivation and with that quality (Fig. 4).

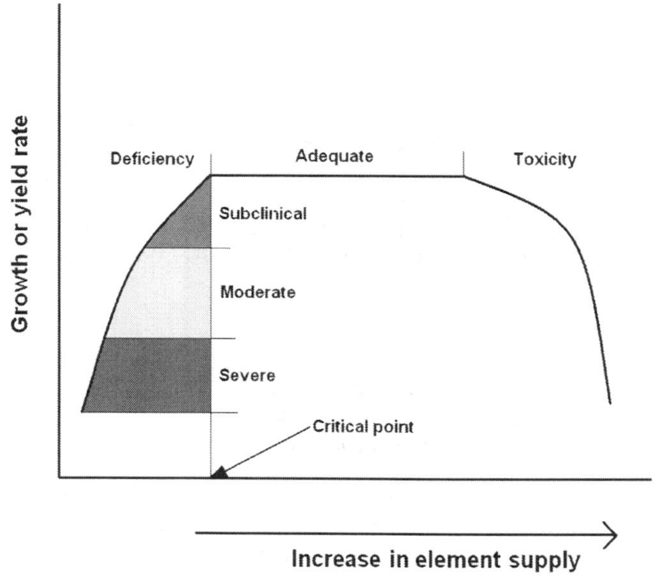

Fig. 4. Effects of copper availability (deficient, sufficient and excessive) on growth and yield of plant

The toxicity caused by copper is a rare phenomena in horticulture, especially where the control of soil pH is controlled and regulated. The most frequent symptoms on plants in terms of excessive copper concentrations are inadequate, obstructed shoot and root growth and vigour, whereas also chlorosis frequently appears on leaves. The excess of copper in soil could be controlled with dosing of lime in soils to increase of pH over 6.5. The effect can be intensified by adding of Fe (iron) because of its antagonism relationship with Cu. The Cu concentrations over 20 mg kg^{-1} in dry matter of plant tissues (parts), and over 30 mg kg^{-1} in soils can cause toxicity, dependent on plant species, age and part (organ), cultivation practices, environmental conditions and the condition of the plant (infections). The resistance to Cu toxicity may include immobilization or exclusion of the metal in/from root tissues, formation of complexes among organic acids and proteins, but also adaptation of enzyme mechanisms (Lepp, 1981; Woolhouse and Walker 1981).

4. Copper in agricultural use: Sustainable horticulture

First records of the agricultural use of Cu compounds date back to 1761, when the discovery of the antibacterial effects of copper sulphate preparations used on seed grains set up further standards in cultivation practices for the following decades.

The most important breakthrough of copper use in viticulture though was undoubtedly in 1880, when the French scientist Pierre-Marie Alexis Millardet from the Bordeaux district in France chanced to notice that those vines, which had been daubed with a paste of copper sulphate and lime in water, in order to make the grapes unattractive to passers-bys and as well as animals, appeared less affected by downy mildew. Only five years later in 1885 Millardet announced his discovery to the world, a cure for the dreaded mildew through the application of the mixture of copper sulphate, lime and water, up to the present day called the Bordeaux mixture. Additionally to the Bordeaux mixture, but including copper sulphate and sodium carbonate (soda crystals) the so called Burgundy mixture appeared few years later.

At that time the Bordeaux and Burgundy mixtures became indispensable fungicides against various fungus diseases of plants, where the prevention enhanced with the proper application, means an appropriate timing and correct use of the fungicide. Consequently, as standing for a successful plant protection method up to the present days, many thousands of tons of copper are used annually in agriculture all over the globe.

The days of prosperity of the production of fungicides based on copper compounds were in the middle of 20th century when many different chemical combinations with copper were applied. In the last decades and these days pharmaceutical corporations have been fabricating copper based fungicides in soluble forms of sulphates, oxychlorides, acetates, carbonates, oleates, silicates, ohydroxides etc. Most compounds of copper adopt the oxidation states Cu^{+} and Cu^{2+}, respectively called cuprous and cupric. Their efficiency against fungal and bacterial infections is mostly reflected in their capability to retain on the plant surface, but not by the number of applications or the concentration of the agens in the fungicide. Copper is also biostatic which means that bacteria cannot grow on surfaces treated with it. Copper barriers can range from copper tape made as a slug barrier to more decorative hammered copper sheeting. Any copper that leaches into the soil around the plants is non-toxic and will not adversely affect the plants or people consuming the vegetables and fruits.

Copper plays an important role not only in pathogen controlling but also in soil fertilization, animal nutrition etc. Excessively long copper use in plant protection but also in terms of fertilisation in modern agriculture lead to copper accumulation in soils, where copper surplus is considered if the soils contain more than two parts per million available copper, bearing in mind the physiological and biochemical needs of the plants. On the other hand the same soil conditions in context of copper deficiency can also affect nutritional statuses of animals and consequently their life expectancy.

4.1 Copper as fungicide

First uses of fungicide reach back 400 years ago where simple water solutions (salts) were frequently used for cereal seed treatments. Not earlier than at the beginning of 20th Century more complex organic chemical solutions were introduced (Bordeaux and Burgundy mixture etc.). Nowadays a considerable number of phytochemical companies offer numerous and different classes of fungicides, where copper fungicides according to their antifungal and antibacterial effects play an important role. The efficiency of copper fungicides, particularly aggressive in moist media, is caused by denaturation of protein structures (secondary and tertiary) of fungi and bacteria, and consequently the interruption of their functions. Copper forms compounds in the oxidation states of +1 (cuprous) and +2 (cupric), but trivalent copper in aqueous solution splits rapidly.

Classification of copper fungicide according to chemical structure:

- copper sulphate (cupric sulphate; $CuSO_4$),
- copper acetate (cupric acetate; $Cu(OAc)_2$),
- cuprous oxide (copper(I)oxide; Cu_2O),
- cupric chloride (copper(II) chloride; $CuCl_2$),
- copper oxychloride (copper(II) oxychloride;$CuCl_2 \cdot 3Cu(OH)_2$),
- cuprous chloride (copper(I) chloride; CuCl),
- cupric nitrate (copper(II) nitrate; $Cu(NO_3)_2$),
- copper cyanide (copper(I) cyanide; CuCN),
- copper soap (solution of copper octanoate),
- copper naphthenate.

4.1.1 Copper sulphate (cupric sulphate)

Copper(II) sulphate is a very versatile chemical compound with a wide use in industry but also in horticulture where it exists in a various forms that differs in degree of hydration. Cupric sulphate is the most frequently used form of Cu in pesticides against many diseases listed in Table 3. It is produced industrially by treating copper metal or its oxides with sulfuric acid, or with placing of copper in a solution of nitric or sulfuric acids as well as in water. The chemical form copper sulphate pentahydrate ($CuSO_4 \times 5H_2O$), a fungicide, mixed with hydrate lime ($Ca(OH)_2$), is called Bordeaux mixture or with sodium carbonate ($Na_2(CO)_3$) Burgundy mixture. Both Cu solutions are used more than 200 years yet, especially in fungi disease control in horticulture (vineyards, orchards etc.).

Bordeaux mixture is produced by dissolving copper sulphate and hydrated lime in water, mostly in a concentration of 1 %; in relation 1:1:100 means 1 kg of copper sulphate and 1 kg of hydrated lime is dissolved in 100 litres of water. Copper sulphate contains approx. 25 %

copper as metal, therefore the content of Cu in 1 % Bordeaux mixture varies between 0 and 25 %. On the other hand 1 % Burgundy mixture is made by dissolving 1 kg of copper sulphate in 50 litres of water and separately 125 kg washing soda (or 0.475 kg soda ash) in 50 litres of water. After that the soda solution has to be slowly added to the copper sulphate solution with stirring. Both mixtures have been effective in controlling most of the fungi diseases of plants, applied in a time span of 10 to 20 days, depending on fungi species, plant, weather conditions, age and phenological stage of plants etc. According to the manner of Cu use it is important that first spraying is done before the diseases is expected and continued at regular intervals throughout the susceptible period. The protective coating of Cu mixtures prevents spores germination and consequently their damages caused by penetration through tissues of unprotected parts of plants. An effective spraying scheme has to ensure permanent copper coating over susceptible plant parts, what demands accuracy of the timing of the spraying, especially at precipitation (> 30 mm) and at intensive fruit growing (fruit enlargement).

Plant	**Disease and Pathogen**
Fruits	
Almond	Blossom wilt (*Sclerotinia laxa; Sclerotinia fructigena*)
	Leaf curl (*Taphrina deformans*)
	Rust (*Puccinia pruni-spinosae*)
	Shot hole (*Clasterosporium carpophilum*)
Apple	Bitter rot (*Glomerella cingulata*)
	Black rot (*Physalospora obtusa*)
	Blossom wilt (*Sclerotinia laxa*)
	Blotch (*Phyllosticta solitaria*)
	Canker (*Nectria galligena*)
	Fireblight (*Erwinia amylovora*)
	Pink disease (*Corticium salmonicolor*)
	Scab (*Venturia inaequalis*)
Apricot	Blossom wilt (*Sclerotinia laxa and Sclerotinia fructigena*)
	Rust (*Puccinia pruni-spinosae*)
	Shot hole (*Clasterosporium carpophilum*)
Avocado	Anthracnose (Black spot) (*Glomerella cingulata*)
	Bacterial rot (*Pseudomonas syringae*)
	Fruit spot (*Cercospora purpurea*)
	Scab (*Sphaceloma perseae*)
Banana	Black rot (Die back) (*Botryodiplodia theobromae*)
	Helminthosporiosis (*Helminthosporium* sp.)
	Sigatoka disease (Leaf spot) (*Mycosphaerella musicola*)
Blackberry	Cane spot (*Elsinoe veneta*)
Blueberry	Fruit rot (*Sclerotinia vaccinii-corymbosi*)
	Leaf rust (*Pucciniastrum myrtilli*)
	Powdery mildew (*Microsphaera alni* var. *vaccinii*)

Plant	Disease and Pathogen
Cherry	Bacterial canker (*Pseudomonas mors-prunorum*)
	Bitter rot (*Glomerella cingulata*)
	Brown rot (Blossom wilt) (*Sclerotinia laxa* and *Sclerotinia fructigena*)
	Leaf scorch (*Gnomonia erythrostoma*)
	Leaf spot (*Coccomyces hiemalis*)
	Scab (*Venturia cerasi*)
	Shot hole (*Clasterosporium carpophilum*)
Chestnut	Blight (*Endothia parasitica*)
	Ink disease (*Phytophthora cambivora*)
Citrus	Anthracnose (Wither tip) (*Gloeosporium limetticola*)
	Black pit (*Pseudomonas syringae*)
	Black spot (*Phoma citricarpa*)
	Brown rot (*Phytophthora* spp.)
	Canker (*Xanthomonas citri*)
	Mal secco (*Deuterophoma tracheiphila*)
	Melanose (*Diaporthe citri*)
	Scab (*Elsinoe fawcetti*)
	Septoria spot (*Septoria depressa*)
	Sooty blotch (*Leptothyrium pomi*)
	Sooty mould (*Aithaloderma citri*)
	Thread blight (*Corticium koleroga*)
Currant (Ribes)	Leaf spot (*Mycosphaerella grossulariae* and *Mycosphaerella ribis*)
	Leaf spot (*Pseudopeziza ribis*)
Fig	Blight (*Phizoctonia microsclerotia*)
	Canker (*Phomopsis cinerescens*)
	Leaf fall and Fruit rot (*Cercospora bolleana*)
	Rust (*Cerotelium fici*)
	Thread blight (*Corticium koleroga*)
Fruit trees	Crown gall (*Bacterium tumefaciens*)
Gooseberry	American mildew (*Sphaerotheca mors-uvae*)
	Cluster cup rust (*Puccinia pringshemiana*)
	Die back (*Botrytis cinerea*)
	Leaf spot (*Mycosphaerella grossulariae*)
Mango	Anthracnose (*Colletotrichum gloeosporioides*)
	Anthracnose (*Gloeosporium mangiferae*)
	Bacterial black spot (*Erwinia mangiferae*)
	Powdery mildew (*Oidium mangiferae*)
	Red rust (*Cephaleuros virescens*)
	Scab (*Elsinoe mangiferae*)

Plant	Disease and Pathogen
Nectarine	Blossom wilt (*Sclerotinia laxa* and *Sclerotinia fructigena*)
	Leaf curl (*Taphrina deformans*)
	Rust (*Puccinia pruni-spinosae*)
	Shot hole (*Clasterosporium carpophilum*)
Olive	Leaf spot (*Cycloconium oleaginum*)
Passion fruit	Brown spot (*Alternaria passiflorae*)
	Grease spot (*Pseudomonas passiflorae*)
Peach	Blossom wilt (*Sclerotinia laxa* and *Sclerotinia fructigena*)
	Leaf curl (*Taphrina deformans*)
	Rust (*Puccinia pruni-spinosae*)
	Shot hole (*Clasterosporium carpophilum*)
Pear	Bitter rot (*Glomerella cingulata*)
	Firebiiglit (*Erwinia amylovora*)
	Leaf spot (Leaf speck) (*Mycosphaerella sentina*)
	Scab (*Venturia pirina*)
	Scab (America) (*Cladosporium effusum*)
	Thread blight (*Corticium koleroga*)
Plum	Bacterial canker (*Pseudomonas mors-prunorum*)
	Bacterial spot (*Xanthomonas pruni*)
	Black rot (*Dibotryon morbosum*)
	Blossom wilt (*Sclerotinia laxa*)
	Brown rot (*Sclerotinia fructigena*)
	Rust (*Puccinia pruni-spinosae*)
	Shot hole (*Clasterosporium carpophilum*)
	Watery rot (Pocket plums) (*Taphrina pruni*)
	Wilt (*Pseudomonas prunicola*)
Raspberry	Cane spot (Anthracnose) (*Elsinoe veneta*)
	Cane wilt (*Leptosphaeria coniothyrium*)
	Spur blight (*Didymella applanata*)
Strawberry	Leaf spot (*Mycosphaerella fragariae*)
Grapevine	"Coitre" (*Coniothyrium diplodiella*)
	Angular leaf spot (*Mycosphaerella angulata*)
	Anthracnose (*Elsinoe ampelina*)
	Crown gall (*Bacterium tumefaciens*)
	Bitter rot (*Melanconium fuligineum*)
	Black rot (*Guignardia bidwellii*)
	Downy mildew (*Plasmopara viticola*)
	Leaf spot (*Isariopsis fuckelli*)
	Powdery mildew (*Uncinula necator*)
	Totbrenner (*Pseudopeziza tracheiphila*)

Plant	Disease and Pathogen
Vine (Sultana)	Sooty dew (*Exosporium sultanae*)
Walnut	Anthracnose (Blotch) (*Gnomonia leptostyla*)
	Downy leaf spot (*Microstroma juglandis*)
	Blight (*Xanthomonas juglandis*)
	Ring spot (*Ascochyta juglandis*)
Vegetable and others	
Artichoke (Globe)	*Ramularia cynarae*
Asparagus	Rust (*Puccinia asparagi*)
Bean (Broad)	Chocolate spot (*Botrytis cinerea*)
	Leaf spot (*Asochyta pisi*)
	Rust (*Uromyces fabae*)
Bean (French and Runner)	Anthracnose (*Colletotrichum lindemuthianum*)
	Common blight (*Xanthomonas phaseoli*)
	Halo blight (*Pseudomonas medicaginis* var. *phaseolicola*)
	Powdery mildew (*Erysiphe polygoni*)
	Rust (*Uromyces appendiculatus*)
Brassicas	Black leg (Canker) (*Phoma lingam*)
	Black rot (*Xanthomonas campestris*)
	Damping off (*Oipidium brassicae*)
	Downy mildew (*Peronospora parasitica*)
Carrot	Bacterial soft rot (*Bacterium carotovorum*)
	Blight (*Alternaria dauci*)
	Leaf spot (*Cercospora carotae*)
Cucurbits	Anthracnose (*Colletotrichum lagenarium*)
	Black rot (*Mycosphaerella citrullina*)
	Downy mildew (*Pseudoperonospora cubensis*)
	Leaf blight (*Alternaria cucumerina*)
	Powdery mildew (*Erysiphe cichoracearum*)
	Stem end rot (*Physalospora rhodina*)
	Wet rot (*Choanephora cucurbitarum*)
	Wilt (*Erwinia tracheiphila*)
Grasses	Brown patch of lawns (*Rhizoctonia* and *Holminthosporium* spp.)
	Red thread (*Corticium fusiforme*)
	Snow mould (*Calonectria graminicola*)
	Stripe smut (*Ustilago striiformis*)

Plant	Disease and Pathogen
Lettuce	Downy mildew (*Bremia lactucae*)
	Ring spot (*Marssonina panattoniana*)
Onion	Downy mildew (*Peronospora destructor*)
Persimmon	Canker (*Phomopsis diospyri*)
Potato	Blight (Late blight) (*Phytophthora infestans*)
	Dry rot (*Sclerotium rolfsii*)
	Early blight (*Alternaria solani*)
	Grey mould (*Botrytis cinerea*)
Rose	Black spot (*Diplocarpon rosae*)
	Downy mildew (*Peronospora sparsa*)
	Leaf spot (Anthracnose) (*Sphaceloma rosarum*)
	Mildew (*Sphaerotheca pannosa*)
	Rust (*Phragmidium mucronatum*)
Spinach	Leaf spot (*Heterosporium variabile*)
	Downy mildew (*Peronospora effusa*)
Sugar beet	Leaf spot (*Cercospora beticola*)
	Downy mildew (*Peronospora schactii*)
Sunflower	Rust (*Puccinia helianthi*)
	Wilt (*Sclerotinia sclerotiorum*)
Sweet potato	Wilt (*Fusarium* spp.)
Tea	Black rot (Die back) (*Botryodiplodia theobromae*)
	Blister blight (*Exobasidium vexans*)
	Grey blight (*Pestalotia theae*)
	Red rust (*Cephaleuros niycoidea*)
Tomato	Anthracnose (*Colletotrichum phomoides*)
	Bacterial spot (*Xanthomonas vesicatoria*)
	Blight (Late blight) (*Phytophthora infestans*)
	Early blight (*Alternaria solani*)
	Foot rot (*Phytophthora cryptogea*)
	Fruit rot (*Didymella lycopersici*)
	Fruit rot (*Phytophthora capsici*)
	Grey leaf spot (*Stemphylium solani*)
	Leaf mould (*Cladosporium fulvum*)
	Leaf spot (*Septoria lycopersici*)
	Mildew (*Leveilluia taurica*)

Table 3. List of diseases and pathogens at some horticultural plants suitable to control with copper fungicides

4.1.2 Copper acetates

Copper(II) acetate is also known as cupric acetate with the chemical formula $Cu(OAc)_2$, where OAc^- stands for acetate ($CH_3CO_2^-$). Since ancient times copper acetates of some form have been used as fungicides (greenish colour), whereas nowadays more frequently as reagents for syntheses of many inorganic and organic compounds. Copper acetate is odourless and efflorescent. It is soluble in alcohol and slightly soluble in ether and glycerol. The commercially available copper acetates are these in hydrated forms ($Cu(C_2H_3O_2)_2\ H_2O$); Mr = 199.67 g mol^{-1}) which contain one molecule of water for each atom of Cu. The production of copper acetates first appeared in France, where the term of "verdigris" (*verte grez*) was applied. At grape remaking and wine production copper metal sheets were used, and when the surface of copper sheets exposed to the air for some days, they had become coated with verdigris, which were removed, and then replaced. The obtained product was known as blue verdigris and in a concentration of 1 kg of basic copper acetate in 500 litres of water; it was used as a fungicide. It contains approx. 32 % of copper and crystallises with 1 mol crystal water. Nowadays it is prepared by dissolving cupric oxide or a copper carbonate in acetic acid; however it can be produced also by treating a copper sulphate solution with lead acetate or by the reaction of acetic acid and copper with subsequent crystallization.

Until the 19th century, verdigris coincidentally happened to be the most vibrant green pigment available and was therefore frequently used in painting. Copper acetates of different chemical compositions make up pigments, colours from green to green-blue, blue-green, and finally blue. Reactions with copper acetate vary among substances such as the following: copper acetates dissolve in mineral acid, alkalis convert them into blue copper hydroxide, oils, resins and proteins react to form green transparent copper oleates, resinates, and proteinates.

Neutral Verdigris is neutral copper acetate which occurs when basic acetates are dissolved in acetic acid, or when basic verdigris is ground up with strong acidic acid. Basic verdigris forms from the combination of air, water vapour, acetic acid vapour, and copper or copper alloy mix. It forms a solid of blue, or blue-green material. It is often made up of fine needles. The chemical formula for basic verdigris can include all of the following (Kühn, 1997):

$$[Cu(CH_3COO_2]_2.\ Cu(OH)_2.\ 5H_2O \text{ (blue)}$$
$$Cu(CH_3COO)_2.\ Cu(OH)_2.\ 5H_2O \text{ (blue)}$$
$$Cu(CH_3COO)_2.\ [Cu(OH)_2]_2 \text{ (blue)}$$
$$Cu(CH_3COO)_2.\ [Cu(OH)_2]_3.\ 2H_2O \text{ (green)}$$

Synonyms of cupric acetate are acetate cuivrique (DOT French); acetato cuprico (DOT Spanish); acetic acid cupric salt; acetic acid, cupric salt; copper acetate; copper acetate ($Cu(C_2H_3O_2)_2$); copper acetate (Cu ($MCO_2)_2$); copper acetate ($Cu(OAc)_2$); copper diacetate; copper (II) acetate; copper (II) diacetate; copper (II) acetate; crystallized verdigris; crystals of venus; cupric acetate monoydrate; cupric diacetate; neytral verdigris; and venus copper.

4.1.3 Cuprous oxide

Cuprous oxide is an inorganic compound of two known forms: copper(I) oxide (Cu_2O; Mr = 143.08 g mol^{-1}), in nature present as cuprite, dark red crystals or granular powder, formed on copper by heat, and cupric oxide copper(II) oxide (CuO; Mr = 79.55 g mol^{-1}), a black-brown crystal or amorphous powder. It is used in making fibres and ceramics and for welding fluxes. Cu_2O presents one of the principal oxides of copper. The most common preparation of cuprous oxide is by oxidation of copper metal $4\ Cu + O_2 \rightarrow 2\ Cu_2O$; $2\ Cu + O_2 \rightarrow 2\ CuO$, commercially however also by reduction of copper(II) solutions with sulfur dioxide, the final product of which is reddish mineral cuprite. Cuprous oxide is commonly used as a pigment (colouring of porcelain and glass), a fungicide (seed dressings) and an antirust protection agent for marine paints. Available on the market are copper(II) oxides with a copper content of around 78 %.

Synonyms of cuprous oxide are yellow cuprocide; red copper oxide; dicopper monoxide; dicopper oxide; brown copper oxide; copper hemioxide; Copper nordox; copper protoxide; copper suboxide; cuprite; cuprocide; fungimar; dikupferoxid (German); óxido de dicobre (Spanish); oxyde de dicuivre (French).

4.1.4 Cupric chloride

Copper(II) chloride is a light brown solid chemical compound with the formula $CuCl_2$ and has the potential of slowly absorbing moisture and forming a blue-green dihydrate., The aqueous solution prepared from copper(II) chloride contains a range of copper(II) complexes depending on concentration, temperature, and the presence of additional chloride ions. Copper(II) chloride occurs as a very rare mineral in nature, tolbachite and the dihydrate eriochalcite, more common however are mixed oxyhydroxide-chlorides, like atacamite $Cu_2(OH)_3Cl$. There are few preparations of cupric chloride known, used as fungicide in agriculture as well:

a. chlorination of copper: $Cu + Cl_2 + 2\ H_2O \rightarrow CuCl_2(H_2O)_2$,
b. treatment of Cu hydroxide, oxide or Cu(II) carbonate with hydrochloric acid,
c. anhydrous $CuCl_2$ prepared directly by the union of copper and chlorine and
d. by crystallization of torrid dilute hydrochloric acid, cooling in $CaCl_2$-ice bath.

$CuCl_2$ is a yellowish to brown, deliquescent powder soluble in water, alcohol and ammonium chloride, used as a mordant in dyeing and printing textiles. CuCl consists of fine grey-black pearls with size of a few hundred μm and a copper content of 64 %. Copper(II) chloride dihydrate ($CuCl_2$ x $2H_2O$) is built up of blue-green crystals, soluble in water and has a copper content of approx. 37 %. Anhydrous copper(II) chloride is a brown crystal powder, soluble in water and highly hygroscopic, with a copper content of approx. 47 %.

Synonyms of cupric chloride; copper(II) chloride; dichlorocopper, Kupferdichlorid (German); dicloruro de cobre (Spanish); dichlorure de cuivre (French).

4.1.5 Cuprous chloride

Copper(I) chloride, known also as lower chloride of copper with the formula CuCl (M_r = 99.03 g mol^{-1}) or Cu_2Cl_2 (M_r = 198.05 g mol^{-1}), is a white solid substance partially soluble in water, but totally in concentrated hydrochloric acid. In middle of 17th century cuprous chloride was first produced by Robert Boyle from mercury(II) chloride and metal Cu: $HgCl_2 + 2\ Cu \rightarrow 2\ CuCl + Hg$. Later Proust J.L. prepared CuCl by heating $CuCl_2$ at red heat in absence of air, causing it to lose half of its combined chlorine, followed by removing residual $CuCl_2$ by rinsing with water, and by the application which was widely used for heating and lighting. During the 19th and early 20th Centuries the acidic solution of CuCl was formerly used for analysis of carbon monoxide content in gases, for example in Hempel's gas apparatus. The moist powder's exposure to air and sunlight, results in a color change to yellow, violet and blue-black. The main use of copper(I) chloride is in phytochemistry as a precursor to the fungicide copper oxychloride (dicopper chloride trihydroxide; $Cu_2(OH)_3Cl$) green crystalline solid, largely stable in neutral media, but decomposes by warming in alkaline media, yielding oxides, virtually insoluble in water and organic solvents, soluble in mineral acids yielding the corresponding copper salts. For this purpose aqueous copper(I) chloride is generated by comproportionation and later air-oxidized:

$$Cu + CuCl_2 \rightarrow 2\ CuCl$$
$$6\ CuCl + 3/2\ O_2 + 3\ H_2O \rightarrow 2\ Cu_3Cl_2(OH)_4 + CuCl_2$$

Synonyms of cuprous chloride are copper chloride; copper monochloride; chlorid medny; copper(1+) chloride; cuprous monochloride; dicopper dichloride; Kupferchlorid (German); cloruro de cobre (Spanish); chlorure de cuivre (French).

4.1.6 Copper oxychloride

Copper oxychloride is a basic copper chloride with the formula $CuCl_2 \times 3Cu(OH)_2$, a green powder used as a blue colour agent and as a fungicide (form as powder, wettable powders and pastes) that controls a wide range of fungal and bacterial diseases of fruits, vegetables and ornamentals. Usually it is manufactured either by the reaction of hydrochloric acid and copper metal or by the air oxidation of cuprous chloride suspensions. It usually contains approx. 57 % of copper and is not soluble in water, but in various acids. Beside its use as fungicide (Table 4) it is also applied as a compound of herbicides and insecticides.

Synonyms of copper oxychloride are copper chloride mixture with copper oxide, hydrate; dicopper chloride trihydroxide; cupric oxide chloride; copper(II) oxychloride; copper oxychloride; vitigran blue; Dikupferchloridtrihydroxid (German); trihidroxicloruro de dicobre (Spanish); trihydroxychlorure de dicuivre (French); tribasic copper chloride; copper chloroxide; copper(II) chloride hydroxide.

4.1.7 Cupric nitrate

Copper(II) nitrate is also known as copper nitrate, its chemical formula is $Cu(NO_3)_2$ (M_r = 232.60 g mol^{-1}). In anhydrous form it is blue coloured, in crystalline and it is used for formulation of fungicides and herbicides. The production of cupric nitrate follows the processes underneath:

a. treating metal Cu with N_2O_4 ($Cu + 2\ N_2O_4 \rightarrow Cu(NO_3)_2 + 2\ NO$),
b. hydrolysis of the anhydrous material (preparation of copper nitrate hydrate) and
c. treating copper metal with an aqueous solution of silver nitrate or concentrated nitric acid ($Cu + 4\ HNO_3 \rightarrow Cu(NO_3)_2 + 2\ H_2O + 2\ NO_2$)

Copper nitrate hydrate ($Cu(NO_3)_2 \cdot nH_2O$) appears either as a green powder or blue crystallised, it is soluble in water, used in electroplating copper on iron, as a catalyst and nitrating agent in organic reactions, fungicides and wood preservatives and as a pigment for ceramics. Copper(II) nitrate trihydrate ($Cu(NO_3)_2 \times 3H_2O$) is a frequent crystalline product with a copper content of around 26 %, consisting of rather large blue-green crystals.

Plant	Disease	Application
Pipfruit, Stonefruit	Black spot, Fire blight, European canker, Leaf curl, Shot hole (die-back), Bacterial spot, Stonefruit blast	Bud burst and green tip (Sept.), leaf fall (May) and winter dormancy.
Citrus, Passionfruit	Verrucosis, Brown rot, Melanose, Black spot, Phytophthora blight	Petal fall and at 3-4 weekly intervals until harvest.
Grapes, Berryfruits	Downy mildew, Leaf spots, Rust	Bud burst to harvest at 14 day intervals. Further applications would be necessary if conditions favour infection.
Roses, Ornamentals	Black spot, Downy mildew, Leaf spots, Fire blight	Bud burst and green tip (Sept), leaf fall (May) and winter dormancy.
Beans, Peas	Seed rots, Anthracnose, Bacterial brown spot, Rust, Rust blight	Dust seed thoroughly prior to sowing. Bud burst and green tip.
Broccoli, Carrots, Cucumber, Lettuce, Zucchini	Anthracnose, Leaf spots, Early and late blight, Bacterial blight	Apply when disease first appears. Repeat at 7 – 14 day intervals, whilst conditions favour infection.
Tomato	Anthraconse, Bacterial speck, Bacterial spot, Late blight, Septoria leaf spot	Apply when disease first appears. Repeat at 7 – 14 day intervals, whilst conditions favour infection.
Walnut	Walnut Blight	Apply at least three sprays at 7 – 10 day intervals. Further applications would be necessary if conditions favour infection.
Ornamentals (flowers and shrubs)	Fungal leaf spots, Downy Mildew	Apply when disease first appears. Repeat at 7-14 day intervals as required. Small scale phytotoxicity tests are recommended as some varieties may be sensitive under certain conditions.
Red beet	Downy Mildew, Rust	Apply at 10 to 14 day intervals, from the seedling stage until maturity, while conditions allow infection.
Strawberries	Leaf Spot, Leaf Scorch	Apply at 10 – 14 day intervals in wet weather or if conditions favour infection.

Table 4. The plants and diseases where the application of copper oxychloride was effective

Synonyms of cupric nitrare are cupric nitrate hemipentahydrate; nitric acid, copper (II) salt, hydrate (2:5); copper II nitrate hemihydrate; Kupferdinitrat (German); dinitrato de cobre (Spanish); dinitrate de cuivre (French).

4.1.8 Copper cyanide

Copper(I) cyanide as an inorganic compound and has the chemical formula CuCN, due to the presence of Cu(II) impurities it can be green, it is a useful in electroplating copper, furthermore it can also be applied as a reagent in the preparation of nitriles. It is insoluble in water but rapidly dissolves in solutions containing CN^- to form $[Cu(CN)_3]^{2-}$ and $[Cu(CN)_4]^{3-}$. CuCN, a white crystalline poisonous powder, is produced by the reaction of cuprous chloride and sodium cyanide and used mainly in electroplating, due to its ability to form complex cyanides. It contains approx. 71 % of copper and is produced as follows:

a. by the reduction of copper(II) sulfate with sodium bisulphite at 60 °C, followed by the addition of sodium cyanide to precipitate pure LT-CuCN as a pale yellow powder ($2\ CuSO_4 + NaHSO_3 + H_2O + 2\ NaCN \rightarrow 2\ CuCN + 3\ NaHSO_4$). By the addition of sodium bisulphite the copper sulphate solution becomes green, at that point sodium cyanide should be added.
b. by treating copper(II) sulfate with sodium cynide in a redox reaction, copper(I) cyanide forms together with cyanogen ($2\ CuSO_4 + 4\ NaCN \rightarrow 2\ CuCN + (CN)_2 + 2\ Na_2SO_4$)

It is used as a fumigant in agriculture. The principal use of hydrogen cyanide is in the manufacture process of acrylates, synthetic fibres, plastics and cyanide salts and pesticides. Cyanide salts are utilized in metal cleaning, gardening, invarious organic reactions in manufacture production. It is also used for the production of monomers (e.g. acrylates) as well as an ingredient of fumigants and pesticides. Copper compounds form a protective barrier on the plant surface and thereby prevent fungi from entering the plant host. The copper compounds as non-systemic fungicides operate as Bordeaux mixture, cupric hydroxide, copper arsenate, copper carbonate, cuprous oxide, copper oxychloride etc.

Synonyms of copper cyanide are cianuro de cobre (Spain); Kupfercyanid (Germany); cyanure de cuivre (France).

4.1.9 Copper naphthenate

Copper naphthenate is a copper salt of naphthenic acid, which is a complex natural mixture of fatty acids, by-product of petroleum refining and it takes part in variable compositions (contaminants, inert, and by-products). Naphthenates are mainly applied for industrial use, including the oriduction of synthetic detergents, lubricants, corrosion inhibitors, fuel and lubricating oil additives, wood preservations, insecticides, fungicides, acaricides, wetting agents as well as oil drying agents used in painting and wood surface treatment. A typical copper naphthenate product appears as a green liquid with about 19% copper naphthenate and 81% unlisted ingredients. The cyclopentylacetic acid, alkyl-substituted cyclopentylacetic acids, fused chains of cyclopentylacetic acids, cyclohexylacetic acids, cyclopentanoic acids, and various low-molecular-weight fatty acids all represent frequent constituents of naphthenic acids.

14

Use of Cu Fungicides in Vineyards and Olive Groves

Elda Vitanovic
Institute for Adriatic Crops and Karst Reclamation
Croatia

1. Introduction

Losses caused by pests, diseases and weeds on all agriculture crops in Europe are considerably heavy (28.8 %). They can be reduced in different ways: by law regulations, professional set up of orchards, breading less sensitive or resistant crops, different technical measures of production, mechanical, physical, biological and chemical measures. The use of pesticides to control microbial, fungal and insect plant pests has long been a feature of conventional agricultural practice and their use has made it possible to increase crop yields and food production. Many of these pesticides have toxic effects that are not confined to their target species. Their application may have negative impact on organisms that benefit a wider agro ecosystem and their use may result in an increased accumulation of heavy metals in the soil. Even if just in traces, heavy metals are the primary sign of soil and groundwater contamination. There are various causes that lead to the pollution of agricultural soils and the problem of soil contamination with heavy metals is a central and current issue in modern ecology.

Fungicide use is the most important component of pest and disease control programs in vine and olive production systems. This is because some fungal diseases have a potential to destroy horticultural crops and make them unsalable. The practical and economic problems for producers are more acute in organic production systems than in the conventional ones, because the use of fungicides in organic production is much more limited. Whilst several synthetic active ingredients are available in the conventional production, these are not allowed in organic agriculture, except for certain copper products, the use of which is considered to be traditional organic practice. In most countries copper fungicides can be used in organic crop production.

Copper fungicides have been used in pome and stone fruit orchards and vineyards for more than 100 years. The most common fungal diseases controlled by copper fungicides in vineyards are *Plasmopara viticola* (B. and C.) Berl. and De Toni and *Phomopsis viticola* Sacc. Copper fungicides such as Bordeaux mixture (a complex of copper sulphate and lime) has been used in viticulture as a plant protection product against the stated fungal diseases since the 18th century. This was the first fungicide to be used on a large scale worldwide. Even today, the only fungicides allowed under organic standards and effective against *Plasmopara viticola* are based on copper hydroxide and copper sulphate. Moreover, other copper compounds have been introduced, including copper carbonate, copper ammonium carbonate, copper hydroxide, copper oxide, copper oxychloride, copper oxychloride sulphate, etc. However, their long-term application and subsequent wash-off from the treated plants have resulted into an extensive copper accumulation in vineyard soils. According to the information gathered to date, a long-term use of copper fungicides in viticulture results in the ingression of significant quantities of copper, which remain in the surface soil layer at 0 - 0.2 m, as has been verified by a number of researchers. The bulk of copper accumulated in leaves and soil after the treatment of the vine with copper fungicides returns to the surface layer of soil through tillage or the biological cycle. Copper can simultaneously be both a micronutrient and a toxic element, depending on its concentration in the soil. In the soil copper is bound to organic matter, to Fe and Mn oxides, adsorbed to clay surface, it is present in the matrix of primary silicate minerals, in secondary minerals or within amorphous matter. The sum of it all can be defined as total copper in soil. Determination of the total content of metals in soils is an important step in estimating the hazards to the vital roles of soil in the ecosystem, and also in comparison with the quality standards in terms of the effects of pollution and sustainability of the system. From the ecotoxicological aspect, it is equally important to determine the bioavailability of copper accumulated in vineyards. Copper availability to biota and its mobility are the most important factors for soil environment. Copper bioavailability is influenced not only by physical and chemical properties of the soil, but also by environmental factors such as climate, biological population, and type and source of contaminants. Copper is toxic for soil organisms and plants, expecialy copper contents as high as those reported in vineyard soils. Even low concentrations of copper in soil may result in long-term effects including reduced microbial and earthworm activity and subsequent loss of fertility. Humans are exposed to copper from many sources. 75 to 99% of total copper intake is from food. Possible undesirable effects of copper fungicides on the health of workers exposed to the chemicals and consumers of crop products treated with them are a major concern. In humans, acute ingestion of copper sulphate may cause gastrointestinal injury, haemolysis, methemoglobinemia, hepatorenal failure, shock, or even death.

In olive orchards, olive leaf spot disease is caused by fungus *Spilocea oleaginea* Cast. Today, olive leaf spot is a significant and serious problem in almost all our olive orchards, including those with organic production. It adversely affects fertility of infected trees, and its recurrence year after year causes degradation of whole olive trees, particularly the young ones. Olive leaf spot is readily controlled by copper fungicides. For effective olive protection, several applications are necessary in one year. Concentration of applied copper fungicides must be strictly under control because of possible copper residues in olive fruits and consequently in oil, which is restricted by law. In years with particularly warm and rainy autumns, one treatment with copper fungicides in autumn is not enough, but it is necessary to perform at least three treatments with copper fungicides. Undoubtedly, increasing the number of treatments in autumn renders it impossible to fully observe all the regulations. On the other hand, if the regulations are fully observed, the question arises whether it is actually possible to adequately protect olive groves against this unpleasant and rising disease at all.

2. Copper fungicides in vineyards and olive groves

The contamination of agricultural soils with inorganic (copper-based) and organic pesticides, including their residues, presents a major environmental and toxicological concern. Agricultural soils are particularly exposed to excessive contamination by heavy metals, the reasons being traffic, households and anthropogenic impact. Anthropogenic impact is especially conspicuous in vineyard soils, orchards and gardens. The use of copper fungicides is the most important component of disease control programs in vine and olive production systems. This is because fungal diseases, such as *Plasmopara viticola, Phomopsis viticola* and *Spilocea oleaginea,* have the potential to destroy vineyards and olive orchards. Due to a devasting and lasting effect fungal diseases can have on horticultural crops, the use of copper fungicides is considered best practice in preventative agrochemical spray programs (McConnell et al., 2003). This involves many fungicide applications in the course of one year. A regular use of fungicides can potentially carry a risk to the environment, particulaty if residues are retained in soil or transferred into water (Kookana et al., 1998; Wightwick & Allison, 2007; Komarek et al., 2010). Concern has been caused by the long-term use of copper fungicides, which can result in an accumulation of copper in the surface layer of the soil (Wightwick et al., 2006; Komarek et al., 2010). This has a negativ effect on soil organisms and carries a potential risk to long-term fertility of the soil (Wightwick et al., 2008; Komarek et al., 2010). Regarding the environmental and toxicological hazards associated with the extensive use of fungicides, the choice of fungicides should be performed carefully according to the physico-chemical properties of the soils and climatic and hydrogeological characteristics of the physico-growing regions. In vine and olive growing production systems, a balance needs to be found between controlling fungal disease risks on crops and protecting agro ecosystems. Copper residues resulting from copper fungicide applications in vineyards and olive orchards have affected the key soil health indicators. To protect soil health, alternatives to copper for disease control will need to be developed, along with remedial technologies to reduce copper contamination in soils. Contamination with metals and organic pollutants, together with erosion and tillage, reduces the quality of the soils and poses a serious environmental and toxicological threat. Vineyard soils are usually highly degraded soils in terms of biochemical properties (Miguens et al., 2007) and are therefore more suspectible to contamination. During the last few decades, some European vineyards have been abandoned, mostly those situated on steep slopes, wich has led to intensive soil erosion and subsequent dispersion of the pollutants into the environment (Novoa-Munoz et al., 2007; Fernandez-Calvino et al., 2008).

Ever since the 18th century copper fungicides have been used in viticulture as plant-protecting preparations against fungal diseases (Merry et al., 1983). As in the past, so too today, a part of the fungicides used for the protection of vines are based on copper as active ingredient. In Croatia, the permission for the application in viticulture, for the control of fungal diseases, has been given to the following preparations: copper sulfate, AI - copper (I) oxide, AI - copper oxychloride, AI - copper hydroxide, AI - copper-hydroxide-calcium sulfate complex, AI - copper-hydroxide - calcium-chloride complex, AI - combination of copper and organic fungicides, and AI – combination of copper and mineral oils.

The number of treatments with copper fungicides is estimated to be extremely high, 8-14, (Gracanin, 1947; Flores Velez, 1996; M. Romic et al., 2001). One author presents data on as many as thirty treatments with copper fungicides a year to protect vineyards from deseases. Today, however, a satisfactory protection can be achieved in 4-6 treatments (Cvjetkovic, 1996). This presents not only economy, but also a significant ecological achievement. In their research some authors state that from 2 to 5 $kgha^{-1}$ of copper is introduced in one vegetativ year (Besnard et al., 2001; M. Romic et al., 2001). Some authors state that from 0 to 7,5 $kgha^{-1}$ of copper is introduced in a year (Parat et al., 2002), while others report data showing that twenty years ago up to 15 $kgha^{-1}$ of copper were put in vineyard soils in a year (Delas & Juste, 1975). As mentioned above, the highest total copper concentration is retained in the surface layer of soil, specifically up to 15 cm of depth. In terms of environmental protection, the question of toxic effect of the accumulated total copper is raised. Hence, in the past ten or so years, numerous studies have been carried out on the concentration of total copper in vineyard soils after a long-term application of copper fungicides. Some results of maximum recorded concentrations of total copper in surface layers of vineyard soils in the world are as follows: South Italy - 75 $mgkg^{-1}$ and North Italy - 297 $mgkg^{-1}$ (Deluisa et al., 1996); Greece - 100 $mgkg^{-1}$ (Parat et al., 2002); Moldova - 230 $mgkg^{-1}$ (Delas & Juste, 1975); Australia - 250 $mgkg^{-1}$ (Pietrzak & McPhail, 2004); South France - 250 $mgkg^{-1}$ (Brun et al., 1998); Bordeaux (France) - 1500 $mgkg^{-1}$ (Flores Velez et al., 1996); South Brazil - 3200 $mgkg^{-1}$ (Mirlean et al., 2007); India - 131 $mgkg^{-1}$ (Prasad et al., 1984); New Zealand - 304 $mgkg^{-1}$ (Morgan & Taylor, 2004).

Arable land usually contains between 5 and 30 $mgCukg^{-1}$ of soil, while in treated vineyards the total copper concentration can range even from 100 to 1500 $mgkg^{-1}$ (Drouineau & Mazoyer, 1962; Delas, 1963; Geoffrion, 1975; Deluisa et al., 1996; Flores Velez et al., 1996; Besnard et al., 1999). In more recent research, some authors state that total copper concentrations in vineyard soils are much lower (200 to 500 $mgkg^{-1}$) (Brun et al., 2003).

The total copper concentration in the north-western Croatia ranges from 5 to 248 $mgkg^{-1}$ with the median of 26 $mgkg^{-1}$. Concentrations higher than the permitted values have been recorded in two sampling locations and are most probably of the anthropogenic origin. High values have been measured in soils subjected to intensive wine-growing, i.e. in soils contaminated by copper sulfate (Miko et al., 2000). The stated author indicates that the total copper concentration in the analysed soils of carbonate terrains range from 6 to 923 $mgkg^{-1}$.

Research results of foreign authors also indicate an increase in the total copper concentration in soil after a long-term application of copper fungicides in viticulture. In the last century, the total copper concentrations of 1500 to 3000 $kgha^{-1}$ were the result of plant-protecting preparations used in viticulture (Geoffrion, 1975). Furthermore, the author presents supporting results of his own research, ranging from 870 to 1870 $kgha^{-1}$ of total copper in vineyard soils under research. Results of several studies show that the applied copper remains in the soil, as it binds tightly to organic matter, clay minerals or to Fe, Al and Mn oxides (Stevenson & Cole 1999). Results of vineyard soils research in eastern China showed an increase of 34.2 $mgkg^{-1}$ in total copper concentration in the surface layer of the soils, after a ten-year application of Bordeaux mixture (Li, 1994). Research of vineyard soils of the French part of the Mediterranean, established that total copper concentrations vary from 31 $mgkg^{-1}$ to 250 $mgkg^{-1}$ compared to the soils of woodland areas, where total copper concentrations ranged from 14 $mgkg^{-1}$ to 29 $mgkg^{-1}$ (Brun et al., 1998). In more recent research, some authors state that total copper concentrations in arable land range from 5 to 30 $mgkg^{-1}$, while in most vineyards they amount to 200 to 500 $mgkg^{-1}$, which they bring into connection with the use of copper fungicides (Brun et al., 2003). Research of agricultural soils of central and eastern Giorgia established values of total copper concentrations five to ten times higher than the ones permitted (the highest value amounted to 1023 $mgkg^{-1}$). By comparison to other profiles, the authors concluded that the results obtained were also connected with the application of copper sulfates used in viticulture (Narimanidze & Bruckner, 1999).

As a consequence of soil erosion, total copper gets run off the surfaces under research, and consequently the leaching of copper into deeper layers of soil is limited (Besnard et al., 2001). A high concentration of copper can be run off by surface waters in very contaminated soils. This has been substantiated by the research of many other authors (Albaladejo et al., 1995; Ribolzi et al., 2002).

The relatively long resistence time of copper in top soils, largely related to the high affinity of copper for organic matter and hydrous oxides, means that long-term accumulation of copper is likely. The accumulation of copper in top soils also corresponds to the zone in the soil profile of greatist bilogical activity. Detrimental affects o elevated copper concentrations upon mycorrhizal associations (Georgieva et al., 2002), microbial populations and function (Dumestre et al., 1999) and a range of mesofauna (Paoletti et al., 1998) have been documented.

Besides vineyards, copper fungicides have also been used in hop fields (Schramel et al., 2000; Komarek et al., 2009), apple (W. Li et al., 2005), avocado orchards (Van Zweiten et al., 2004) and during the cultivation of tomatoes and potatoes (Adriano, 2001).

2.1 Behavior of copper in vineyard and olive grove soils

The levels of copper in soil averagely vary from 5 $mgkg^{-1}$ to 50 $mgkg^{-1}$. Copper belongs to a group of heavy metals which are adsorbed tightly onto soil colloids, binding to them as Cu2+ cation. The stated metal derives from primary minerals, where it is found in a univalent form, and after the disintegration it oxidizes in Cu^{2+}. In the soil it forms very stable and complex compounds with organic acids, semi-disintegrated or humified organic matter. Thus bound, it is sparsely accessible to plants and therefore its deficiency occurs more often in very humified soils, due to the "organic" fixation. In the research of the mobility of copper within the soil profile it has been concluded that the translocation of the total copper in soil occurs in both directions, specifically in the form of Cu complex, usually with amino acids, resulting in significant amounts being contained in plant roots (Vukadinovic & Loncaric, 1998).

Copper in soil is found: bound to organic matter of the soil, adsorbed to clay surface, bound to Fe and Mn oxides, present in the structure of primary silicate minerals, present in secondary minerals or within amorphous matter. The sum of all the above can be defined as total copper in soil (Chaignon et al., 2002; Parat et al., 2002). Such strong sorption/complexation properties make it one of the least mobile metals in soils. However, metals of antropogenic origin present in general a greater mobility in soil comparatively to a natural origin where the metals are strongly associated with soil components (Baize, 1997).

Copper is bound to the adsorption soil complex in the form of Cu^{2+} or CuOH. Since this bond is very tight, plants have difficulties using this part of copper. For the removal of copper ions from the adsorption complex, H^+ ions are the most effective. Furthermore, the amount of total copper in soil depends primarily on the parental rocks from which the soil has developed. An important source of copper is the mineral chalcopyrite ($CuFeS_2$). In the forms of CuS or Cu_2S it is found in marshy soils. Acidic rocks such as granite contain around 10-100 $mgkg^{-1}$ of Cu, while basic rocks contain a somewhat higher amount. Hardly soluble phosphates, carbonates and copper sulfides can also be found in the soil. At a higher content of organic matter, an intensified nutrient fixation occurs. Accessibility of the copper thus bound varies, sometimes the bond is so tight that plants are not able to use copper, while sometimes they are. Deficiency of total copper in soil most often occurs precisely in soils rich in organic matter (Anic, 1973). The same author states that apart from the pH reaction of the soil, the concentration of Al ions and calcium in soil is important as well. Although a very low concentration of copper is needed for it to function in the soil, it is still often deficient. The surplus of total copper is more frequent in acidic soils, as well as in orchards and vineyards after a long-term application of Bordeaux mixture ($Ca(OH)_2 + CuSO_4$) and other plant-protecting preparations based on copper as the active ingredient. Due to its poor mobility, total copper is accumulated in the surface layer. Although the accumulation results in a higher concentration of total copper, toxicity usually does not occur, as it quickly transforms in forms less accessible to the plant.

Plants receive copper mostly in ionic form, and it returns into the soil through harvest residues; this is why surface horizons (15-20 cm) are often richer in total copper than deeper mineral layers (Gracanin, 1947). This has been confirmed by the research of other authors as well (Ribolzi et al., 2002). According to the stated author, total copper is particularly deficient in some heath, acidic mineral and many marshy soils in Europe, America, Australia, the Filipines... The author notes that Maquenne and Demoussy conducted research of French heath soils containing the lowest concentration of total copper, as low as 2 mg in 1 kg of soil. He also reports data on the presence of total copper in the main types of soil in the former states of the USSR. It was thus established that red soils abound the most in total copper, and are followed by chernozemic soils and finally by very podzolic sandy soils, while in the peat and marshy soils the concentration of total copper oscillates.

In sandy soils more than 3% of total copper concentration is accounted for by exchangeable copper, of which exchangeable copper bound to nitrates accounts for 1%, and free Cu accounts for 2-9%. The rest of exchangeable copper is bound to organic matter in the soil (Temminghoff et al., 1994).

The research led the stated authors to the conclusion that the concentration of total copper is inversely proportional to the soil depth (up to 60 cm). They noted the highest concentration level of total copper in the surface layer, at 0 to 3 cm of depth (Besnard et al., 2001).

2.2 Bioavailability of copper in vineyerd and olive grove soils

Factors affecting the distribution and migration of heavy metals in the profiles of agricultural soils are as follows:

- soil type, despite its morphogenetic features being disrupted by tillage
- change in the sequence of genetic horizons and active depth of the profile
- amount of organic matter and pH reaction
- consumption process and accumulation of clay, Fe and Mn oxide
- whether the element is of geogenous, pedogenous or anthropogenous origin.

The process of intake of copper through plant roots is an active process and it is believed that a specific transmitter exists. In terms of intake, it is competed by Mn, Fe and Zn, and it has also been noted that a good supply in plants of nitrogen and phosphorous often results in copper deficiency (Vukadinovic & Loncaric, 1998). The stated authors point out that the availability of total copper is considerably influenced by pH reaction of the soil, and that its availability increases with the acidity. Total copper binds more tightly to organic matter of the soil than other micronutrients (e.g. Zn^{2+}, Mn^{2+}), which is why Cu-organic complexes play an important role in copper mobility regulation and availability in the soil (M. Romic & D. Romic, 1998). The stated authors point out that in fluvial and alluvial soil the redistribution of copper within fractions occurs relatively quickly, it is not retained in the exchangeable fraction, which considerably decreases the risk of its mobility and inclusion into the food chain. Apart from organic matter, soil carbonates proved to be another important factor controlling copper mobility in soils. Activity of copper in calcareous soils is to a great extent controlled by the surface precipitation of CuCO3 (Besnard et al., 2001; Ponizovsky et al., 2007). This is especially important in alkaline soils containing high concentrations of carbonates, which is the case for many vineyards. Results of the research on the relation between pH reaction of the soil and availability of total copper show that increase in the pH value of the soil causes increase in the amount of the bound copper in soil, reducing its mobility in the process (McLaren & Crowford, 1973). A finding that some authors point out as one of the greatest is that increase in bioavailable copper concentration is proportional to the increase in pH reaction of the acidic soils rhizosphere (Chaignon et al., 2002; Parat et al., 2002). A high total copper concentration in calcareous soils is caused by deficiency of other heavy metals, such as Fe and Zn, due to their antagonistic interrelation (Chaignon et al., 2001). Equal results were obtained in their research of carbonate soils. They conclude that plants take in copper, as well as other metals (Cd), more intensely from contaminated soils poor in iron and zinc.

Many authors have also researched the correlation between total copper and other metals in soils. In the research of agricultural soils in Zagreb and the surrounding area, it has been established that the content of lead shows a good correlation with the content of copper, zinc and cadmium ($r>0.43$) (M. Romic & D. Romic, 1998). The same results were obtained by some foreign authors while doing research on soils in Georgia (Narimanidze & Bruckner, 1999).

Total copper has a high positive correlation with the amount of organic matter in soil, while total zinc does not show correlation with any of the soil features, nor with the elements encompassed by the research (Cu, Fe, Ni, Cd, Cr) (M. Romic, 2002) .

Iron is the main factor responsible for the accumulation of total copper in the clay fraction of the soil (Parat et al., 2002).

2.3 Physiological role of copper in plants

In 1931 Anna L. Sommer established that copper was one of the first microelements found to be essential for the growth and development of plants. (Anic, 1973).

Physiological role of copper is very important, as it is an integral part or activator of many enzymes which participate in oxidation processes. It affects protein synthesis, it stabilizes chlorophyll molecules and participates in the synthesis of anthocyanins. It is included in the structure of plastocianin, cyto-chrom-oxidase c (transport of electrons), phenol oxidase (oxidation of phenol into quinone), lactase and phenolase, hidroxilase (translocation of phenyl alanine into tyrosine), oxygenase, ascorbic acid oxidase, superoxyde-dismuthase, several amino-oxidase, galacto-oxidase etc. Copper has a distinct affinity to protein structure, and consequently 70% of copper in plants is bound to proteins in chloroplasts, where they act as stabilisers, especially of chlorophyll. Furthermore, it has a significant role in the metabolism of nitrogen compounds, as it regulates the binding of ammonium keto acids, affects the synthesis of nucleic acids, bacterial leg-hemoglobin, the metabolism of carbohydrates, the formation of pollen and plant fertility, it increases resistance to low temperatures etc. (Vukadinovic & Loncaric, 1998).

The authors point out that copper toxicity is manifested with the reduced growth of roots and shoots, with the chlorosis of older leaves and dark-red margin necrosis.

The critical lower limit of this element ranges from 1 $\mu gCug^{-1}$ to 5 $\mu gCug^{-1}$ of dry matter, and depends on many factors, such as: variety, plant organ, plant development level. The upper toxicity limit of the stated element in the leaf ranges from 20 $\mu gCug^{-1}$ to 30 $\mu gCug^{-1}$ of dry matter (Marschner, 1995). The author adds that increased copper concentrations in the plant are connected with the application of plant-protecting products, giving as example vineyards and copper fungicides used for disease control.

The highest total copper concentrations are found at the depth of up to 15 cm (Brun et al., 1998), in which soil layer plant roots are found, therefore the authors conclude that plants are directly exposed to high levels of copper contamination. The same authors indicate that Cu concentrations in roots are a good indicator of Cu bioavailability in soils (Brun, et al., 2001; Chopin, et al., 2008). However, this time-consuming approach is not suitable for routine analyses. Furthermore, it should be pointed out that Cu uptake by roots is species-dependent and influenced by root type and size (Brun, et al., 2008; Chopin, et al., 2008).

2.4 Copper content in grapes, must and wine

Considering the high total copper concentrations accumulated in the surface layer of vineyard soils, due to a long-term use of plant-protecting products based on copper as active ingredient, many questions arise. One of them is whether increased concentrations of the above-mentioned are found in grapes, must and wine, and, if so, whether they can at all be connected to the concentrations found in soil. The answer to this question is comprehensive and deserves a separate research. Since this is not the subject mater of this paper, only some literature data will be briefly presented. The purpose of this subtitle is to give a wider introduction into the issue of copper contamination of soils, as well as to give an insight into the possible consequences of this problem on grapes, must and wine, i.e. on food products. It is important to point out the origin of copper in wine, coming from three sources, which are as follows: plant-protecting products based on copper as active ingredient, winemaking equipment in cellars, and addition of Cu-salt for the elimination of H_2S from wine. The highest concentration of copper taken in from soil is accumulated in the plant root (Chaignon et al., 2002).

Copper is found in rigid parts of the cluster (seeds, skin, stem, etc.), and through processing it enters must and wine (Radovanovic, 1970). The same author presents data on copper concentration in must, which ranges from 0.2 mgL^{-1} to 4.0 mgL^{-1} (an average of 2.0 mgL^{-1}), and on copper concentration in wine, which ranges from 0.1 mgL^{-1} to 5.0 mgL^{-1}. Must always contains a relatively high copper concentration, around 5.0 mgL^{-1} (Ribereau-Gayon et al., 2000). The authors come to a conclusion that the largest part of copper in must originates from copper fungicides (copper-sulfate) used in vineyards for disease control. The same conclusion has been reached by some other researchers (Puig-Deu et al., 1994). The above-mentioned authors report that new wines contain as little as 0.3 mgL^{-1} to 0.4 mgL^{-1}, which is significantly less than the maximum level of copper concentration in wine permitted by the EU regulations, which amounts to 1.0 mgL^{-1}. Croatia has the same permitted level of copper concentration, prescribed by the Wine Act (Croatian Official Gazette, 96/1996) and the related regulations. Copper concentrations in Californian wines are as follows: white wines 0.13 mgL^{-1}, rose wines 0.16 mgL^{-1}, red wines 0.17 mgL^{-1}, and sparkling wines 0.07 mgL^{-1} (Boulton et al., 1996). According to the results of some research, wine contains from 30 μgL^{-1} to 1500 μgL^{-1} of copper, which is an average of 0.2 mgL^{-1} (Margalit, 1996). The author points out that copper concentrations of around 4.0 mgL^{-1} are dangerous for wine, and states that wine with concentrations from 0.5 $mgCuL^{-1}$ to 1.0 $mgCuL^{-1}$ is safe. Copper concentration in wine of 1.0 mgL^{-1} is detected by the senses (Zoecklein et al., 1995).

2.5 Contamination of different types of vineyard soils with copper in this part of Mediterranean region

2.5.1 Different types of vineyard soils

Vine is a very old field crop, native of the Mediterranean countries. Its beginnings date from the times of the Ancient Egypt, Greece and Rome. In the time of the Roman Empire cultivation of vine spread fast to Croatia (Licul and Premuzic, 1993.). From the time between the two World Wars until the middle of the last century, conditions for the development of viticulture were not very favorable. It was not until after that period that a rapid development of viticulture started, through the introduction of new varieties and rootstocks, and through the work on the clonal selection and hybridisation. Viticulture is a very important agricultural field in our country, as it makes more than a tenth of the overall agricultural production value. Its importance is also evident from the fact that, apart from the wine-growers whose aim is production for the market, there are tens of thousands of other wine-growers, who are either subsistence wine-growers or they grow vine as a hobby in small vineyards which are not even listed (Mirosevic, 1996).

This part of Mediterranean region consists of a number of sub-regions, while each sub-region comprises a number of vineyard areas. Following an extensive analysis of the region, sampling was made at four different locations, which had been selected by taking into account differences between respective types of soil typical in the area under research. Areas with anthropogenic colluvial soils were selected as first location. The secound location comprised areas with anthropogenic soils on flysch, areas with anthropogenic soils on terra rossa were selected for third location, while anthropogenic terrace soils on cretaceous limestones were selected as fourth location. The age of the vineyards (40 - 70 years old) had an important role in the selection, due to long-term use of copper fungicides. In order to establish the so-called "background" concentration, woodland soils were selected, as no copper-based plant-protection products had ever been used on them.

Anthropogenic colluvial soils, on flysch and terra rossa, were made by the work of man – by land clearing, digging, terracing, and fertilisation, with the aim to increase their fertility and to protect them from erosion. Viticultural area with anthropogenic terrace soils is defined as hilly and rolling. The parent rock consists of deposits of cretaceous formation, specifically limestone, particularly developed in coastal areas of wine-growing hills. In the past, this area was relatively poorly populated, man's influence was minimal, and only an insignificant amount of land was cultivated. In the course of historic development the number of inhabitants increased, and so did the land cultivation, and people started making terraces.

Characteristics of anthropogenic colluvial soils vary greatly depending on the origin of quaternary deposit, depth, properties of substrate onto which colluvium is deposited, as well as hydrological and geomorphological and lithological conditions of the location in which colluvium is accumulated (Males et al., 1998). On the basis of the results of chemical analyses, it has been determined that the anthropogenic colluvial soils are alkaline reactions (pH 7.30-7.73), the alkalinity increasing with depth, although insignificantly. This type of soils are highly carbonate (26.4-76.8%)and limy (5.79-20.07%) and poorly humic (1.10-11.12%). It has been established that they are moderately to well-supplied with nitrogen. Anthropogenic colluvial soils have shown poor to good supply level of potassium, while they are very poorly to poorly-supplied with phosphorus. Loam texture dominates in these soils. The soils are skeletal, their water capacity is low, while the air capacity is high (Vitanovic et al., 2010a).

Significant variations of the properties of anthropogenic soils on flysch are a result of the lithological complexity of flysch (Males et al., 1998). Anthropogenic soils on flysch are alkaline reactions (pH 7.16-7.44) and their alkalinity increases with depth, although insignificantly. These soils are very carbonate (8.2-62.0%), very limy (3.24-29.98%) and poorly humic (1.84-5.71%). In the researched soils the supply level of nitrogen is good, the supply level of potassium is poor to good, while phosphorous is very poorly to poorly present.

This type of soil shows great variability of the mechanical composition. It has been established that these soils are as skeletal as anthropogenic colluvial soils. Generally, chemical and hydrophysical properties of this type of soil are totally converse in comparison to other soils (Vitanovic et al., 2010a).

Anthropogenic soils on terra rossa are deep, with a specific red color (Males et al., 1998). This type of soil has alkaline reaction (pH 6.81-7.26) which increases with depth. Anthropogenic soils on terra rossa are, unlike the above described soil types, less carbonate (0.4-26.0%). Moreover, they have lower content of total lime (0-9.16%), i.e. they are not limy. However, the analysis has determined that they are considerably humic (1.63-11.47%) and well-supplied with total nitrogen, potassium and phosphorous. The textures are clay loam and silty clay loam. It has been determined that these soils are very skeletal. Generally, this type of soil, in addition to a heavy texture composition, has favorable hydrophysical properties (Vitanovic et al., 2010a).

A characteristic of antropogenic terrace soils is a shallow A-horizon of small-grained soil, located above the parent rocks (Males et al., 1998). Anthropogenic terrace soils are alkaline reactions (pH 6.77-7.29) and their alkalinity increases with depth, however, they have a weaker alkaline reaction than the above described soils. These soils have a low content of total carbonates (1.2-48.0%) and active lime (0-25.56%), but they are very highly humic (3.50-11.24%). They are very rich in nitrogen and very well-supplied with potassium and phosphorous, the content of which increases with depth. An analysis of the mechanical composition indicates that the soils are comprised of loam and silt loam. The soils are skeletal, mainly medium gravelly to cobbly, the water capacity is low, while their air capacity is high. Their water permeability is high. The structure of soils in these vineyards is mainly stable and well-defined (Vitanovic et al., 2010a).

2.5.2 Copper introduction by copper fungicides and total copper content in different types of vineyard soils

Bordeaux mixture, copper-hydroxide-Ca-chloride complex and copper (I) oxide are most frequently used copper fungicides in this part of Mediterranean region. Each of them contains different quantities of active ingredients, and is applied in different concentrations. The total quantity of copper introduced into one hectare of vineyards during one vegetative year has been calculated. The result shows that every vegetative year 2.90 $kgCuha^{-1}$ is introduced in antropogenic colluvial soil, antropogenic soil on flysch and on terra rossa. Copper fungicides used in antropogenic terrace soils on cretaceous limestones introduce 4.20 $kgCuha^{-1}$ every vegetative year (Vitanovic et al., 2010a).

The total copper concentrations in anthropogenic colluvial soils ranged between 70.50 $mgkg^{-1}$ and 181.62 $mgkg^{-1}$, while concentrations from 21.85 $mgkg^{-1}$ to 49.05 $mgkg^{-1}$ were recorded in control areas. Anthropogenic soils on flysch contained from 163.68 $mgCukg^{-1}$ to 302.05 $mgCukg^{-1}$, while the concentrations of this metal in control areas ranged from 44.42 $mgkg^{-1}$ to 124.77 $mgkg^{-1}$. The total copper concentrations in antropogenic soils on terra rossa ranged from 113.46 $mgkg^{-1}$ to 252.89 $mgkg^{-1}$ in vineyard soils, whereas in control areas they ranged from 52.03 $mgkg^{-1}$ to 290.11 $mgkg^{-1}$. Anthropogenic terrace soils on cretaceous limestones contained from 138.79 $mgkg^{-1}$ to 625.79 $mgkg^{-1}$. Concentrations of this metal in control area soils varied from 45.94 $mgkg^{-1}$ to 140.01 $mgkg^{-1}$. What is evident from the above-mentioned data is that total copper concentrations in all types of vineyard soils were higher than in the control soils. According to the research results, vineyard soils of this part of the Mediteranean contain from 70.50 $mgkg^{-1}$ to 625.79 $mgkg^{-1}$ of total copper, while concentrations of the metal in control areas are quite lower (21.85 $mgkg^{-1}$ - 290.11 $mgkg^{-1}$) (Vitanovic et al., 2010a; 2010b).Results of total copper concentrations in all researched soils show a significant difference in concentrations of total copper between the vineyard and control areas. Significantly higher concentrations of this metal were identified in vineyard soils. Based on the obtained results, it can be concluded that total copper accumulates in the surface layer of vineyard soils due to long-term use of copper fungicides. The results of total copper concentrations in various types of researched soils also indicate a considerable difference in total copper concentrations (with 95% certainty) between colluvial anthropogenic soils and anthropogenic terrace soils on cretaceous limestones. Significantly higher concentrations of this metal were identified in anthropogenic terrace soils on cretaceous limestones. The reason for higher concentrations of total copper in heavier soils can be found in stronger bonding of copper with particles of heavier soils. In such soils copper is known to leach more slowly and in lesser quantities into lower layers, while its leaching is more excessive and faster in lighter soils. There are no significant differences in concentrations of the metal under research between anthropogenic soils on flysch and terra rossa and other researched anthropogenic soils (Vitanovic et al., 2010a; 2010b).

Considering the average concentrations of the metal under research, anthropogenic colluvial soils and anthropogenic soils of terra rossa were contaminated with copper, while anthropogenic soils on flysch and anthropogenic terrace soils on cretaceous limestones were polluted with researched metal. Human environment (food, drink, air) is either directly or indirectly connected with soil. Its quality is directly dependant on soil properties. The results of this research are very important, since they are the basis for defining soils as polluted, which will irreversibly remain such, and contaminated, in which the contamination can still be reduced (Vitanovic et al., 2010a).

2.6 Olive leaf spot - The biggest problem in most Mediterranean olive groves

Olive leaf spot, also known as peacock spot, is caused by the fungus *Spilocaea oleaginea,* which attacks the olive exclusively. The Frenchman Castagne described the disease for the first time as early as 1845, under the name *Cycloconium oleaginum.* Today, it is present throughout the world, in all olive-growing regions. The disease is common worldwide and causes serious problems in cooler olive-growing regions, with yield losses estimated to be as high as 20% (Wilson & Miller, 1949). Experts regard it as one of the most widespread and dangerous olive fungal diseases (Obonar et al., 2008). It has been known in this part of the Adriatic coast since the 8th century.

Since the disease most often attacks leaves, it is on them that the symptoms are most conspicuous (Photo 1).

Photo 1. Olive leaf spot

2.6.1 Symptoms and damages of disease

The infected leaves have spots on them which resemble spots on peacock's tail feathers, hence the name peacock spot. In the initial phase of infection dark-green oily spots, difficult to detect, appear on the upper leaf surface. The spots gradually turn yellow, with a discernible yellow-brown halo around them. In this later phase of infection the spots are very conspicuous. As the disease develops, the spots turn dark-brown, their number increases, and they cover more and more surface of the infected leaves. In the final phase of infection the leaf skin detaches from the epidermis, causing the spots to turn white (Photo 2). In rare cases, the symptoms of the disease also occur on the undersides of leaves, as well as on stems and fruits.

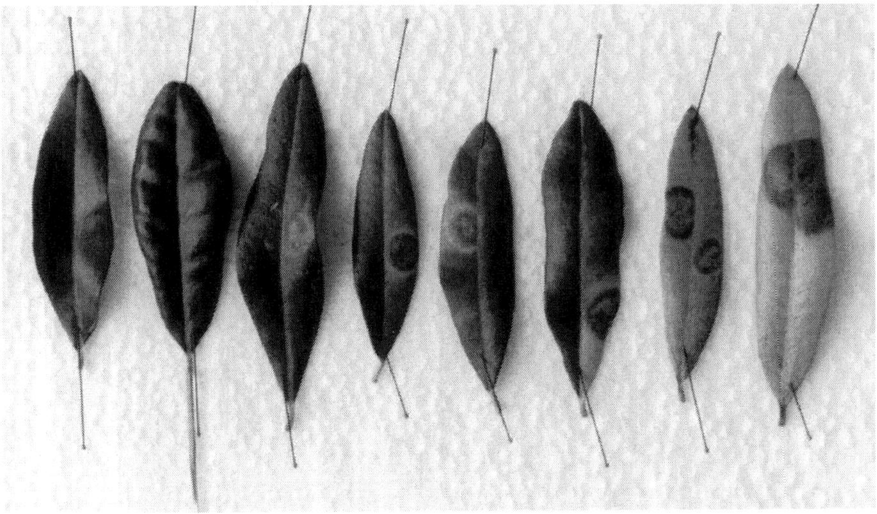

Photo 2. Symptoms of the disease

Damage caused by olive leaf spot is manifested in yield reduction of infected trees, as a consequence of a great loss in leaf mass. The fungus penetrates below the leaf epidermis, spreading into and in between cells. It feeds itself and releases toxins which disturb the creation of starch, necessary for normal plant functions, which consequently leads to the falling off of the sick leaves. Since the leaves fall after the flower buds have been formed, they become insufficient for an adequate diet of flowers, which consequently remain under grown and do not bear fruit. In the case of a serious autumn and subsequently a serious spring infection, leaves fall off massively. Consequently, the trees become partially or completely defoliated. In older olive grove reoccurrence of the disease for several consecutive years can cause defoliation and destruction of individual trees (Photo 3). The damage is greater if the disease spreads in a young olive grove, as then all olive trees can be destroyed (Miller, 1949; Azeri, 1993; Katalinic et al., 2009). Fruit infection can cause unacceptable blemishes on table olives, and when it occurs on oil-producing cultivars, infection may cause a delay in ripening and a decrease in oil yields (Verona & Gambogi, 1964).

2.7 Use of copper fungicides against olive leaf spot and possible copper traces in olive oil

Application of copper-based fungicides is the main method of olive leaf spot control throughout olive-growing regions of the world (Teviotdale et al., 1989; Graniti, 1993; Katalinic et al. 2009). For effective olive protection, several applications of copper fungicides are necessary in one year. For a long time it was thought that olive leaf spot infection could be prevented by two applications of copper preparations, i.e. by one treatment in autumn (September - October) and another one in spring (March - April). The autumn treatment, before the rains begin, is necessary because the rainy period creates conditions for primary infections. This reduces the possibility of spring infections by 87% to 95% (Obonar et al., 2008; Katalinic et al. 2009). Timing of the fungicide applications is vital for effective control of olive leaf spot (Graniti, 1993). In the Mediterranean region, fungicides are usually applied before the onset of the main infection periods, which often coincide with the main shoot-growth seasons (spring and/or autumn) (Prota, 1995). In organic production, for protection of olives against olive leaf spot only the use of copper fungicides is allowed. Concentration of applied preparations must be strictly under control because of possible copper residues in olive fruits and consequently in oil, which is restricted by law. In years with particularly warm and rainy autumns, one treatment with copper fungicides in autumn is not enough, but at least three treatments with copper fungicides are necessary. However, here a problem is encountered. To control olive leaf spot, preparations with increased concentrations are used, and in Croatia treatments with copper fungicides are allowed up to 56 days prior to harvest. In countries members of the EU copper fungicides are allowed up to 21 days prior to harvest. Under local conditions, harvest begins in early November, which means that control should be completed as early as the beginning of September. In years with the aforementioned conditions this is not possible, due to great quantities of precipitation and high temperatures during the whole autumn period, which is optimal for the development and spreading of the disease. The stated conditions require an increased number of treatments. However, in olive groves with organic production the application of only 3 kg of copper per hectare is permitted per year (Croatian Official gazette, 12/2001, 91/2001). The law also provides the maximum quantity allowed in table olives and olive oils, which is 0.40 $mgCukg^{-1}$. Undoubtedly, increasing the number of treatments in autumn renders it impossible to fully observe all the regulations. On the other hand, if the regulations are fully observed, the question arises whether it is actually possible to adequately protect olive groves against this unpleasant and spreading disease at all. In Mediterranean regions, such as Italy and Spain, three fungicide applications (end of winter, end of spring and late autumn) are recommended (Graniti, 1993). In Californian olive groves a single spray of copper-sulphate (Bordeaux mixture) gave satisfactory control of olive leaf spot when applied prior to rains in autumn or early winter (Wilson & Miller, 1949). Some authors also reported that one annual application of copper-based fungicides in late autumn, before the rainy period, effectively controlled olive leaf spot under low disease attack in Californian olive-growing region (Teviotdale et al., 1989). In New Zealand olive groves, multiple applications of copper and/or systemic fungicides are commonly used annually, but often fail to provide effective control, resulting in very high disease levels in some regions (MacDonald et al., 2000).

Olive leaf spot is a disease which has caused death of an increasing number of olive groves in almost all olive-growing areas in recent years. Urged by this problem we set up an experiment in order to find out if several autumn applications of copper fungicides would leave copper residues in olive oil. The experiment was conducted in 2009 and 2010 in a 25-year-old olive grove. In both years olives were treated 3 times in the autumn, every 15 days, with copper fungicides (copper-oxychloride) in the concentration of 0.3%. Copper-oxychloride is permitted for use even in organic production. Due to its good qualities it is especially used for the protection of quality vine varieties, and its application in olive-growing is of a particular importance. Due to its short waiting period (21 days from last spray to harvest), in European countries it is used to fight olive leaf spot in all the applications. Each year of the research 12 olive trees were randomly selected, 4 trees for each treatment combination (four trees - one treatment, four trees - two treatments and four trees - three treatments). The first treatment was performed on 15th September 2009 on all the 12 trees. The second treatment was done on 1st October 2009 on 8 trees, and the third was performed on 15th October 2009 on the last 4 trees. The treatments in 2010 were performed in the same way, on 14th September, 1st October and 15th October 2010. After the harvest, the fruits were not washed and were processed in a mini oil mill of the type Abencor mc2 - water bath with adjusted conditions (temp. < 27 °C and kneading time around 40mins, centrifuge 3000 r.p.m.). The copper residues in olive oil were determined by flame atomic absorption spectrometry (SpectrAA Varian 220). The results of this research show that in both years no traces of copper above the permitted limit (0.13 - 0.26 $mgkg^{-1}$) were found in any of the samples.

3. Conclusion

Owing to the importance of fungal diseases in the production of grapes, olives and olive oil, as well as to a great number of people engaged in the stated productions, it is necessary to continually improve their knowledge of fungal diseases and measures of their control. The importance of this knowledge is also accentuated by the fact that numerous chemical products of higher or lower toxicity for people and environment are used in the protection of vines and olives. It is, therefore, necessary to have a particularly good knowledge of the qualities of products for the protection of the stated crops, as well as of the correct application procedures. Even if just in traces, heavy metals are the primary sign of the soil and groundwater contamination. Polluted agricultural soils present a serious agro-environmental concern. Soil health is fundamental for the sustainability of agriculture. The benefits of having healthy soils are numerous. In healthy soils microorganisms play a major role, especially in nutrient cycling, making many nutrients available to plants. Contamination of vineyards and olive groves with copper compromises soil health and opposes the ideals of IPM and organic production. Copper fungicides have been used in vineyards and olive groves for protection from different fungal diseases for a long time. Many researchers indicate that the use of copper fungicides has led to an increase of total copper in surface layer of vineyard soils. Therefore, the selection of fungicides should be performed carefully, bearing in mind agricultural practices, physico-chemical characteristics of the soils and climatic and hydrogeological regimes in individual vine and olive growing regions. Pest control such as IPM and pest management in organic production, which manage pest damage with the least possible hazard to humans and the environment, should be encouraged. Our research has shown that application of copper fungicides to vineyards has resulted in the accumulation of copper residues in surface soils. Concentrations of up to 625.79 mg/kg were found in antropogenic terrace soil on cretaceous limestones. The reason of significant quantities of heavy metals remaining in the surface layer of vineyard soils lies in the long-term use of copper fungicides and the number of treatments per one vegetative year.The research has shown that repeated treatment of olive groves with copper fungicides in the period prior to the harvest is very efficient in olive leaf spot control. The large number of treatments has been efficient in the control of the disease, so that after processing of the treated olives no traces of copper in oil above the permitted limits were found.

On the basis of all the research and findings to date, as the positive so too the negative ones, on the use of copper fungicides in viticulture and olive-growing, it will be necessary in the future to supplement the knowledge of all the agricultural producers and thus enhance the efficiency of plant-protection measures and reduce the threat for people, useful organisms and environment.

4. References

Adriano, D.C. (2001). Trace element sin the terrestrial environments: biogeochemestry, bioavilability, and risks of metals. Springer-Verlag, New Yourk-Berlin-Heidelberg-Tokyo

Albaladejo, J.; Ortiz, R.; Guillen, F.; Alvarez, J.; Martinez-Mena, M. & Castillo, V. (1995). Erodibility of agricultural soils in the semiarid mediterranean area of Spain. Arid Soil Reserch and Rehabilitation. Vol. 9. (n.d. 1995), pp. 219-226, ISSN 0890-3069

Anic, J. (1973). Ishrana bilja. Faculty of Agriculture, Zagreb, Croatia

Azeri, T. (1993). Research on olive leaf spot, olive knot and Verticillium wilt of olive in Turkey. EPPO Bulletin, Vol. 23, (n.d. 1993), pp. 437-440, ISSN 0250-8052

Baize, D. (January 1997). Teneurs en elements traces metalliques dans les sols (France). INRA Editions, ISNB 978-2-7380-0747-6, Paris, France

Besnard, E.; Chenu, C. & Robert, M. (1999). Distribution of copper in vineyard soils, as influenced by organic amendments. Symposium J09-Environmental Pollution and Natural Backgrounds. In: Journal of Conference Abstracts Cambridge Publications. 01.02.2010, Available from http://www.the-conference.com/JConfAbs/4/564.html

Besnard, E.; Chenu, C. & Robert, M. (2001). Influence of organic amendments on copper distribution among particle-size and density fractions in Champagne vineyards soils. Environmental Pollutions. Vol. 112, No. 3, (n.d. 2001), pp. 329-337, ISSN 0269-7491

Boulton, R.B. (1996). Principle and practices of vinemaking, The Chapman & Hall, ISBN 9781441951908, University of California, Davis, USA

Brun, L.A.; Maillet, J.; Richarte, J.; Herrmann, P. & Rémy, J.C. (1998). Relationships between extractable copper, soil properties and copper uptake by wild plants in vineyard soils. Environmental Pollution. Vol. 102, No. 2-3, (n.d. 1998), pp. 151-161, ISSN 0269-7491

Brun, L.A.; Maillet, J.; Hinsinger, P. & Pepin, M. (2001) Evaluation of copper availability to plants in copper-contaminated vineyard soils. Environmental Pollution. Vol. 111, (n.d. 2001), pp. 293-303, ISSN 0269-7491

Brun, L.A.; Le Corff, J. & Maillet, J. (2003). Effects of elevated soil copper on phenology, growth and reproduction of five ruderal plant species. Environmental Pollution. Vol. 122, No. 3, (n.d. 2003), pp. 361-368, ISSN 0269-7491

Chaignon, V.; Sanchez-Neira, I.; Jaillard, B. & Hinsinger, P. (2001). Relation between copper phytoavailablity and chemical properties of 29 contaminated soils from a vine-growing area. Proceedings of the 6th International Conference on the Biogeochemistry of Trace Elaments, Guelph, Ontario, Canada July-August 2001

Chaignon, V.; Sanchez-Neira, I.; Herrmann, P.; Jaillard, B. & Hinsinger, P. (2002). Copper bioavailability and extractability as related to chemical properties of contaminated soils from a vine-growing area. Environmental Pollution. Vol. 123, (n.d. 2002), pp. 229-228, ISSN 0269-7491

Chopin, E.I.B.; Marin, B.; Mkoungafoko, R.; Rigaux, A., Hopgood, M.J. & Delannoy, E. (2008). Factors affecting distribution and mobility of trace elements (Cu, Pb, Zn) in a perennial grapevine (Vitis vinifera L.) in the Chapmagne region of France. Environmental Pollution. Vol. 156, (n.d. 2008), pp. 1092-1098, ISSN 0269-7491

Croatian Official Gazette (1996). Regulations on Wine Production, No. 96/1996

Croatian Official Gazette (2001). Organic Agriculture and Organic Food Production Act, No. 12/2001

Croatian Official Gazette (2001). Regulations on Organic Plant Production, No. 91/2001

Cvjetkovic, B. (1996). Plamenjaca vinove loze. Croatian Journal of Plant Protection. Vol. 5, pp. 241-244, ISNN 1332-9545

Delas, J. (1963). La toxicité du cuivre accumulé dans les sols. Agrochimica. Vol. 7, (n.d. 1963), pp. 258-288, ISSN 0002-1857

Delas, J. & Juste, C. (1975). Quelques problemes poses par les sols viticoles acides. Connaissance de la Vigne et du Vin. Vol. 2, (n.d. 1975), pp. 67-80, ISSN 0010-597X

Deluisa, A.; Giandon, P.; Aichner, M.; Bortolami, P.; Bruna, L.; Lupetti, A.; Nardelli, F. & Stringari, G. (1996). Copper pollution in Italian vineyard soils. Communications in Soil Science and Plant Analysis. Vol. 27, (n.d. 1996), pp. 1537-1548, ISSN 1532-2416

Drouineau, G. & Mazoyer, R. (1962). Contribution á l'étude de la toxicité du cuivre dans les sols. Annales Agronomiques. Vol. 13, No. 1, (n.d. 1962), pp. 31-53, ISSN 0003-3839

Dumestre, A.; Sauve, S.; McBride, M.; Baveye P. & Berthelin, J. (1999). Copper speciation and microbial activity in long term contaminated soils. Archives of Environmental Contamination and Toxicology, Vol. 36, (n.d. 1999), pp. 124-131, ISSN 0090-4341

Fernandez-Calvino, D.; Pateiro-Moure, M.; Lopez-Periago, E.; Arias-Estevez, M. & Novoa-Munoz, J.C. (2008). Copper distribution and acid-base mobilization in vineyard soils and sediments from Galicia (NW Spain). European Journal of Soil Science, Vol. 59, (n.d. 2008), pp. 315-326, ISSN 1351-0754

Flores Velez, L.M.; Ducaroir, J.; Jaunet, A.M. & Robert, M. (1996). Study of the distribution of copper in an acid sandy vineyard soil by three different methods. European Journal of Soil Science. Vol. 47, (December 1996), pp. 523-532, ISSN 1351-0754

Geoffrion, R. (1975). L'altération des terres á vigne par une longue répétition des traitements a base de cuivre et de soufre. Phytoma; Défense des Cultures. Vol. 267, (n.d. 1975), pp. 14-16, ISSN 0370-2723

Georgieva S.S.; McGrath, S.P.; Hooper D.J. & Chambers, B.S. (2002). Nematode communities under stress: the long term effects of heavy metals in soil treated with sewage sludge. Applied Soil Ecology, Vol. 20, (n.d. 2002), pp. 27-42, ISSN 0929-1393

Gracanin, M. (1947). Pedologija, II dio - Fiziologija bilja. Poljoprivredni nakladni zavod, Zagreb, Croatia

Graniti, A. (1993). Olive scab: review. EPPO Bulletin, Vol. 23, (n.d. 1993), pp. 774-777, ISSN 0250-8052

Katalinic, M.; Kacic, S. & Vitanovic, E. (2009). Stetnici i bolesti masline (first edition). Agroknjiga, ISBN 9789537246198, Split, Croatia

Komarek, M.; Vanek, A.; Chrastny, V.; Szakova, J.; Kubova, K. & Drahota, P. (2009). Retention of copper originating from different fungicides in contrasting soil types. Journal of Hazardous Materials. Vol. 166, (n.d. 2009), pp. 1395-1402, ISSN 0304-3894

Komarek, M.; Cadkova, E.; Chrastny, V.; Bordas, F. & Bollinger, J.C. (2010). Contamination of vineyard soils with fungicides: A review of environmental and toxicological aspects. Environment International. Vol. 36, (n.d. 2010), pp. 138 - 151, ISSN 0160-4120

Kookana, RS.; Baskaran, S. & Naidu, R. (1998). Pesticide fate and behaviour in Australian soils in relation to contamination and management of soil and water: a review. Australian Journal of Soil Research. Vol. 36, No. 5, (n.d. 1998), pp. 715 - 764, ISSN 1838-675X

Li, R.N. (1994). Effect of long-term applications of copper on soil and grape copper (Vitis vinifera). Canadian Journal of Soil Science. Vol. 74, No. 3, (August 1994), pp. 345-347, ISSN 0008-4271

Li, W.; Zhang, M. & Shu, H. (2005). Distribution and fractionation of copper in soils of apple orchards. Environmental Science and Pollution. Vol. 12, (n.d. 2005), pp. 168-172, ISSN 0944-1344

Licul, R. & Premuzic, D. (1993). Praktično vinogradarstvo i podrumarstvo. Nakladni zavod Znanje, ISBN 953-6124-09-2, Zagreb, Croatia

MacDonald, A.J.; Walter, M.; Trought, M.; Frampton, C.M. & Burnip, G. (2000). Survy of olive leaf spot in New Zealand. New Zealand Plant Protection, Vol. 53, (n.d. 2000), pp. 126-132, ISSN 1175-9003

Males, P.; Milos, B.; Pezo, I. & Bubic, J. (1998). Selection of grape rootstock for Kaštela fields, Proceedings of the Kaštela - Croatia cradle, Kastel Stari, Croatia, September-October 1998

Margalit, Y. (January 1996). Concept in wine chemestry, Wine Appreciation Guild, ISBN 978-0932664914, San Francisco, USA

Marschner, H. (1995). Mineral nutrition of higher plants, Academic Press INC., ISNB 012473541X, San Diego, CA, USA

Mc Laren, R.G. & Crawford, D.V. (1973). Studies on soil copper. II. The specific adsorption of copper by soils. Journa of Soil Science. Vol. 24, (n.d. 1973), pp. 443-452, ISSN 1365-2389

McConnell, S.; Wightwick, A.; Smith, T. & Porteous, C. (2003). Code of Environmental Best Practice for Viticulture - Sunraysia region. Environmental best practices Volume 1 - Environmental best practices. Department of Primary Industries. ISBN 1-74106-638-7, Victoria, Australia

Merry R.H.; Tiller, K.G. & Alston, A.M. (1983). Accumulation of copper, lead and arsenic in some Australian orchard soils. Australian Journal of Soil Research, Vol. 21, (n.d. 1983), pp. 549-561, ISSN 1838-675X

Miguens, T.; Leiros, C.; Gil-Sotres, F. & Trasar-Cepeda, C. (2007). Biochemical properties of vineyard soils in Galicia, Spain. Science of Total Environment. Vol. 378, (n.d. 2007), pp. 218-222, ISSN 0048-9697

Miko, S.; Peh, Z.; Halamic, J.; Prohic, E.; Galovic, L. & Sparica, M. (2000). Geochemical mapping in Croatia - state of the art. Proceedings of Secound Croatian Geology Congress, Cavtat, Croatia, May 2000

Miller, H.N. (1949). Development of the leaf spot fungus in the olive leaf. Phytopathology, Vol. 39, (n.d. 1949), pp. 403-410, ISSN 0031-949X

Mirlean, N.; Roisenberg, A. & Chies, J.O. (2007). Metal contamination of vineyard soils in wet subtropics (southern Brazil). Environmental Pollution. Vol. 149, No. 1, (n.d. 2007), pp. 10-17, ISSN 0269-7491

Mirosevic, N. (1996). Vinogradarstvo (first edition). Nakladni zavod Globus, ISBN 953-167-084-6, Zagreb, Croatia

Morgan, R.K. & Taylor, E. (2004). Copper accumulation in vineyard soils in New Zealand. Journal of Environmental Sciences. Vol. 1, No. 2, (n.d. 2004), pp. 139-167, ISSN 1001-0742

Narimanidze, E. & Bruckner, H. (1999). Survey on the metal contamination of agricultural soil in Georgia. Land Degradation & Development. Vol. 10, No. 5, (September-October 1999), pp. 467-588, ISSN 1085-3278

Novoa-Munoz, J.C.; Queijeiro, J.M.G.; Blanco-Ward, D.; Alvarez-Olleros, C.; Martinez-Cortizas, A. & Garcia-Rodeja, E. (2007). Total copper content and its distribution in acid vineyard soils developed from granitic rocks. Science of Total Environment. Vol. 378, (n.d. 2007), pp. 23-27, ISSN 0048-9697

Obonar, F.O.; Jaspers, M.V.; Jones, E.E. & Walter, M. (2008). Greenhouse and field evaluation of fungicides for control of olive leaf spot in New Zealand. Crop Protection, Vol. 27, No. 10, (October 2008), pp. 1335-1342, ISSN 0261-2194

Paoletti, M.G.; Sommaggio, D.; Favretto, M.R.; Petruzzelli, G.; Pezzarossa, B. & Barbafieri M. (1998). Earthworms as useful bioindicators of agroecosystem sustainability in orchards and vineyards with different inputs. Applied Soil Ecology, Vol. 10, (n.d. 1998), pp. 137-150, ISSN 0929-1393

Parat, C.; Chaussod, R.; Leveque, J.; Dousset, S. & Andreux, F. (2002). The relationship between copper accumulated in vineyard calcareous soils and soil organic matter and iron. European Journal of Soil Science. Vol. 53, No. 4, (December 2002), pp. 663-669, ISSN 1351-0754

Pietrzak, U. & McPhail, D.C. (2004). Copper accumulation, distribution and fractionation in vineyard soils of Victoria, Australia. Geodarma. Vol. 122, (n.d. 2004), pp. 151-166, ISSN 0016-7061

Ponizovsky, A.A.; Allen, H.E. & Ackerman, A.J. (2007). Copper activity in soil solutions of calcereous soils. Environmental Pollution. Vol. 145, (n.d. 2007), pp. 1-6, ISSN 0269-7491

Prasad, B.R.; Basavaiah, S.; Subba Rao, A. & Subba Rao, I.V. (1984). Forms of copper in soils of grape orchards. Journal of the Indian Society of Soil Science. Vol. 32, (n.d. 1984), pp. 318-322, ISSN 0019-638X

Prota, U. (1995). Le malattiedell'olivo. Informatore Fitopatologico, Vol. 45, (n.d. 1995), pp. 16-26, ISSN 0020-0735

Puig-Deu, M.; Lamuela-Raventos, R.M.; Buxaderas, S. & Torre-Boronat, C. (1994). Determination of copper and iron in must: Comparison of wet and dry ashing. American Journal of Enology and Viticulture. Vol. 45, No. 1., (n.d. 1994), pp. 25-28, ISSN 002-9254

Radovanovic, V. (1970). Tehnologija vina, Građevinska knjiga, Beograd, Serbia

Ribéreau-Gayon, P.; Glories, Y.; Maujean, A. & Dubourdieu, D. (2000). The Chemistry of Wine Stabilization and Treatments. In: Handbook of Enology, Volume 2, 86-97, John Wiley & Sons Ltd., ISBN 987-0-470-01037-2, Chichester, England

Ribolzi, O.; Valles, V.; Gomez, L. & Voltz, M. (2002). Speciation ond origin of particulate copper in runoff water from a Mediterranean vineyard catchment. Environmental Pollution. Vol. 117, No. 2, (August 2002), pp. 261-271, ISSN 0269-7491

Romic, M. & Romić, D. (1998). Sadržaj olova kadmija, cinka i bakra u poljoprivrednim tlima Zagreba i okolice. Agriculturae Conspectus Scientificus. Vol. 63, No. 3, (n.d. 1998), pp. 147-154, ISSN 1331-7768

Romic, M.; Romic, D. & Kraljickovic, J. (2001). Bakar u vinogradarskim tlima, In: Land management and soil protection for future generations, Z. Ratz, (Ed.), 1-77, Croatian Society of Soil Science, Zagreb, Croatia

Romic, M. (2002). Sadržaj, oblici i preraspodjela imisije teških kovina u poljoprivrednim tlima šireg područja Zagreba. PhD disertation. Faculty of Agriculture University of Zagreb, Zagreb, Croatia.

Schramel, O.; Michalke, B. & Kettrup, A. (2000). Study of copper distribution in contaminated soils of hop fields by single and sequential extraction procedures. Science of Total Environment. Vol. 263, (n.d. 2000), pp. 11-22, ISSN 0048-9697

Stevenson, F.J. & Cole, M.A. (March 22. 1999). Cycle of Soil: Carbon, Nitrogen, Phosphorus, Sulfur, Micronutrients , John Wiley & Sons, ISNB 978-0471320715, New York, USA

Temminghoff, E.J.M.; Vanderzee, S.E.A.T.M. & Keizer, M.G. (1994). The influence of pH on the desorption and speciation of copper in a sandy soil. Soil Science. Vol. 158, No. 6, (December 1994), pp. 398-408, ISSN 0038-075X

Treviotdale, B.L.; Sibbett, G.S. & Harper, D.H. (1989). Control of olive leaf spot by copper fungicides. Journal of Applied Agricurtural Research, Vol. 4, (n.d. 1989), pp. 185-189, ISSN 2006-750X

Van Zwieten, L.; Rust, J.; Kingston, T.; Merrington, G. & Morris, S. (2004). Influence of copper fungicide residues on occurrence of earthworms in avocado orchard soils. Sciencie of Total Environment. Vol. 329, (n.d. 2004), pp. 29-41, ISSN 0048-9697

Verona, O. & Gambogi, P. (1964). On the characteristics of oil produced by olives attacked by Cycloconium oleaginum. Agricoltura Itaiana., Vol. 64, (n.d. 1964), pp. 1135-1139

Vitanovic, E.; Vidacek, Z.; Katalinic, M.; Kacic, S. & Milos, B. (2010a). Copper in surface layer of Croatian vineyard soils. Journal of Food Agriculture and Environment, Vol. 8, No. 1, (January 2010), pp. 268-275, ISSN 1459-0255

Vitanovic, E.; Katalinic, M.; Kacic, S.; Zanetic, M.; Milos, B. & Radinovic, S. (2010b). The content and correlation of copper, iron and zinc in the surface layer of the Croatian vineyard soils. Journal of Food, Agriculture and Environment, Vol. 8, No. 3&4, (July-October 2010), pp. 1134-1140, ISSN 1459-0255

Vukadinovic, V. & Loncaric, Z. (1998). Ishrana bilja. Faculty of Agriculture, ISBN953-6331-04-7, Osijek, Croatia

Wightwick, A.M.; Mollah, M.; Smith, J. & MacGregor A. (2006). Sampling considerations for surveying copper concentrations in Australian vineyard soils. Australian Journal o Soil Research, Vol. 44, No. 7, (October 2006), pp. 711-717, ISSN 1838-675X

Wightwick, A.M. & Allinson, G. (2007). Pesticide residues in Victorian waterways: a review. Australian Journal of Ecotoxicology. Vol. 13, (n.d. 2007), pp. 91 - 112, ISSN 1323-3475

Wightwick, A.M.; Mollah, M.R.; Partington, D.L. & Allison, G. (2008). Copper fungicide residues in Australian vineyard soils. Journal of Agricultural and Food Chemestry. Vol. 56, (2008), pp. 2457-246, ISSN 0021-8561

Wilson, E.E. & Miller H.N. (1949). Olive leaf spot and its control with fungicides. Hilgardia, Vol. 19, (n.d. 1949), pp. 1-24, ISSN 0073-2230

Zoecklein, B.W. (December 31. 1995). Wine analysis and production. Springer, ISBN 978-0412982415, University of California, Davis, USA